Laser Ignition of Energetic Materials

Laser Ignition of Energetic Materials

S Rafi Ahmad
Michael Cartwright

WILEY

Registered office: John Wiley & Sons, Ltd, The Atrium, Southern Gate, Chichester, West Sussex, PO19 8SQ, UK

Editorial offices: 9600 Garsington Road, Oxford, OX4 2DQ, UK
The Atrium, Southern Gate, Chichester, West Sussex, PO19 8SQ, UK
111 River Street, Hoboken, NJ 07030-5774, USA

For details of our global editorial offices, for customer services and for information about how to apply for permission to reuse the copyright material in this book please see our website at www.wiley.com/wiley-blackwell.

Library of Congress Cataloging-in-Publication Data

Ahmad, Rafi, 1944–
Laser ignition of energetic materials / Dr S Rafi Ahmad and Dr Michael Cartwright.
pages cm
Includes bibliographical references and index.
ISBN (cloth : alk. paper) 1. Explosives. 2. Detonators. 3. Lasers–Industrial applications. 4. Heat–Transmission. I. Cartwright, Michael, 1940– II. Title.
TP293.A37 2014
662′.2–dc23

2014012438

A catalogue record for this book is available from the British Library.

ISBN: 9780470975985 (Cloth)

Wiley also publishes its books in a variety of electronic formats. Some content that appears in print may not be available in electronic books.

Cover images courtesy of iStockPhoto
Cover design by Andrew Ward

Set in 10/12pt Times by Aptara Inc., New Delhi, India.
Printed and bound in Malaysia by Vivar Printing Sdn Bhd

1 2015

Dedicated to our spouses Angela & Estelle for their support, patience and encouragement during the preparation of the book.

Contents

About the Authors

Dr S Rafi Ahmad

Dr Ahmad founded the Centre of Applied Laser Spectroscopy within Cranfield University (CU) in 1988 and led it until he retired in 2012. He received the degree of D.Phil. from the University of Oxford (UK) in 1972 on his thesis on 'Laser Interaction with Solid Materials'. During his career as a scientist within the Ministry of Defence (UK), and later as an academic at CU, the scope of his research areas expanded to include, among many diverse topics, laser ignition of energetic materials and laser-induced processing of natural and synthetic polymers for biomedical applications. His research was funded by many national and international bodies and he was the principal investigator and the coordinator of a number of EU funded projects involving many partners from EU countries in research topics including Jute modification (INCO-DC) and plastic identification (BE-7148). He also served as the UK representative in the management committee of the EU's EULASENET network and one of the co-coordinators of the COST-G7 action on artwork preservation using lasers. Dr Ahmad retired from Cranfield University in 2012. He has supervised 11 PhD students and published 62 peer reviewed papers and other books.

Dr Michael Cartwright

After graduating in chemistry from London in 1963, Michael Cartwright gained industrial experience in the nuclear industry at Windscale and pharmaceutical research in London, before moving to Bath University as a researcher in nuclear chemistry. He was awarded a D.Phil by London University in 1974 for a thesis on 'Radiation Damage in Solids', based on research work performed at Bath University. He then proceeded to do research in inorganic thermo-chemistry and organo-metallic coordination complex chemistry. His interest expanded towards structural chemistry and X-ray diffraction which resulted in later years in his inauguration of the University's first single crystal four circle structure determination facility.

He moved to Cranfield University, at RMCS Shrivenham, in 1986, to lecture and perform research in energetic materials. His major interest was the relationship between molecular structure and explosive sensitivity, which developed into a series of lectures for the M.Sc in Explosives Ordnance Engineering. He was co-founder of the M.Sc. Forensic Engineering and Science course and he helped to develop the university's interests in the environmental impact of explosives, particularly in waste waters from manufacturing plants and land contamination at munitions disposal sites. He helped to design the inter-site master's courses on environmental diagnostics and waste water chemistry. Energetic materials research was funded by the MOD, government agencies, research organisation and private companies. He represented CU on the SCC of the MOD, devising test methods for assessment of accidental initiation risks for energetic materials. He also represented CU on several NATO organisations, examining various aspects of energetic material science. He retired from Cranfield in 2009.

Preface

The practical laser was invented by T. H. Maiman at the Hughes Research Laboratories in the USA, way back in 1960. It was then hailed as 'the tool looking for applications'. In no time, the tool found applications in almost all fields of science and technology. The headline defence application was in a 'Star Wars' anti-missile system but, not surprisingly, within a couple of years, research and development on its applications in the defence industry, particularly for high explosives ignition/initiation, got under way. Due to the 'Cold War' prevailing at the time, most of this research was shrouded in secrecy. However, for a variety of reasons, there has been a long pause in tangible developments in this field until recently. Current emphasis on the safety of energetic materials during manufacture, storage, use and transportation, has prompted a spate of research activities throughout the industrial world on the synthesis and ignition initiation of high-performance munitions. These must have whole-life cost-effectiveness, through-life safety and end-of-life environmentally friendly disposal options. These aspects are the objectives of research and development in this field, and the book aims to elucidate the background and the current state of the art in the field of laser initiation.

The book starts with a brief chronological resume of the invention, development and the use of materials generally termed 'explosives'. This is intended as a purely historical background introduction and is compiled from various sources. An extensive review of the research and development in the application of lasers for ignition/initiation in energetic materials, identifying some of the critical parameters involved, is provided in Chapter 2. This includes a number of references, in addition to a bibliography of recent relevant publications.

Since the book topic encompasses two very different fields of science and technology, these are, for completeness and convenience of the readers, elaborated in Chapters 3 and 4. Chapter 3 provides the basic science behind the technologies, manufacture and general properties of lasers, while Chapter 4 provides a background to the general properties and synthesis of energetic materials. This includes the essential components, both as mixtures of fuels and oxidiser and single energetic molecules, with a chemical classification of these latter materials. The contents are considered to be adequate background for researchers in this field. There are also references and bibliography for the inquisitive reader. Note that further information on these topics is readily available in a number of open literature sources.

Chapter 5 examines the limitations of the current materials and methods of improving safety, for example, with plastic bonded explosives, PBXs and so on. Consideration is also given to the synthesis of new explosives, an active field of research and development. Some of these newer materials are less environmentally toxic. It was therefore considered prudent to include a chapter reviewing these aspects and, in particular, high nitrogen materials, since some of these materials may find future applications in laser ignition.

Fundamental processes associated with the decomposition of energetic materials, ranging from simple burning through deflagration to detonation, are discussed in Chapter 6, along with the effects of explosives in terms of shock pressure and explosive power. Additional methods of improving explosive power are discussed. A brief appendix details some of the methods used for measuring velocity of the shock wave. Chapter 7 examines the energetic changes associated with the initiation process and the currently used techniques for the initiation of energetic materials, with only brief reference to the use of optical or laser systems. Classification of explosives by ease of initiation and the use of explosives trains to minimise hazards are considered, along with the basic properties of current initiatory primers. For both general safety and for safe ignition using lasers, a synopsis of the development of alternative primary explosives is presented in Chapter 8. Some of the materials discussed show particular sensitivity to laser radiation and have high explosive performance, sometimes in excess of existing high explosives.

The theoretical basis of laser interaction with energetic materials involves both optical and thermal properties of materials and both these aspects are covered in Chapters 9 and 10. Chapter 11 provides a synopsis of practical research conducted in this field, mainly citing examples of work carried out at the authors' laboratories. Finally, a general conclusion of the work conducted so far in this field, and the future prospects and direction of research, is included in Chapter 12.

Acknowledgements

One of the authors wishes to acknowledge the assistance of Mr Edwin Billiet and Dr X Fang for assistance with laboratory work for most of the data presented in the book and also for drawing some of the figures. The authors also acknowledge Cranfield University (Defence Academy) for providing the environment and opportunity to conduct tests and evaluations which are now in the public domain.

1

Historical Background

1.1 Introduction

Historically, mankind has tried to dominate both fellow human beings and other animals for as long as humans have been around. Some of this domination was achieved by killing other species. This had two aspects; survival and providing food.

Survival was dictated by the fact that many animals regarded humans as excellent sources of food and were quite capable of killing humans. Humans could have two approaches; avoid areas known to contain threatening species or produce devices – weapons – which would enable humans to kill the threatening animals. Humans then developed a taste for the flesh of some of the animals they had killed, thus increasing the sources of food available. As the human population increased, conflict between humans for food and territory increased, and so humans started to fight amongst themselves. By using weapons, humans could overcome physical disadvantages, and the optimum situation was to be able to kill your opponent before they could kill you.

The sword and lance effectively extended the human arm and kept your opponent at bay but, as lances became longer and longer, they became more unwieldy. A remote killing weapon was required. Simple javelins, which could be thrown at the opposition, extended the distance between opponents but required considerable physical stature and skill to achieve the correct flight trajectory for the javelin. Therefore, in order to overcome human physical limitations, mechanical advantage devices were used. The earliest weapons for remote killing were simple slings. These could carry a stone and were capable of accelerating it to high velocity by spinning the sling in a circle. When one of the supporting thongs was released, the stone would travel in an almost straight line from the point of release. Impact of the stone with an animal or human was capable of killing or injuring the animal.

With the development of wood manufacturing skills, bows and arrows became individual weapons or, when grouped together became a lethal hail of arrows which did not depend on the individual accuracy of the archer. The longbow was the ultimate in these weapons. Improved performance came when mankind developed stored energy devices, such as

Laser Ignition of Energetic Materials, First Edition. S Rafi Ahmad and Michael Cartwright.

Figure 1.1 Small-scale basic ballista. Reproduced with permission from Cranfield University © 2014.

the ballista and crossbows, both of which stored mechanical energy in wooden elements but required winding up before loading the stone or arrow projectile. These overcame the limitations of physical stature required to effectively use the longbow. The ballista, Figure 1.1, was also used to fire barrels of burning oil at the enemy when they had formed shield walls against arrows. The oil container burst on impact and was one of the first deployments of pyrotechnics weapons.

1.2 The Gunpowder Era

Meanwhile, the Chinese were developing the first chemical explosive gunpowder. The earliest record of this was around 800 AD. Initially, the mixture was for use as a medicine but, as with all good inventions, serendipity intervened and a batch of the medicine fell on to the fire over which it was been cooked; it very rapidly burnt with a flash, smoke and rushing sound. The potential for this was recognized, and the Chinese started to use the mixture as a propellant for their lances/javelins. When attached to the normal throwing spear, these early rockets could extend the useful range of the javelin by as much as a factor of two.

It took about 400 years for the technology to appear in Europe, when a cleric Roger Bacon was credited with discovering the properties of gunpowder. He was so afraid of its properties that he hid the details of the composition in code in religious manuscripts. The recognition of its propellant properties resulted in the manufacture of muzzle-loaded cannons.

An idea of the chronology of the development of the science is given in Table 1.1 on page 3.

1.3 Cannons, Muskets and Rockets

The barrels of the first cannon systems were simple wooden devices made from hollowed-out tree trunks, which were wrapped with wet ropes for added strength. The development of bronze and cast iron technology led to the production of iron-barrelled guns, such as

Table 1.1 *Some significant discoveries in the history of explosives.*

Explosive	Credited to	Nationality	Date
Gunpowder	{Anon	(Probably Chinese)	Before 1000
	{R. Bacon	English	c.1246
First battlefield cannon		Italian	c 1326
Crecy bombard		English	1346
Hand cannon		Italian	c 1364
Leonardo's mortar	Leonardo da Vinci	Italian	c 1483
Mercury fulminate	Kunckel	German	c.1690
Picric acid[1]	Woulff	German	1771
Mercury fulminate percussion cap	Forsyth	Scottish	1825
Nitrocellulose[2]	{Pelouze	French	1838
	{Schonbein	German	1845
Nitroglycerine	Sobrero	Italian	1846
TNT	Wilbrand	German	1863
The fulminate detonator	Nobel	Swedish	1865
Dynamite	Nobel	Swedish	1867
Ammonium nitrate mixtures	Ohlsson & Norrbin	Swedish	1867
Tetryl	Mertens	German	1877
N.C. propellants[3]	{Schultze	German	1864
	{Vieille	French	1884
Ballistite	Nobel	Swedish	1883
Cordite	Abel & Dewar	British	1889
Lead azide	Curtius	German	1890
PETN	Rheinisch-Westfaelische Sprengstoff A.G	German	1894
RDX	Henning (patented by	German	1899
	Herz)	German	1920
NTO	von Manchot and Noll	German	1905
Tetrazene	Hoffman & Roth	German	1910
HMX			
Slurry explosives	Cook	USA	1957
Emulsion explosives			
PBX			

[1] Explosive properties of Picric acid were not investigated for a further 100 years.
[2] Pelouze produced NC but did not understand the chemistry whereas Schonbein correctly identified the chemistry and made some propellant uses.
[3] Schultze produced the first successful powdered NC propellants and Vielle was credited with the first NC propellants for rifled barrel guns.

the Bombard, used at the battle of Crecy in 1346 (shown in Figure 1.2). This weapon used solid projectiles in the form of either suitable stone or cast metal (e.g. iron) spheres. The development of these weapons resulted in the foundation of the Board of Ordinance in 1414. The operators of these weapons were known as Bombardiers – a term still used for an artilleryman with the rank of corporal in the British Army.

In the fifteenth century, cannon were also deployed at sea on warships and these enabled the opposition to be destroyed at distance without needing to engage in hand to hand combat. A number of cannons were deployed along each side of the ship, and a broadside

Figure 1.2 *Crecy bombard 1346. Reproduced with permission from Cranfield University © 2014.*

could be loosed at the opposition. Typical cannons are shown in Figure 1.3, which displays a typical army cannon in the foreground and a naval cannon in the background.

The naval cannon was mounted on a four-wheeled trolley rather than the two wheels of the army. This provided better stability onboard a ship in heavy seas. The iron guns were made of a number of staves, or bars, of iron which were formed into a cylinder around a mandrel. Collars and hoops of wrought iron were heated and slipped over the cylinder. As these cooled, they contracted to form a reinforced tube. Surprisingly, breech-loaded cannon (often regarded as a modern invention, introduced when screw cutting technology was developed during the Industrial Revolution) were available in the early fifteenth century. The early systems used a simple hollow steel tube mounted on a wooden trough, with a space between the end of the metal tube and the end of the wooden support. A closed metal cup containing the propellant charge was then inserted into the gap and rammed into the rear open end of the barrel. The system was then sealed by inserting a wooden plug behind the charge-bearing cup to ensure the cup remained in place when the gun was fired. The components are displayed in Figure 1.4.

Figure 1.3 *Showing a field gun in the foreground and a naval gun in the background. Reproduced with permission from The Mary Rose Trust. © The Mary Rose Trust, 2014.*

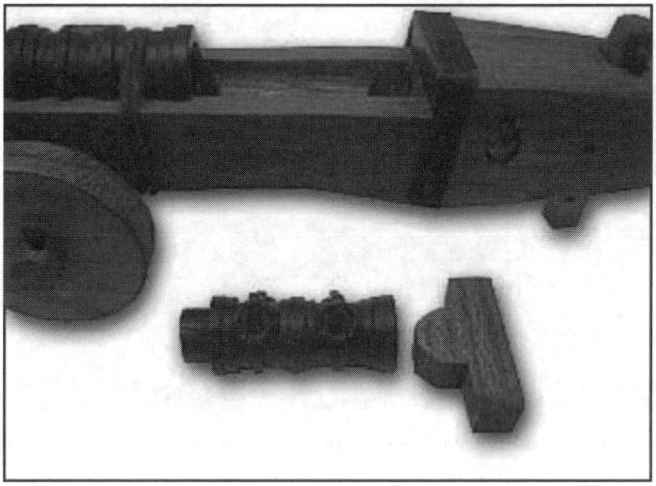

Figure 1.4 *Showing active region of a breech-loaded cannon as used on the* Mary Rose. *Reproduced with permission from The Mary Rose Trust. © The Mary Rose Trust, 2014.*

Although the breech was by no means gas-tight, these guns were powerful enough to inflict heavy damage on another ship at close range. This has been demonstrated in field trials using a replica iron gun. Later breech-loaded cannons were cast metal closed tubes, but with a slot in the side into which the charge could be inserted and sealed in position again with a wooden plug. These were more durable than the wooden supported systems. The beauty of the system was that each gun had a number of charge cups which could be filled with gunpowder prior to engagement, so that a higher rate of fire could be produced. Also, swabbing out of the gun between shots was not so crucial, since the next charge did not have to be loaded down a barrel which could contain hot residues, liable to ignite the charge before the gun was ready. This, again, increased the rate of fire. Notice that the ignition of the charge was by transmission of an externally generated flame through a touch hole. This hole allowed some of the burning gases to be vented, thus reducing the gas propulsion of the projectile and also blowing hot embers back towards the gun crew.

As well as the standard stone or iron shot, these guns could fire a vicious Tudor version of canister shot consisting of chopped flint stones. These were particularly useful when the opposing forces had a number of soldiers exposed on an open deck.

Gunpowder was the propellant in cannon for the next 400 years, but it was the longbow that was responsible for English superiority at the battle of Agincourt 1425. Cannons were mainly used as siege weapons, and they required skilled operators to achieve success with the minimum number of shots. Several options were proposed by Leonardo Da Vinci to alleviate some of the requirement for skilled operators, and he proposed a multi-barrelled system, as shown in Figure 1.5, which provides sketches from his work book.

One of the most fearsome siege weapons was the Dardanelles Gun. This was a cast bronze weapon weighing over 18 tonnes, and it was five metres long when assembled from its two parts, which screwed together. ‘Superguns’ were not a modern invention, but only became practical after the Industrial Revolution. The solid cannonball weighed about 400 kg could penetrate over one metre thick walls with an effective range of almost a

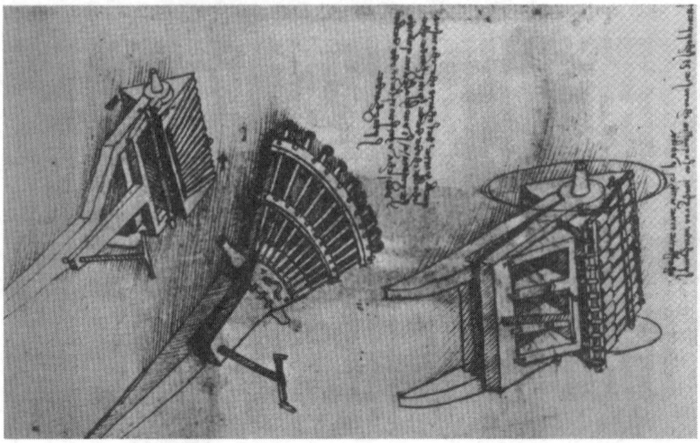

Figure 1.5 *Leonardo Da Vinci's sketches for multi-barrelled cannon systems.*

kilometre. Leonardo also proposed mortar-type guns for siege work on the principle that, if it was not possible to penetrate the wall, fire over it. Two mortars from his original designs are depicted in Figure 1.6. They could both fire single or multiple flint stone solid shot. Note on one of the mortars the sophistication of the range/angular adjustment and also the rudiments of sighting instruments

This attention to cannons resulted in the creation of the first body of specialist artillerymen, The Guild of St George (1537), which became Honourable Artillery Company in 1668. The gunpowder was of variable performance because of three factors. Solids had to be ground to the correct size and then correctly mixed in the correct proportions, which was, according to the early Chinese formulation, equal parts of brimstone, (sulfur), charcoal (carbon) and saltpetre (potassium nitrate). Small particle size materials, which produced

Figure 1.6 *Leonardo Da Vinci-designed siege mortars.*

the optimum gun performance, were very dusty, and hence the millers worked outdoors whenever possible. Note that the ingredients were ground separately, since it was discovered early in the use of gunpowder that grinding all the ingredients together resulted in ignitions, with serious consequences. It was only with the development of the 'Corning process' that the dust problem was reduced. In corning, the finely divided powders are mixed with water to a paste, which is spread out and dried to a cake. The cake is then broken into smaller fragments (≈ 1–2 mm diameter), which can easily be filled into bags to load into cannons.

Because of the difficulties recruiting skilled artillerymen, alternative siege weapons were developed. The Petard was essentially a barrel of gunpowder fitted with a slow match fuse and was carried into battle by a Petardier. At some suitable stage, he would rush forward and place the Petard against the wooden gates of the fort, light the slow match and retreat as fast as possible. If the slow match was badly made, it would burn rapidly, ignite the charge before the Petardier had retreated and he could be killed by the gunpowder explosion. Hence the expression 'Hoist by ones own Petard'.

The first explosive charges fired from guns were containers of gunpowder fitted with a slow fuse which was ignited by the burning gunpowder propellant charge. This was again a very hazardous operation, since most of the containers did not survive the ignition of the gunpowder propelling charge, so they were effectively an added propellant system with no useful performance. Thus they were very unpopular with artillerymen.

1.3.1 Musketry

Over a period of time, the gunpowder system was adapted to personal usage in the form of the musket. Again, this was muzzle-loaded and fired solid shot from a smooth-bore barrel. The gunpowder charge was poured into the barrel from a powder horn, and packed down using a cotton wad to retain the powder and a ramrod to force the charge into a compact charge before inserting the ball shot. The shot would be also rammed in, to prevent it from rolling out if the gun was pointed downwards. The internal charge was set off by igniting a small gunpowder charge contained in a cup on the outside of the barrel, and the burning process was transferred through a touch hole into the interior. This firing hole was a source of reduced performance, since some of the gases produced inside the barrel could vent to the outside, thus reducing the gas pressure driving the ball.

Because the escaping gases could burn the firer, some of the heavier muskets were mounted on a crude stand so that the weapon could be fired at arm's length, reducing the hazard to the operator. The firing charge was set off by a range of devices. The first initiation device was a glowing length of burning rope, which was applied either manually or by a rotating arm carrying the 'slow match' – a 'matchlock' device. A series of alternatives were developed, including a flint being struck against a hard surface, causing a spark which ignited the gunpowder in the pan or cup – 'the flintlock'. The basic flintlock and the adapted wheel lock are shown in Figure 1.7.

If the material in the cup ignited, but failed to set of the main propellant charge, then it gave rise to the term 'flash in the pan' for a trivial output of no use to the firer. This situation left the gun in a dangerous condition which required careful re-priming. A later stage development came in 1805 when another clergyman, Alexander Forsyth, produced a metal cap which contained a sensitive composition, mercury fulminate. The cap covered the touch hole, as shown in the lower image of Figure 1.7, and was initiated by a mechanical

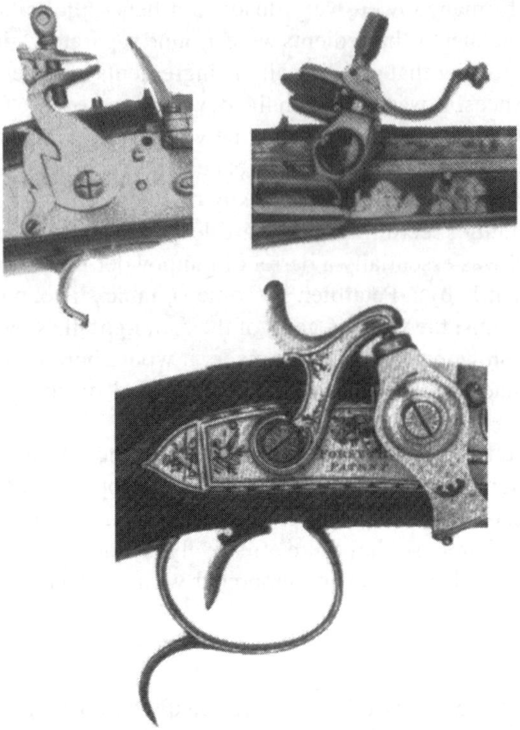

Figure 1.7 *Showing ignition systems in musketry. Top left: original flintlock. Top right: wheel lock. Bottom: first percussion cap system. Reproduced with permission from Cranfield University © 2014.*

blow delivered by a hammer striking the outside of the cup. The mercury fulminate reliably delivered an explosion to set off the charge at the rear of the barrel. This dramatically reduced the firing time between the operation of the trigger and the ejection of the projectile. The cup provided a partial seal against gas venting through the fire hole, thus increasing the gas pressure driving the musket ball. The cap system also increased the kickback on the weapon, since momentum must be conserved during the firing. If the bullet exited the barrel with greater velocity and momentum, then the rearwards velocity of the gun also increased.

Note that, at this stage, the projectiles were still solid, with no explosive properties, and that the wadding was important because the shot was a poor fit in the barrel, due to manufacturing problems. After the English Civil war in the mid-seventeenth century, which led to the establishment of a permanent army under the control of Parliament, the development of the musket accelerated. Also, in 1671, the government purchased the Woolwich site for their armaments, and this became Royal Arsenal 1805. This site provided fireworks for the celebration which accompanied Handel's Fireworks Music.

One example of the devastating performance achieved with gunpowder muskets was the 'Brown Bess' used by the English army at the battle of Culloden in 1746, when it was trying to subdue the Scottish rebels. The solid, ≈ 20 mm diameter, musket ball was lethal to 100 m in the hands of a skilled musketman, and a volley of shots could decimate an advancing force of soldiers armed with swords and lances. Typical shields of the time, a metal skin on a wooden backboard, offered little protection and, as recent trials have demonstrated, the same musket ball could pass through the shields and bodies of two soldiers standing one behind the other.

Prior to 1759, the Board of Ordinance did not make gunpowder but used the East India Company as suppliers, and they often sourced their supplies from the French Arsenal in Paris. Antoine Lavoisier, of combustion theory fame, was the administrator, and he ensured that good quality control was applied at all stages – hence, French gunpowder was the best available in Europe. In 1759, the Board of Ordinance purchased Faversham Kent powder mills. The factory blew up 1781, and the Board then bought the Waltham Abbey Powder mills in 1787 to continue the supply. Around this time, another cleric, Bishop Watson, obtained charcoal from distilling wood in closed vessels. This increased purity charcoal gave $\approx 70\%$ improvement in performance. The explosion at the Waltham Abbey Powder mills in 1843, which resulted in many fatalities, was one of the driving forces for the development of alternatives to gunpowder and the first efforts with nitrocellulose. Modern gunpowder mills are subject to very stringent safety measures.

1.3.2 Rocketry

Some military men used rockets, with cutting blades attached to the front, essentially to increase the range of the weapon over the musket and thus keeping the musketeers out of musket range. However, their flight was very unpredictable because of the uneven burning of the gunpowder, and also steering was purely by having the centre of mass behind the centre of thrust, which state was achieved by attaching the rocket to the front of a long, solid wooden stick. This stick added considerable mass to the projectile, reducing its range. Most of these devices flew along spiral paths. The conflict in India in the eighteenth century saw the Indian forces using batteries of rockets in the battle of Guntar against the advancing English infantry to great effect, both from a lethal viewpoint but also from a psychological viewpoint, since the rockets made screaming sounds as they passed through the air.

After the battle of Seringapatam in 1799, Congreve developed rockets and explosive missiles with ranges of up to 3,500 metres. In 1806, these were used to attack a French flotilla in Boulogne, destroying many of the ships and blocking the harbour. Similar devices were used at the battles of Copenhagen and Walcheren, with similar results. The inaccuracies of the rocket system did not matter; because of the extended target area and the density of the targets, they were bound to hit *something*, even if it was not what was aimed for. At the battle of Leipzig in 1813, rockets were used by the army. Congreve's rockets were used by British Army until 1865, when Boxter invented an improved type. In 1845, Hale invented fin-stabilized rockets, avoiding the useless mass of the guidance stick.

1.4 Explosive Warheads

The first steps along the path for explosive warheads came with the development of the grenade. This was a container, usually of wood or metal, which contained a gunpowder charge, with or without some added musket balls and a length of burning fuze, initially a slow match. As the two sides involved in the conflict approached each other, grenadiers would light their fuzes, charge towards their opponents and hurl the grenades into the ranks of the opposition. Sometimes, the thrower would make use of the old sling technology to

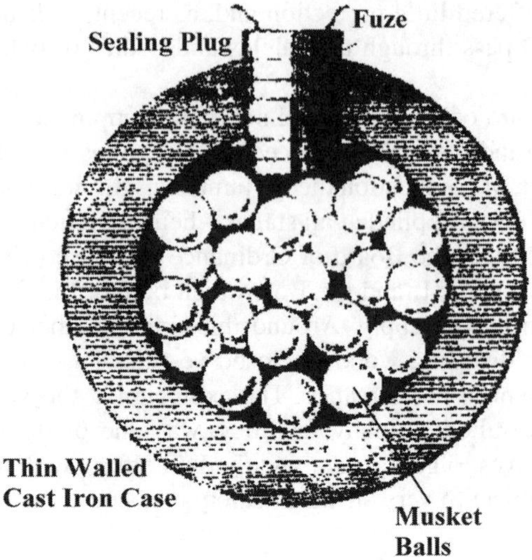

Figure 1.8 Shrapnel's exploding cannon shell ≈ 1780. Reproduced with permission from Cranfield University © 2014.

increase the effective range over which the projectile could be hurled. The hazard of this job was again reflected in the recognition of the Grenadiers regiments being the leading soldiers in any parade, standing to the right of the line. There was a small financial reward to grenadiers, but it did not compensate for the high grenadier mortality. Opposition riflemen could easily identify the grenadiers and concentrate their firepower on them.

At this stage, two developments occurred which had a significant effect on the effectiveness of both guns and rockets. Colonel Shrapnel designed a cast iron hollow sphere which was packed with musket balls and gunpowder, with the filling hole being sealed by a pyrotechnic delay device, ignited by the output from the gunpowder propelling charge. The system is shown in Figure 1.8.

The system originally blew out the plug and the musket balls through the plug hole, giving a shotgun effect, but the later system was designed to shatter the case and scatter case and musket balls in all directions. If the delay element was correctly assembled, the gunpowder contained in the shell was initiated by the delay just as the projectile arrived at the opposition, explosively bursting the casing and thus showering the opposition with the musket balls. If the delay operated while the projectile was still in the barrel, then it behaved rather like a large-scale modern-day shotgun. The modern term 'shrapnel' is applied to any fragment from the warhead.

Congreve applied the same strategy to his rocket systems, using a similar warhead, as shown diagrammatically in Figure 1.9. One advantage of this system was that, because the delay was at the end of the rocket motor, the problem of sealing the delay against propellant gas pressure was minimized. Also, the rocket was down range before the warhead exploded if the fuse was badly fitting. The technology was very imprecise and only rarely did they achieve their design operating conditions. Nevertheless, the foundations were laid for further developments in the field of explosives. The development of rifled gun barrels

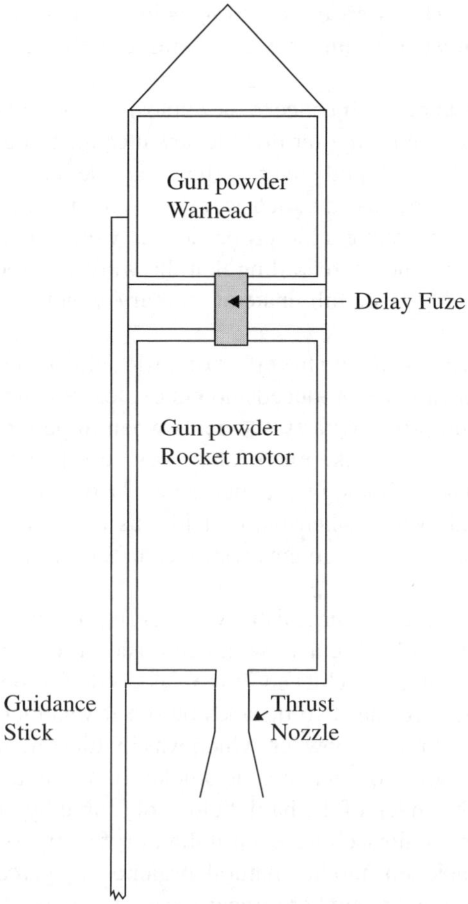

Figure 1.9 *Congreve's exploding warhead rocket.*

led to the demise of unstable rocket systems until the Second World War, when the use of anti-tank rockets by both infantrymen and aircraft used double-base propellants and were fin-stabilized. Now they are the principal air-launched weapon for both ground attack and aircraft self-defence, as well as ground-to-air defence. Naval Maroons are modern variations of the Congreve system.

1.5 Explosives Science

While all these developments in gun and rocket technology were taking place, the science of chemistry was quietly developing in some surprising directions. Instead of trying to turn everything into gold, as per the alchemist's dream, some people started to look at the changes which could be induced in the world around them. Metals which could be simply manufactured by roasting on fires with or without the assistance of charcoal had been around for some considerable period of time, and some chemists were starting to look at non-metallic materials. The discovery of strong acids such as sulfuric, hydrochloric and nitric acids led to a number of chemists experimenting with these materials to see what happened.

The first such experiment of significance to the explosives community was that performed by Kunckel, in Germany around the turn of the seventeenth century. He wondered what would be the effect of mixing 'liquid silver' – the metal mercury – with nitric acid in the presence of alcohol, which he had distilled from fermented wines. He obtained a white powder which showed many remarkable properties, as well as exploding at the slightest touch. He shelved the experiments, recording that the work was too dangerous for him to pursue. He had inadvertently made fulminate of mercury, which we now know as mercury isocyanate.

Later on, Woulff experimented with the effect of nitric acid on one of the products he had isolated from coal tar – phenol. He produced another explosive material, picric acid, trinitro phenol, which again could produce nasty accidents when impure or in contact with some metals, notably the lead used to make many laboratory vessels before the development of suitable glass devices. The technology was put aside for over one hundred years, before resurfacing during World War I, when picric acid was an alternative filling to TNT in munitions until its inherent safety problems resurfaced. Today, the lead salts of picric acid are used in initiating systems.

By the early nineteenth century, chemistry was developing at a very rapid pace. Early experiments on the reaction of nitric acid with cellulosic fibres were undertaken, and the first preparations of nitrocellulose achieved. It took a decade for the science involved to be correctly enunciated, and a further two decades before it was identified and trialled as a gun propellant replacement for gunpowder, which was inefficient, hazardous to handle and produced a significant quantity of unavoidable smoke. In a famous painting depicting the battle of Waterloo in 1815, much of the battlefield is obscured by the gunpowder smoke.

Shortly after the work on nitrocellulose, an Italian Professor, Ascanio Sobrero, worked on the effect of nitric acid on another natural organic oil, glycerol. He manufactured nitroglycerine and, having had a number of near misses with his students, he put the work to one side. In the meantime, civil engineers were starting to tunnel through rock to build railways, and the gunpowder available was only marginally better than men with picks and shovels on the hard granites encountered. Thus, for the first time in almost a thousand years, non-military people were driving the development of explosive technology. It was found that nitroglycerine was a powerful explosive, capable of shattering the hardest rocks. Unfortunately, it was also prone to accidental ignition, with fatal consequences. Immanuel Nobel and his son Alfred started to manufacture nitroglycerine at their plant in Helenebourg. Their success as suppliers of blasting explosives to the civil engineering industry was blighted by two problems:

1. It was very prone to accidental initiation.
2. It could be very unreliable when used in boreholes.

The first problem came to a head in 1864, when an explosion destroyed the Nobels' manufacturing plant and the second problem indicated the solution to the first problem. The liquid nitroglycerine could be absorbed into porous rocks, thus reducing its effectiveness in bore holes. The Nobels recognized that other nitroglycerine-absorbing materials, such as certain types of diatomaceous earths (e.g. Keiselgur and also Fuller's earth) produced materials which still had an explosive performance but were less hazardous in operations. These were identified as Guhr dynamites. nitroglycerine, when absorbed on these materials, was then comparatively safe to handle, but now was difficult to set off or initiate. Simple match fuzes no longer worked, so the Nobels, in 1865, turned to the work of Kunckel with mercury fulminate and used this compound in copper tubes to initiate their Guhr dynamites. They had developed the world's first detonator device.

The Nobels also discovered that nitroglycerine could react with nitrocellulose NC to produce a jelly-like material, which had lost the sensitivity of nitroglycerine but was still sufficiently explosive to be useful in rock blasting and gave reliable performance. The term gelatine dynamites was coined for these materials. It was also in the 1860s that nitroglycerine was trialled as the first replacement for gunpowder. The development of engineering skills produced higher-precision gun barrels and also the opening and closing breech devices, enabling a change from muzzle-loading systems to breech-loading and, with the development of the rifled barrel to spin stabilize the projectile, a step up in the performance of artillery was produced.

Whilst Nobels were working with nitroglycerine, Wilbrand was experimenting with concentrated nitric sulfuric acid mixtures on a number of hydrocarbons, the most important of which was toluene. As a result, he discovered trinitrotoluene or TNT – a most important explosive material, mainly because of its ease of handling. When melted in hot water, Liquid TNT could be readily cast into shells and be allowed to set. The problem of it cracking and other filling defects took a while to recognize and treat. The other advantage of TNT was that it had a good safety record, because it was difficult to set off accidentally. However, there was a downside to TNT which took longer to recognize. It is very toxic, and has appreciable vapour pressure when heated at its melting point. As a result, a number of people died during the early part of the twentieth century due to TNT poisoning. The environmental impact of the manufacturing process was also quite a problem, due to the red water produced during the purification process. This was left in lagoons to age, but bacteria struggle to digest TNT at above 200 ppm in solution – far below the red water nitrated toluene levels.

Further developments in chemistry lead to the production of further explosive materials, all of which gave enhanced performance. Lead azide became the detonator of choice, replacing mercury fulminate, and PETN (see appendix for it systematic name) was a booster designed to give reliable initiation of the high explosives. During the two world wars, chemical engineering technology was developed and applied to enable mass production of explosives – mainly TNT. In 1914, world wide production of TNT was $\approx$ 100 tons per annum, but by 1917 this had increased to $\approx$ 750 000 tons per annum. The Great War also accelerated the search for new, more powerful explosives, such as RDX and HMX. During the Second World War, rockets powered by NC-based propellants came into use again, either as area weapons (exemplified by the German multi-launch Nebbleweffer rocket battery) or air-launched systems used against armoured vehicles. Also, the first infantry, hand-launched, rocket-powered anti-armour weapons appeared. Following the two world wars, attention turned back to civilian demands for safer explosives to be used in quarrying and civil engineering.

The quickest way to dismantle a building is by the use of explosives. Engineers had to store gelatine dynamites in special stores, and the search for alternatives which did not pose such a serious hazard was instigated. Ammonium nitrate (AN)-based systems started to develop, particularly for quarrying operations, where the exposed nature of the sites posed considerable explosive storage problems. AN-based systems had the advantage of being cheap, since they were bulk manufactured for agricultural uses as a fertilizer. Simple mixtures with fuel oil, ANFO, which had been discovered about one hundred years previously, were early successes, and later systems such as emulsions, slurries and gels developed during the latter half of the twentieth century. The slurries and gels were often supplied in the form of sausages in plastic skins, of various diameters, which could be readily stored and packed into boreholes in mines and quarries. Also, some of the systems could be manufactured on site on demand. As a result, quarries did not have to store explosive compounds.

This is not to say that AN is not explosive, but that it requires special conditions and initiation with a booster before it will explode. This fact was very dramatically demonstrated at Oppau in Germany. An attempt to beak up the cake formed on the surface of a pile of AN stored outside with an explosive charge resulted in the complete detonation of the whole pile, causing many fatalities and devastating damage for several miles.

The drive to improved performance and safety has led to a number of newer explosive materials which will be identified in a later section. Military users have now realized the potential benefits of designing munitions which can not function accidentally but only on specific demand and, as a result the use of laser initiation, has a critical role to play in insensitive munitions (IM). To this purpose, encasing the explosive in a plastic matrix has reduced the occurrences of accidental initiation. Previously, RDX had been desensitized by using cast formulations with TNT, but the limit was 25% TNT, because of the problem of wetting the RDX solid. Even this level required very careful manufacturing methods to exclude trapped air. The other problem was the reduced performance of the mixture due the extreme oxygen deficiency of TNT. Casting the RDX with a polymer binder allowed almost 90% RDX levels. By using energetic binders, the loss of performance from pure RDX and HMX could be minimized. More will be discussed on this area in the a later section. The *in situ* manufacture of the polymer was not without problems, both from the presence of unwanted side reactions but also from the toxicity of some of the cross-linking agents used, for example, di- and tri- isocyanates.

Due to increasing concern for the environment, a search is on for replacements for heavy metal salts such as lead azide and styphnate, and also less toxic high explosives. Nitrotriazalone NTO, which was first prepared and characterized in the early twentieth century, has started to be evaluated. While NTO is marginally lower in performance than RDX, it is much less sensitive than RDX and is easier to manufacture in high purity. It is very soluble in water and, very importantly, its toxicity is several orders of magnitude less than RDX or TNT. The toxic level for animals is $\approx$ 10 g per Kg body mass, and bacteria can readily digest NTO. Other materials are being evaluated and, again, will be discussed in later sections.

Bibliography

Brown, G.I. (1998). *The Big Bang: a History of Explosives*, Sutton Publishing Stroud, U.K., ISBN 0 7509 1878 0.

Bowen, D. (1977). *Encyclopaedia of War Machines*, Octopus Books, London, ISBN 0-7064-0648-6.

Martin, F.W. (1990). *The First Golden Age of Rocketry*, Smithsonian Institute Press, Washington, ISBN 0-87474-987-5.
Kelly, J. (2005). *Gunpowder*, Basic Books, NY, ISBN 0-465-03718-6.
McDonald, G.W. (2010). *Historical Papers on Modern Explosives*, L.A. Verne, TN USA.
Boddu, V. and Redmner, P. (eds) (2010). *Energetic Materials*, CRC press (Taylor and Francis), ISBN 978-1-4398-3513-5.
Pollitzer, P. and Murray, J.S. (eds) (2003). *Energetic Materials*, Part 1 Decomposition Crystal and Molecular properties; Part 2 Detonation and Combustion, Elsevier, Amsterdam, NL.
Klapotke, T.M. (ed.) (2007). *High Energy Density Materials*, Ser Ed D.M. P. Mingos, Springer, Berlin, ISBN 978-3-540-72201-4.
Marinkas, P.L. (ed.) (1996). *Organic Energetic Materials*, Naval Science Publishing Inc, NY, ISBN 1-56072-201-0.
Kaplotke, T.M. (2012). *Chemistry of High Energy Materials*, De Gruyter, Berlin, ISBN 978-3-11-027358-8.
Bailey, A. and Murray, S.G. (1989). *Explosives Propellants and Pyrotechnics*, 2nd edn, Brassey's, London, ISBN 0 08 036249 4.
Akhavan, J. (1988). *Explosives*, 2 edn, RSC, London.
Russell, M.S. (2009). *The Chemistry of Fireworks*, 2nd edn, Royal Society of Chemistry, Cambridge, 978-0-85404-127-5; Kosanke, K., Kosanke, B., Sturmer, B. *et al.* (2004) *Pyrotechnic Chemistry*, J Pyrotechnics Inc., + 1-970-245-0092.
Myler, R., Kohler, J. and Humbive, A. (2002). *Explosives*, 5th edn, Wiley VCH, Weinheim, ISBN 3-527-30267-0.
Shidlovsky, A.A. (1997). *Fundamentals of Pyrotechnics*, Trans from Russian, ISBN 0-929931-13-0.
Asay, B.W. (ed.) (2010). *Shock Wave Science and Technology Reference Library*, vol. **5** Non Shock Initiation of Explosives, Springer Verlag, Berlin, ISBN 978-3-540-87952-7.
Agrawal, J.P. and Hodgson, R.D. (2007). *Organic Chemistry of Explosives*, Wiley, Chichester, UK, ISBN 978-0-470-02967-1.
Olah, G.A. and Squire, D.R. (ed.) (1991). *Chemistry of Energetic Materials*, Academic Press, NY, ISBN 0-12-525440-7.

2

Review of Laser Initiation

2.1 Introduction

The first experiments with laser initiation of energetic materials as an alternative to standard electrical initiation were performed in the early 1960s. This was a direct consequence of the success of laser beams in cutting through steel and other metal sheets by rapidly melting them. Most of the lasers used were in the red to infrared region, since it was thought that direct heating was the optimum method to achieve the rapid combustion needed for typical propellant ignition. Looking at the scattering of laser radiation in Raman spectrometry, the closer the radiation wavelength is to the molecular vibration frequencies, the stronger the inelastic scattering is, and the more energy is transferred to molecular vibrations; the criteria for an increase in temperature leading to ignition.

Later work used lasers to initiate secondary high explosives directly, thus eliminating the need for highly sensitive primary explosives. The experimental arrangement for this study is shown in Figure 2.1. Note the advantage of not having a very sensitive primary explosive, so that the chances of accidental initiation are very much reduced.

Laser-initiated field guns started to be tested by the US army in the early 1990s, with the Nd/Glass laser operating at 1.05 μm being the one of choice because of its small dimensions, ruggedness, long life and low cost. The object of these preliminary trials was to initiate the gunpowder booster directly without using the standard, percussion-sensitive military M82 primer. This wavelength was also very adaptable, since it could be transmitted over considerable distances by high purity silica optical fibres, with very little loss due to radiation absorption. Simple use of splitter systems in the optical fibres enabled a number of initiation points to be simultaneously initiated, giving an optimized burn within a single propellant charge. Fibre optic transmission of the laser beam also enabled multiple charges to be initiated simultaneously from a single firing command. This latter has a distinct attraction for quarrying operations if delay detonators can be initiated.

A number of prototype guns using laser initiation have been demonstrated in the early 21st century, but regular operation has not been implemented. The Crusader SPH XM2001

Laser Ignition of Energetic Materials, First Edition. S Rafi Ahmad and Michael Cartwright.

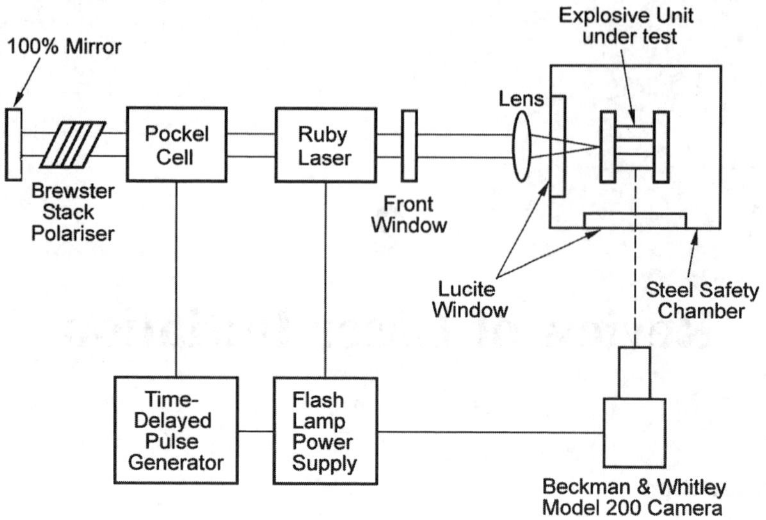

Figure 2.1 *First experiments with laser initiation of secondary explosives. Reprinted with permission from [84]. Copyright 1971, AIP Publishing LLC.*

was a self propelled 155 mm calibre howitzer which used a laser initiation system into a modular charge but, since it was deemed too heavy for field deployment, efforts have been directed to producing a demonstrator laser initiated system that operates on a modified British-designed lightweight 155 mm gun system.

One system in which laser initiation has been developed is in the slapper detonator, shown schematically in Figure 7.23. The high-intensity laser beam impinges on a metal foil, causing it to flash evaporate, rupturing the polymer film and forcing a fragment of the film into a barrel and accelerating the fragment to such a velocity that the kinetic energy delivered to a secondary explosive, typically fine grain HNS, at the other end of the barrel is sufficient to shock-initiate a detonation in the HNS. These systems are common in missile warheads, where reduced probability of accidental initiation is vital, and the cost of this one component is readily absorbed into the overall cost of an expensive weapon. However, it is not a viable proposition for routine deployment of explosive devices. The power required to achieve such a rapid temperature rise in the copper foil is very large compared to the standard thermal ignition system.

An alternative system of laser ignition which is receiving attention is the photochemical approach. High-energy photons in the 200 to 600 nm region have energies which correspond to the energy required to excite electrons in chemical bonds within molecules to either the conduction band or complete ionization levels. These processes can lead to dissociation of the molecule into reactive species by breaking chemical bonds. Because this process is very rapid and releases considerable energy, it should result in an ignition and require overall less energy than the thermally induced ignition. The optimum is when the photon energy and the band gap are identical, when each absorbed photon will excite one molecule, but it is possible for lower-energy photons to bring about the same effect by multi-phonon absorption. The greater the number of photons that have to be absorbed for the reaction to occur, the lower the probability of the event, due to increased probability of the partially excited

Table 2.1 Critical hotspot parameters.

	Critical temperature K (°C)		
Explosive	Hot Spot Radii 10^{-2}	Hot Spot Radii 10^{-3}	Hot Spot Radii 10^{-4}
PETN	623 (350)	713(440)	833(560)
RDX	(385)	(485)	893(620)
Tetryl	698(425)	(570)	1086(813)

state dissipating some of the absorbed energy. With systems which employ photochemical activation, the probability of accidental initiation is reduced.

2.2 Initiation Processes

There has been much recent work utilizing monochromatic light from laser sources to ignite or initiate reactive materials. Ignition processes have been investigated extensively, and most authors have attributed the instigation of reaction as being due to a thermal inhomogeneity called a hot spot. This term was coined, by Bowden and Yoffe [1], and much of the early investigation and analysis of this phenomenon was conducted by them. They suggested a minimum dimension of circa 0.1 μm for a hot spot from which ignition might occur. Additionally, they specified that the temperature should be $\approx$ 700 K or greater and that the source should be maintained for a minimum of $\approx$ 10 μs. They proposed a linking relationship between some of the parameters for hot spots in terms of size, temperature and duration. They produced the Table 2.1, showing the relationship between hot spot dimension and minimum temperatures for ignition effects.

In any event, many such points may exist, but only a few meet the conditions proposed by Bowden and Yoffe to form *critical* hot spots from which an ignition event may originate. These authors also suggested that a bright flash might be sufficient to ignite an explosive. Later work, discussed in the section on novel primers, has produced energetic materials which can be initiated by the flash output from a digital camera. However, generally it was only with the availability of lasers that radiation of sufficient power might be absorbed at a single wavelength to ignite or even initiate explosive materials.

Two approaches have been used to initiate explosives and are indicated in Figure 2.2. In the first, the beam is incident directly upon the target, initiating a self-sustaining burning process which, if conditions are correct, may transition through high speed burning, to deflagration, to detonation.

The development of lasers as laboratory tools occurred comparatively recently, and the early work on initiation was sporadic, investigating only a small range of mechanisms that might operate [2, 3]. It was known that high-intensity flashes from discharge lamps could lead to explosion in sensitive primary materials [4, 5] but not in secondary explosives. The research indicated that laser ignition was possible in many explosives of varying sensitivity, and that direct detonation was seen at high power density; it concluded that no DDT occurred. Later research suggests that, in order to explain the observation that there was a time lag between the start of the laser pulse start and the onset of light emission from

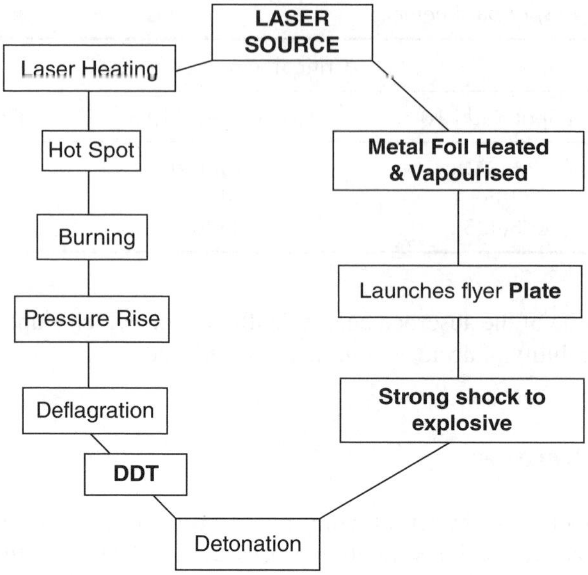

Figure 2.2 *Schematic of alternative methods of laser initiation of energetic materials.*

reaction, a mechanism akin to SDT or DDT may, indeed, be occurring [6]. This time lag was reduced as radiation density was increased.

Several mechanisms have been proposed to explain these effects. Electrical breakdown was one proposal, along with several routes by which a shock wave might be created. One of these was a light-induced 'pressure' which impacted the explosive. In the second, thermal energy was converted into a shock wave. The mechanism operating during burning or deflagration to detonation transition (DDT) was first investigated at the end of the Second World War [7–9] and will be further discussed in later chapters. In the second method, the laser light is fed down an optical fibre, on the end of which has been sputtered a metal film. When the high-power laser is pulsed, the metal in contact with the fibre is vaporized, and the confined plasma drives the remainder of the coating as a flyer plate travelling at high velocity towards the target, impacting on the explosive and delivering a shock which initiates the explosive. The mechanisms of this direct shock initiation to detonation (SDT) [10, 11] will also be discussed in later sections.

The advantages of using a light pulse are many. One of the most appealing practical benefits is that the optical signal, transmitted down an enclosing fibre, is immune to electromagnetic interference, ambient temperature or pressure. The present review aims to summarize the relevant research that may be found in the open literature. The promising nature and performance of the present devices suggests that there may be many applications for such a robust and safe means of triggering an explosive.

There are a number of other aspects of the explosive reaction, in addition to the initiation process, in which laser interaction with solids have been used in explosive science. Raman spectroscopy can use the impinging laser light to examine the chemical composition of the first stages of the thermal decomposition induced by the laser heating and in the deflagration process. Other spectroscopies, at different wavelengths [12–20], are also possible, again providing an insight into the mechanisms operating during the initiation process. There is a range of other experimental investigations where a laser is used as a diagnostic tool in the study of explosives, rather than as a stimulus to initiation. These include studies of deflagration [21] and laser speckle interferometry to determine strain displacements [22,23].

2.3 Initiation by Direct Laser Irradiation

There are many different classes of investigation that have been conducted in this area. Apart from the emphasis on differing behaviour under varying impulses, the materials and their chemical structure have been varied. Some of the work focuses on propellants and pyrotechnics [24–29], while some of the other work looks at the response of primary explosives [30–36] and secondary materials [37–43]. These references examine different aspects of the process, and the variables studied have been identified and discussed. Individual observables will be selected and the relevant findings summarized for variations in some particular parameter.

2.3.1 Laser Power

One of the most important areas for study is the effect of laser power. The cost of laser systems increases with increase in delivered power and also as the size of the electrical power systems required to operate the laser increases. This latter can be offset to some extent against the use of optical fibres for multiple simultaneous initiations of distributed devices. If laser initiation systems are to become more widespread, then the cost must be comparable with existing initiation systems or offer some dramatic improvement in safety and reliability.

Several studies have noted that as the energy or power of the laser pulse is increased, the threshold for ignition is reduced [35,44–46]. Most authors believe that it is power, the rate of delivery of the energy, and not the energy that controls initiation [34], and this supports the minimum hot spot duration proposed by Bowden and Yoffe. It has also been noted that there is a delay time after irradiation before the composition shows evidence of deflagration or detonation [3,30,33,34,47–49].

Both primary and secondary explosives show delayed reaction, and several examples are documented. Even when detonation is the final result, this delay time always appears to be present. As power density increases, the delay time is reduced. Using single crystals of primary explosive lead azide, size $40 \times 200 \times 10$ μm, irradiated with a ruby laser at 694.3 nm, delay times to ignition, as a function of pulse energy, were measured [34] for both a Q switched and non-Q switched arrangement. This enabled precise control of the laser pulse to be established. The delay time observed reduced as both the pulse length increased and the incident beam energy increased. Full detonation was observed to occur in most cases after irradiation, but there always appeared to be some delay observed, and the delay time could only be decreased to a limiting value for the material [50]. This effect is seen for the propagation of deflagrations and for full detonation. In both cases, this may reflect the mechanism by which irradiation leads to the start and build up of reaction. Additionally, it depends upon the pulse length [32,46,51].

There are several features of the processes occurring that have been noted. First, it is power – the rate of energy delivery – and not energy levels that control the initiation. Second, electron microscopy of the crystals indicates that the sites that are initiated appear to correlate with flaws or defects in the crystal. Since a perfect crystal would be transparent to the radiation, this suggests a thermal mechanism for the initiation, implying that energy is absorbed in the material. The absorption must then occur at defects, which might be cracks, clusters or dislocations within the lattice. As well as providing electron transition energies corresponding to the laser energies, these inhomogeneities could also lead to focusing of the laser beam and high energy densities at particular local sites. These may lead to a number of initiation mechanisms [35].

2.3.2 Laser Pulse Duration

Lasers may be operated to give out continuous radiation, but they may also be pulsed for an arbitrary duration, using an incorporated Pockel cell in the laser cavity. Several investigations of the effect of pulse width upon initiation threshold at various wavelengths have been carried out [29, 30, 33, 35, 43, 45, 52, 53]. Despite the range of variables in the pulse characteristics and materials irradiated, the energy required for initiation decreases as pulse length is increased. This implies that it is a thermal mechanism [35] which is ultimately responsible for the reaction.

It is interesting to speculate at what time the reaction starts. When high powers are used, the pulse length is found to exceed the time taken until explosion occurs [54] so that, again, a critical situation must be reached. Where this delay time was measured as a function of pulse length in lead azide [35], it was found that pulses of 80 ns to 800 μs resulted in time delays to reaction of 0.1–1 ms. For short pulses, a critical constant energy threshold was found, while for long pulses, there was one of constant critical power. In some work [30], the effect of multiple pulses was compared with that of a single one. It was found that the threshold for ignition by multiple pulses was around twice that for a single pulse. As a result, the integrated energy content of the pulse must be matched to the kinetics of the relevant reactions in explosive decomposition, as well as to critical bond energies, to ensure that explosion results. A further factor involving the energy dissipation via molecular vibrations is also very important, since this may prevent the critical temperature being achieved.

2.3.3 Absorbing Centres

It might be anticipated from the various analytical spectroscopies that the primary mechanism for material energy absorption of radiation is direct coupling to suitable bonds within the structure. This mechanism may occur by two processes – either by direct ionization of electrons from the bond, typically with UV photons, or by exiting thermal vibrations of a particular bond, usually in the IR spectral region. These mechanisms occurring on a bond length-scale must have precise, tuned, laser wavelength. This contrasts with the situation in which the absorbing centres are in the nano to micron scale, which may be occurring at defects, flaws and impurities within the microstructure to create hot regions within the material. In this case, it might not be necessary to match the laser wavelength directly to a particular molecular bond energy process. It has to be borne in mind that molecules around the edges of the defect may well have different bond lengths and energies compared to the normal ideal lattice situation.

Several works have investigated the addition of such centres into various energetic microstructures [3, 4, 30, 34, 55–58]. The earlier work included irradiation of the primary explosive lead azide with neutrons and fission fragments [4]. Since ignition was not observed, it was concluded that the typical 10 nm flaws induced would not initiate. This observation was cited as evidence that hot spot size must be a minimum dimension of microns. Such a size is consistent with the physical dimensions of inhomogeneities in a crystalline lattice with defects such as flaws, cracks or dislocations, and these have been hypothesized as reaction sites [34, 35]. The mechanism for local explosion at these sites leading to bulk deflagration or detonation is again believed to be one of SDT or DDT.

2.3.4 Pressed Density

In single crystal, inhomogeneities in the bulk have been attributed to crystal imperfections, as described above. However, most operational devices employing reactive materials consist of pressed powdered materials or filled containers, into which molten material has been cast. In the former case, it is only occasionally that a density close to the theoretical maximum density – normally the single crystal density – is achieved. When full density has not been reached, voids exist between grains, and grain boundaries divide them. The presence of heating centres and voids around the site of a potential hot spot reduces the temperature and pressure that may exist, thereby increasing the initiation threshold level. The net result of this effect is that the performance of explosive samples is influenced by the mechanical preparation conditions to which the device is subjected. As with other properties, such as the initiation pressure sensitivity, increased pressed density (with consequent decreased porosity and pore volume) leads to reduced sensitivity to laser-induced initiation. There have been several systematic investigations of such effects [2, 3, 30, 32, 43, 50, 59–61].

Other physical features of the pressing process can have additional influence on the absorption of radiation. The surface of a loosely pressed material, for example, generally has an optically dull appearance, due to the poor packing of grains and their random orientations. This leads to increased scattering from the surface, thus reducing the beam intensity for the absorption process. However, increased pressing pressure results in a density approaching that of the single crystal value, and the surface beneath the drift becomes more reflective, particularly if the die surfaces are highly polished – a phenomenon utilized in IR spectroscopy with KBr pellets. Thus, the surface scattering is decreased and the initiation threshold lowered, which counteracts the increase in threshold brought about by the elimination of crystal flaws. In this respect, the physical roughness of the surface and the wavelength of the radiation will play their role in determining the relative importance of such factors, with shorter wavelengths being more susceptible to such effects.

In one of the more complete studies [50], pure samples of PETN, HNS, HMX and also with added graphite (5–10%) were subjected to different variations in a number of experimental parameters, including the pressed density of the targets. The variation in required power density for prompt detonation of the explosives was a function of pressed density when irradiated at 308 nm with an excimer laser. HNS could not be detonated at any of the powers available up to the laser's maximum power of 5 GW cm^2. The optical reflection was determined to be similar at this wavelength for each material. The authors concluded that the laser, at these energies, initiates by energy deposition, leading to bond breakage at these wavelengths, but the specific bond ruptured was not identified.

2.3.5 Strength of Confining Container

The confinement of explosives generally has a number of effects on the reaction process. The reaction is generally easier to initiate and propagate. For a detonating system, the shock reflection at the wall reduces the critical diameter and strengthens the running shock fronts up to the maximum VOD. The stronger the wall, the higher the percentage of the shockwave energy that is reflected from the wall. For a burning system, once reaction starts, any hot gaseous reaction products cannot escape and are confined close to the burning surface, reducing the heat loss and thus increasing the temperature and also the pressure. Both effects increase the rate of reaction, according to Vielle's Law (see Chapter 6), increasing the temperature and pressure build-up even further.

Confinement can take one of three forms to generate such effects. In the first, the explosive may simply be pressed into a metallic pipe with thick walls, and this has been found to sensitize PETN, RDX and HNS when confined within aluminium pipes [43]. Thinner walled steel pipes also increase the sensitivity of the explosives.

The second form of confinement may be regarded as self- or internal confinement provided by a thick cylinder of energetic material around the reaction site. Several workers have noted that the confinement provided by the charge dimensions affects the measured threshold but they have found this effect to be more prominent for longer wavelengths rather than short wavelengths [41, 43].

A third confinement may be to place the explosive beneath a cover plate which is transparent to the laser radiation that is incident upon the target [43, 50, 61]. When reactions occur near to the surface, gaseous products are confined but, when they are occurring at depth, the cover has a minimal effect [62]. On the other hand, gaseous vapours of the material irradiated can no longer escape from around the surface, which may not favour ignition [63]. An equivalent effect to the latter is to apply a high, inert gas pressure to the periphery of the explosives target. Since the ignition of the explosive itself causes local sites to produce gaseous products, the net effect of high external pressure is, generally, to lower ignition thresholds. If external pressurization is by gases that may be involved in the reaction, then the scheme is complicated even further by the pressurizing medium.

An explanation of these various pressure considerations may be found in consideration of the Beer-Lambert law, which shows that the depth of penetration is proportional to the density of the absorbing species. A study [64] has shown the effects of irradiating RDX at 10 and 30 atmospheres pressure surrounding the explosive. The incident wavelength was varied from 9 to 11 μm, and the ignition energy measured in each case. With high confining pressure, the energy rises with the absorption depth. However, with lower confinement, there is a region below 20 μm where the threshold increases sharply – whereas, above this value, the increase seems independent of confining pressure. This behaviour is explained by noting that the reaction zone is squeezed down into closer contact with the surface by the high pressures, leading to a lowering of the ignition thresholds. Even though the wavelength is in the infrared, there is still a surface effect in this case. The implication of the above is that the radiation wavelength must be selected to ensure that the absorption depth is sufficient to render the gas products and external pressure irrelevant. It should be remembered that these effects are more important for long wavelengths with short penetration depths than for short wavelengths.

2.3.6 Material Ageing

Materials that have been stored for lengths of time, or thermally cycled, have been known to have differing properties from newly-produced ones, as a result of chemical or physical changes that might have occurred during storage [65]. The problem is most frequently encountered with nitroester-based propellant formulations, which have a specified shelf life, but has also been noted in some primaries [30] and secondaries [66]. Aged composite samples have been examined microstructurally and show a fibrous structure which mutates over time. Individual crystallites grow, but the primary effect is an increase in the porosity noted in the sample with time [28]. The net result of this change upon performance is determined to be an increased energy flux necessary to initiate aged material. Similarly, it has been noted that the primary, lead azide, becomes less sensitive as it is aged [30]. This may be due to a combination of hydrolysis and formation of basic azide and lead carbonate if the material is exposed to atmospheric CO_2. Both of these chemical reactions reduce the sensitivity and the explosive performance of lead azide formulations. A number of the more recently trialled energetic materials suffer from similar problems and these will be discussed in later chapters.

2.3.7 Laser-Induced Electrical Response

Aduev *et al.* [48] investigated the chemical and physical effects of laser-induced initiation and noted that electrical conductivity of the solid was also changing in the laser-irradiated lead azide. They applied a 1.06 μm radiation in a 30 ps pulse to a target fitted with current sensors, photodetectors and acoustic transducers. The laser pulse incident on the crystal was recorded and, 70 ns later, a current started to flow, indicating that the material became conducting. A further 50–70 ns after this, a detonation was recorded running through the lead azide primary explosive. Thus, the irradiation of the lead azide appeared to have altered the band structure of the solid significantly, with a multiplication of electronic excitations that caused it to conduct electric current. Similar reorganizations of the electron energy levels appear to have additionally caused the onset of explosion in this primary. This initiation has recently been attributed to the migration of irradiation-induced excitons in the azide ion, leading to reaction between adjacent azide ions and the metal cation [67]. The mechanism involves the overlap of non-bonded orbitals on adjacent azide ions, and also suitable orbitals on the metal cation, facilitating the exciton electron transfer process. The order in which this sequence of observable phenomena occurs is still under investigation, since the change in electrical conductivity precedes detonation in this case, and both processes are delayed relative to the incident near the infrared pulse.

2.4 Laser-Driven Flyer Plate Initiations

The alternative method of initiating an explosive target is to use the energy created by vaporizing a metallic coating to drive a thin, flying plate to impact upon it [68–75]. This is the principal behind the 'slapper detonator'. The wave induced in the target by flyer plate impact will cause an inert shock to transition into a reactive detonation front [11]. The physical and chemical phenomena associated with the impact of flyer plates upon targets and the shock wave introduced have been studied for a considerable period of time [76]. The variables that control their loading include the physical properties of the impacting plate, its velocity and the planarity of the impact [77]. In the case of most investigations of loading by plate impact, the flyer is millimetres thick, which induces a pulse of length several microseconds in duration. This pulse length is determined by the transit time of a shock wave through the target and the following release back into the target from the rear surface.

Additionally, the plates are typically 50 mm or greater in diameter, which means that lateral release waves take some time to reach the centre of the loaded area, again implying several microseconds of loading time. However, the laser-driven plates, typically 5 μm thick, are of the order of 1 mm in diameter [74]. Thus, typical loading pulses are only of circa 5 ns duration, which is very short in comparison with flight duration of the flying plates. As a result of the firing pulse, the flying plates have hot vaporized material behind them, which is continually heating them as they travel, leading to an essentially molten plate if they are allowed to travel a sufficient length of time. The flyers thus only travel a few hundred microns before they break up, unless some means of delaying thermal conduction and ablating the discs is found.

Typically, the metal sheets are created by depositing a thin film of metal onto a window or optical fibre end surface. The laser pulse travelling through the transparent medium vaporizes the rear of this metal film and forms plasma, confined by the remaining part of the sheet. The glass or window material is also involved in the process, since it controls heat transfer into the system and also provides a proportion of the plasma. The standard window material adopted for this application has been fused silica but others have been tried. Polycrystalline quartz has been found to be inferior, while polyethylene has shown promise [78] at some of the laser wavelengths. The requirements for the flyer plate are high melting point and low molecular weight. Titanium and aluminium have been selected from the limited number of candidates, as both seem to have similar effectiveness, as judged by their success in initiating a range of compositions [71]. The velocities of the flyers produced, ≈ 3 km s^{-1} (mm μs^{-1}) are higher than those achieved by normal propellant-powered guns, but are slightly inferior to those obtained from electric guns ≈ 5 km s^{-1} [64]. Typical lasers used for these systems produce power densities of 1–20 GW cm^2 and these launch flyers with $\approx 35\%$ efficiency for energy conversion; the remaining energy is dissipated by other mechanisms [69].

These high-power densities and energy losses inevitably result in high temperatures in the plasma behind the flyer and, as a result of the rapid thermal conduction through the system, the flyer is rapidly vaporizing as it travels. This process of ablation has been found to account for ≈ 0.5 μm of the aluminium film [79], suggesting a minimum value for the original metal coating. One means of moderating such effects successfully is to create a composite flyer that has layers of thermal insulators incorporated within its structure. Several composite flyers have been tried, and two of these promising laminates are magnesium/magnesium fluoride/copper, $Mg/MgF_2/Cu$ and aluminium/alumina/aluminium ($A1/A1_2 0_3/Al$) [71, 72]. The plasma persists longer, the broadening decays more slowly for composites and the layered aluminium/alumina/aluminium sequence has been found to give the best performance, due to the alumina acting as the best thermal confinement [71, 78]. As a direct result of the insulating layer of alumina, a strongly absorbing plasma is also produced [77].

Coupling the laser energy into an optical fibre in such a way as to allow high optical power to be transmitted provides several challenges in coupling of the laser output and damage to the optical fibre. The high energy densities involved means that fibres carrying such power levels inevitably show damage. It has been found that larger core diameter fibres have higher damage thresholds than smaller ones. Thus, the damage increases as the fibre diameter decreases. This fibre damage problem is under investigation in the general use of fibres. The plate is accelerated rapidly, achieving 90% of its velocity in two pulse widths [72]. In terms of a device using optical signals to detonate a material, this process offers an optimum means of initiating reaction, while being unaffected by electrical noise [80]. However, the laser also offers a means of probing the mechanisms by which high explosives react, and it also provides further methods for investigating the physical states that exist during the initiation process.

More recent research has continued to elucidate the mechanisms involved under a range of sample and source conditions, as well as the effects of additives distributed in the critical regions in the sample. One other aspect of the research has been to look for materials that have low thresholds for laser initiation at appropriate wavelengths, but which are less sensitive to conventional accidental initiation. In this respect, some transition and heavy metal complexes are particularly attractive, since some of them respond to laser diode output. Typical examples are cobalt complexes with nitrogen-rich ligands such as amines and polynitrosubstituted hetrocyclic ligands [81–83]. The usually involve perchlorates and nitrates as counter ions, but others are now being considered. More detailed discussion of these will be given in the section on novel primary explosives.

2.5 Summary and Research Rationale

The distances separating inhomogeneities in an explosive target are significantly far apart that the variations in laser intensity, which may arise from phenomena such as intensity variations due to interference patterns generated across the beam, are insignificant. Thus, lasers ignite materials and there is a particular threshold for this to occur, which is reduced as energy is increased. There is, however, a time delay between irradiation and light emission, due to reaction. This time reduces with increased power, but never drops to zero.

The wavelength of the radiation controls the depth to which the radiation is absorbed in the material. This follows the Beer-Lambert law relating transmission to concentration of absorbing species. The short wavelengths have greater absorption depths, and the reaction is less dependent upon surface conditions than with longer wavelengths. Clearly, ambient conditions can also influence the reaction scheme. High pressure leads to a reaction zone squeezed down on to the surface, while thermal heating also increases the reaction rate. As pulse length is increased, the energy threshold comes down, which implies that there is a thermal ignition mechanism operating. The pulse leads to higher pressure within the beam, and this leads to chemical reactions releasing energy; longer pulses sustain this process.

If shock waves are generated, impulsive reactions are accelerated and multiple hot spots may be activated. This corresponds more closely with the accepted form of an SDT mechanism and, hence, reaction may not rely solely upon the precise region around which the beam interacts. The creation of hot spots within a large bulk of the irradiated medium may clearly be accelerated by the addition of suitable additives. Their effect, if their absorption properties are suitably chosen relative to the incident laser wavelength, is to lower ignition or initiation thresholds in the target. The material is incorporated at a suitable level in the region of the target which the laser radiation is capable of penetrating.

The vaporization of metallic films, coated onto a window or the end of a polished optical fibre, creates a surface plasma that drives the remainder of the coating at high velocity away from the fibre/window. The impact of this flyer may shock initiate the explosive target at which it is aimed. The run distances are short since the flyer is constantly being vaporized by the hot plasma following behind. However, laminar coatings, including materials of poor thermal conductivity such as alumina, create more stable plates controlling the portions vaporized and those still remaining as solids. Currently, according to recent trials, aluminium with alumina interlayer offer the most successful coatings in terms of performance and ease of production.

The conversion of the laser energy, at its particular wavelength and in a pulse of prescribed length, and its coupling into the explosive target that is to be initiated, depends on the utilization of the appropriate quantum states of the material irradiated. The chemical bond structure of the material is vital as is the thermal and mechanical preparation are the on which the target it is based. The presence of lattice defects and impurities/additives can alter the electron energy states and their distribution in the solid thus modifying the initiation thresholds for the materials. The various mechanisms operating require further investigation before viable devices can be made to challenge conventional systems.

2.5.1 Rationale for Research

The desire to pursue laser initiation is largely based on the requirements to reduce the number of accidental ignitions in weapon systems and also in civil engineering such as demolition and quarrying. From a munitions viewpoint, the methods of achieving this reduced vulnerability falls into a number of categories, depending on the design criteria of the device. Both propellants and high explosives generally require the high sensitivity of a primary explosive to the design firing stimulus. However, the probability of initiation by an accidental stimulus can also be very high. If the primary explosive can be dispensed with, then the safety margin is dramatically improved. The ability of lasers to deliver high energy densities can eliminate the need for hazardous primary explosives.

If the laser is a cheap diode, then attaching the diode directly to the explosive means that the firing signal for the diode has to be transmitted either by wire or tuned radio frequency transmission, both of which can be disorganized. Lasers also offer two mechanisms for remote firing of explosives charges. The most important is via a fibre optic cable, which has a number of advantages over simple metal wire electrical conductors. Electromagnetic fields can induce electrical currents in metal conductors, and these may be sufficient to fire the electric initiation sequence of a device such as the bridge wire, which is electrical heating of a primary explosive.

It is virtually impossible to interfere with the laser system using a fibre optic cable by stray optical signals. Light can not enter the fibre optic from outside for two reasons. The outside of the fibre is coated with an opaque coating and, even if this coating is removed and the silica fibre is exposed, it is not possible for external light to be transmitted by the fibre. The fibre transmits by total internal reflection from the walls at very small angles to reduce the dispersion of the beam. External light radiation would have to also come in at very low angles with respect to the fibre surface, and would thus be totally reflected back externally out of the fibre. By having a system whereby an optical signal other than the firing signal is reflected back down the fibre, there is a ready circuit continuity check. More work on using low-cost laser diodes and optical fibres is required.

A very unusual remote laser initiation is by direct impingement of the laser beam on an uncovered explosive charge, or by burning a way in through the covering on the explosive. This method is only used as a last resort, since the initiation sequence is very unpredictable and is also impossible under circumstances when there is a significant scattering of the laser light by suspended particles in the laser beam. The laser beam has to be focused onto the target, and a miss even by millimetre distances would render the device useless. It looks good in the film industry, but is impractical in real life. There is also a safety issue, since the laser power could induce vision problems if it accidentally hits the eye.

Bibliography

Brish, A.A. (1969) On the Mechanism of initiation of condensed explosives by radiation of an optical Quantum Generator. *Fizika Goreniya i Vzryva*, **5**, 475–480.

Yong, L., Nguyen, T., Waschl, J.A. (1995) Laser ignition of explosives, pyrotechnics and propellants: A review. Report DSTO-TR-0068, DSTO Aeronautical and Maritime Research Lab, Melbourne.

Ilyushin, M.A. and Tselinskiy, I.V. (2000) Use of Laser initiation of power-intensive compounds in science and technology: A Review. *Zhurnal Prikladnoi Khimii*, **73**, 1233–1240.

Bourne, N.K. (2001) On the laser ignition and initiation of explosives. *Proceedings of the Royal Society of London*, **A 457b**, 1401–1426.

Bowden, M.D., Cheeseman, M., Knowles, S.L. and Drake, R.C. (2007) Laser initiation of energetic materials: a historical review. *Proceedings of SPIE*, **6662**, 666208.

Kennedy, J.E. (2010) Spark and Laser Ignition, in: Asay, B.W. (ed.), *Shock Wave Science and Technology Reference Library*, vol. **5**, Non Shock Initiation of Explosives. Springer Verlag, Berlin, pp. 582–605.

Golubev, V.K. (2012) Optical initiation of energetic materials: Recent scientific investigations and technical applications. NTREM Symposium Pardubice, Czech Rep.

References

(1) Bowden, F.P. and Yoffe, A.D. (1949) Hot spots and the initiation of explosion, in *Proc. Third Symp. on Combustion and Flame and Explosion Phenomena*, Williams & Wilkins, Baltimore, MD, pp. 551–560.

(2) Brish, A.A., Galeev, I.A., Zaitsev, B.N. *et al.* (1966) Laser- excited detonation of condensed explosives. *Combustion, Explosion, and Shock Waves*, **2**, 81–82.

(3) Brish, A.A., Galeev, I.A., Zaitsev, B.N. *et al.* (1969) Mechanism of initiation of condensed explosives by laser radiation. *Combustion, Explosion, and Shock Waves*, **5**, 326–328.

(4) Bowden, F.P. and Yoffe, A.D. (1952) *Initiation and Growth of Explosion in Liquids and Solids* (republ. 1985), Cambridge University Press.

(5) Bowden, F.P. and Yoffe, A.D. (1958) *Fast Reactions in Solids*, Butterworth, London.

(6) Paisley, D.L. (1989) Prompt detonation of secondary explosives by laser, in *Proc. Ninth Symposium (Int.) on Detonation*, Office of the Chief of Naval Research, Arlington, VA, pp. 1110–1117.

(7) Ubbelohde, A.R. (1949) Transition from deflagration to detonation: The physico-chemical aspects of stable detonation, in *Proc. Third Symp. on Combustion and Flame and Explosion Phenomena*, Williams & Wilkins, Baltimore, MD, pp. 566–571.

(8) Roth, J. (1951) Experiments on the transition from deflagration to detonation, in *Proc. First ONR Symp. on Detonation*, Office of Naval Research, Washington DC, pp. 57–70 (This article has been republished in 1987).
(9) Macek, A. (1959) Transition from deflagration to detonation in cast explosives. *Journal of Chemical Physics*, **31**, 162–167.
(10) Kendrew, E.L. and Whitbread, E.G. (1960) The transition from shock wave to detonation in 60/40 RDX/TNT, in *Proc. Third ONR Symposium on Detonation*, Office of Naval Research, Washington DC, pp. 574–583.
(11) Jacobs, S.J., Liddiard, T.P. Jr. and Drimmer, B.E. (1963) The shock-to-detonation transition in solid explosives, in *Proc. Ninth Symp. (Int.) on Combustion*, Academic Press, New York, NY, pp. 517–529.
(12) Avouris, P., Bethune, D.S., Lankard, J.R. *et al.* (1981) Time-resolved infrared spectral photography: Study of laser-initiated explosions in RN3. *Journal of Chemical Physics*, **74**, 2304–2312.
(13) Edwards, P., Weaver, D.P. and Campbell, D.H. (1987) Laser-induced fluorescence in high pressure solid propellant flames. *Applied Optics*, **26**, 3496–3509.
(14) Eloy, J.-F. and Delpuech, A. (1988) Experimental study of photon-phonon interactions in an explosive by laser probe mass spectrometry (LPMS-25), in Schmidt, S.C. and Holmes, N.C. (eds) *Shock Waves in Condensed Matter – 1987*, Elsevier, Amsterdam, NL, pp. 557–560.
(15) Leeuw, M.W., Rooijers, A.J.T. and Steen, A.C.v.d. (1989) Fast spectrographic analysis of laser initiated decomposition reactions in explosives, in *Proc. Ninth Symposium (Int.) on Detonation*, Office of the Chief of Naval Research, Arlington, VA, pp. 710–713.
(16) Nilsson, H. and Ostmark, H. (1989) Laser ignition of explosives: Raman spectroscopy of the ignition zone, in *Proc. Ninth Symposium (Int.) on Detonation*, Office of the Chief of Naval Research, Arlington, VA, pp. 1151–1161.
(17) Ostmark, H. and Nilsson, H. (1989) Laser ignition of explosives: A mass spectroscopic study of the pre-ignition reaction zone, in *Proc. Ninth Symposium (Int.) on Detonation*, Office of the Chief of Naval Research, Arlington, VA, pp. 162–171.
(18) Trott, W.M. and Renlund, A.M. (1989) Pulsed-laser-excited Raman spectra of shock- compressed TATB, in *Proc. Ninth Symposium (mt.) on Detonation*, Office of the Chief of Naval Research, Arlington, VA, pp. 153–161.
(19) Ostmark, H., Carison, M. and Ekvall, K. (1996) Concentration and temperature measurements in a laser-induced high explosive ignition zone. Part 1: LW spectroscopy measurements. *Combustion and Flame*, **105**, 381–390.
(20) Ostmark, H., Ekvall, K., Carlson, M., Bergman, H. and Pettersson, A. (1995) Laser ignition of explosives: A LW study of the RDX ignition zone, in Short, J.M. and Tasker, D.G. (eds) *Proc. 10th Int. Detonation Symposium*, Office of Naval Research, Arlington, VA, pp. 555–562.
(21) Andrews, J.R. and Netzer, D.W. (1976) Laser Schlieren for study of solid-propellant deflagration. *AIAA Journal*, **14**, 410–412.
(22) Asay, B.W., Laabs, G.W., Peterson, P.D. and Funk, D.J. (1996) Measurement of strain and temperature fields during dynamic shear of explosives, in Schmidt, S.C. and Tao, W.C. (eds) *Shock Compression of Condensed Matter 1995*, American Institute of Physics, Woodbury, NY, pp. 925–928.
(23) Asay, B.W., Laabs, G.W., Henson, B.F. and Funk, D.J. (1997) Speckle photography during dynamic impact of an energetic material using laser-induced fluorescence. *Journal of Applied Physics*, **82**, 1093–1099.
(24) Saito, T., Shimoda, M., Yamaya, T. and Iwama, A. (1991) Ignition of AP-based composite solid propellants containing nitramines exposed to CO2 laser radiation at subatmospheric pressures. *Combustion and Flame*, **85**, 68–76.
(25) Ostmark, H. and Roman, N. (1993) Laser ignition of pyrotechnic mixtures: Ignition mechanisms. *Journal of Applied Physics*, **73**, 1993–2003.
(26) Haas, Y., ben Eliahu, Y. and Weiner, S. (1994) Infrared laser-induced decomposition of GAP. *Combustion and Flame*, **96**, 212–220.
(27) Haas, Y. and Ben-Eliahu, Y. (1996) Pulsed laser induced decomposition of energetic polymers: Comparison of ultraviolet (355 nm) and infrared (9.3 jim) initiation. *Propellants, Explosives, Pyrotechnics*, **21**, 258–265.

(28) Ulas, A. and Kuo, K.K. (1997) Effect of aging in ignition delay times of a composite solid propellant under CO_2 laser heating. *Combustion Science and Technology*, **127**, 319–331.
(29) Radenac, E., Gilard, P. and Roux, M. (1998) Laser diode ignition of the combustion of pyrotechnic mixtures: Experimental study of the ignition of Zr/Cl04 and ZrIPbCrO4, in *Proc. 29th Int. Ann. Conf of ICT: Energetic Materials Production, Processing and Characterization*, Fraunhofer Institut für Chemische Technologie, Karlsruhe, Germany, pp. paper 40.
(30) Aleksandrov, E.I. and Voznyuk, A.G. (1978) Initiation of lead azide with laser radiation. *Combust. Explos. Shock Waves*, **14**, 480–484.
(31) Aleksandrov, E.K. and Tsipilev, V.P. (1981) Dimensional effect in the initiation of compressed lead azide by single-pulse laser radiation. *Combustion, Explosion, and Shock Waves*, **17**, 550–553.
(32) Aleksandrov, E.I. and Tsipilev, V.P. (1982a) Effect of pressing pressure on the sensitivity of lead azide to the action of laser radiation. *Combustion, Explosion, and Shock Waves*, **18**, 215–218.
(33) Aleksandrov, E.I. and Tsipilev, V.P. (1984) Effect of the pulse length on the sensitivity of lead azide to laser radiation. *Combustion, Explosion, and Shock Waves*, **20**, 690–694.
(34) Hagan, J.T. and Chaudhri, M.M. (1981) Low energy laser initiation of β lead azide. *Journal of Materials Science*, **16**, 2457–2466.
(35) Hagan, J.T. and Chaudhri, M.M. (1983) Low energy laser initiation of single crystals of b-lead azide. 7th Symp. (Inti) on Detonation, pp. 735–744.
(36) Kawakaki, M., Hada, H. and Uchida, H. (1986) Transfer of photoelectrons and photoholes through AgBr/AgCl interface, and relative locations of the energy bands. *Journal of Applied Physics*, **60**, 3945–3953.
(37) Botcher, T.R., Ladouceur, H.D. and Russell, T.P. (1998) Pressure dependent laser induced decomposition of RDX, in Schmidt, S.C., Dandekar, D.P. and Forbes, J.W. (eds) *Shock Compression of Condensed Matter – 1997*, American Institute of Physics, Woodbury, NY, pp. 989–992.
(38) Bykhalo, A.I., Zhuzhukalo, E.V., Kovalskii, N.G. *et al.* (1985) Initiation of PETN by high-power laser radiation. *Combustion, Explosion, and Shock Waves*, **21**, 481–483.
(39) Castille, C., Germain, E. and Belmas, R. (1992) Origine physique des points chauds dans les compositions explosives pressées au TATB. *Propellants, Explosives, Pyrotechnics*, **17**, 249–253.
(40) Harrach, R.J. (1976) Estimates on the ignition of high explosives by laser pulses. *Journal of Applied Physics*, **47**, 2473–2482.
(41) Ng, W.L., Field, J.E. and Hauser, H.M. (1986) Thermal, fracture, and laser-induced decomposition of pentaerythritol tetranitrate. *Journal of Applied Physics*, **59**, 3945–3952.
(42) Ramaswamy, A.L. and Field, J.E. (1996) Laser-induced ignition of single crystals of the secondary explosive cyclomethylene trinitramine. *Journal of Applied Physics*, **79**, 3842–3847.
(43) Renlund, A.M., Stanton, P.L. and Trott, W.M. (1989) Laser initiation of secondary explosives, in *Proc. Ninth Symposium (Int) on Detonation*, Office of the Chief of Naval Research, Arlington, VA, pp. 1118–1127.
(44) Ostmark, H. (1985) Laser as a tool in sensitivity testing of explosives, in Short, J.M. (ed.) *Proc. Eighth Symposium (Int.) on Detonation*, Maryland Naval Surface Weapons Center, White Oak, Silver Spring, MD, pp. 473–484.
(45) Tang, T.B., Chaudhri, M.M. and Rees, C.S. (1987) Decomposition of solid explosives by laser irradiation: a mass spectroscopic study. *Journal of Materials Science*, **22**, 1037–1044.
(46) Ulas, A. and Kuo, K.K. (1997) Effect of aging in ignition delay times of a composite solid propellant under CO2 laser heating. *Combustion Science and Technology*, **127**, 319–331.
(47) Aleksandrov, E.I. and Tsipilev, V.P. (1982b) Influence of the generation regime on the singularities of the size effect in laser initiation of pressed lead azide. *Combustion, Explosion, and Shock Waves*, **18**, 663–665.
(48) Aduev, B.P., Aluker, E.D., Belokurov, G.M. and Krechetov, A.G. (1995) Predetonation conductivity of silver azide. *Journal of Experimental and Theoretical Physics Letters*, **62**, 215–216.
(49) Chernai, A.V., Sobolev, V.V., flyushin, M.A. *et al.* (1996) Mechanism of explosive compositions ignition by laser monopulses. *Zhurnal Fizicheskoi Khimii.*, **15**, 134–139.
(50) Paisley, D.L. (1989) Prompt detonation of secondary explosives by laser, in *Proc. Ninth Symposium (Int.) on Detonation*, Office of the Chief of Naval Research, Arlington, VA, pp. 1110–1117.

(51) Jumper, E.J. (1984) Implications of applying a global energy balance to laser-supported and chemical detonation waves. *Physics of Fluids*, **27**, 2361–2364.
(52) Yang, L.C. (1981) Performance characteristics of a laser initiated microdetonator. *Propellants, Explosives, Pyrotechnics*, **6**, 151–157.
(53) Tasaki, Y., Kurokawa, K., Hattori, K., Sato, T., Mijajima, T. and Takano, M. (1989) Experimental study of laser-initiated detonator. *4th Congres International de Pyrotechnics,* La Grande Motie, France, 225–230.
(54) Ostmark, H. (1985) Laser as a tool in sensitivity testing of explosives, in Short, J.M. (ed.) *Proc. Eighth Symposium (Int.) on Detonation*, Maryland Naval Surface Weapons Center, White Oak, Silver Spring, MD, pp. 473–484.
(55) Aleksandrov, E.I., Sidonskii, O.B. and Tsipilev, V.P. (1991) Influence of combustion in the vicinity of absorbing inclusions on the laser ignition of a condensed medium. *Combustion, Explosion, and Shock Waves*, **27**, 267–272.
(56) Aleksandrov, E.I. and Voznyuk, A.G. (1988) Effect of absorption heterogeneities on the laser triggering of explosive decomposition. *Combustion, Explosion, and Shock Waves*, **24**, 730–733.
(57) Aleksandrov, E.I., Voznyuk, A.G. and Tsipilev, V.P. (1989) Effect of absorbing impurities on explosive initiation by laser light. *Combustion, Explosion, and Shock Waves*, **25**, 1–7.
(58) loffe, V.B., Dolgolaptev, A.V., Aleksandrov, V.E. and Obraztsov, A.P. (1985) Laser pulse ignition of condensed systems containing aluminum. *Combustion, Explosion, and Shock Waves*, **21**, 316–320.
(59) Aleksandrov, V.E., Dolgolaptev, A.V., loffe, V.B. and Levin, B.V. (1985) Inflammation of porous systems by monopulse laser radiation. *Combustion, Explosion, and Shock Waves*, **21**, 54–57.
(60) Zinchenko, A.D., Sdobnov, V.1., Tarzhanov, V.1. *et al.* (1991) Action of a laser on a porous explosive substance, without initiation. *Combustion, Explosion, and Shock Waves*, **27**, 219–222.
(61) Paisley, D.L. (1990) Laser-driven miniature flyer plates for shock initiation of secondary explosives, in Schmidt, S.C., Johnson, J.N. and Davidson, L.W. (eds) *Shock Compression of Condensed Matter - 1989*, Elsevier, Amsterdam, pp. 733–736.
(62) Ostmark, H., Carison, M. and Ekvall, K. (1994) Laser ignition of explosives: Effects of laser wavelength on the threshold ignition energy. *Journal of Energetic Materials*, **12**, 63–83.
(63) Ostmark, H. and Grans, R. (1990) Laser ignition of explosives: Effects of gas pressure on the threshold ignition energy. *Journal of Energetic Materials*, **8**, 308–322.
(64) Ostmark, H., Carison, M. and Ekvall, K. (1994) Laser ignition of explosives: Effects of laser wavelength on the threshold ignition energy. *Journal of Energetic Materials*, **12**, 63–83.
(65) Kelley, F.N. (1969) Solid propellant mechanical properties testing, failure criteria, and aging, in Boyars, C. and Klager, K. (eds) *Propellant Manufacture, Hazards, and Testing*, American Chemical Society, Washington DC, pp. 188–243.
(66) Behrens, R. and Bulusu, S. (1996) Thermal decomposition of HMX: Low temperature reaction kinetics and their use for assessing response in abnormal thermal environments and implications for long-term aging. *Materials Research Society symposia proceedings*, **418**, 119–126.
(67) Cartwright, M. (2012) Correlation of structure and sensitivity in inorganic azides III. A mechanistic interpretation of impact sensitivity dependency on non-bonded nitrogen to nitrogen distance. *Propellants, Explosives, Pyrotechnics*, **37**, 639–646.
(68) Bloom, G., Chau, H., Glaser, R. *et al.* (1984) Improvements in thin-pulse shock initiation threshold measurements, in Asay, J.R., Graham, R.A. and Straub, G.K. (eds) *Shock Waves in Condensed Matter – 1983*, North-Holland, Amsterdam, pp. 535–538.
(69) Trott, W.M. and Meeks, K.D. (1990) Acceleration of thin foil targets using fiber-coupled optical pulses, in Schmidt, S.C., Johnson, J.N. and Davidson, L.W. (eds.) *Shock Compression of Condensed Matter - 1989*, Elsevier, Amsterdam, pp. 997–1000.
(70) Paisley, D.L., Warnes, R.H. and Kopp, R.A. (1992) Laser-driven flat plate impacts to 100 GPa with sub-nanosecond pulse duration and resolution for material property studies, in *Shock Compression of Condensed Matter - 1991* (eds S.C. Schmidt, R.D. Dick, J.W. Forbes and D.G. Tasker), Elsevier, Amsterdam, pp. 825–828.
(71) Frank, A.M. and Trott, W.M. (1996) Investigation of thin laser-driven flyer plates using streak imaging and stop motion microphotography, in Schmidt, S.C. and Tao, W.C. (eds)

Shock Compression of Condensed Matter 1995, American Institute of Physics, Woodbury, NY, pp. 1209–1212.
(72) Farnsworth, A.V. Jr. (1996) Laser acceleration of thin flyers, in Schmidt, S.C. and Tao, W.C. (eds) *Shock Compression of Condensed Matter 1995*, American Institute of Physics, Woodbury, NY, pp. 1225–1228.
(73) Hatt, D.J. and Waschl, J.A. (1996) A study of laser-driven flyer plates, in Schmidt, S.C. and Tao, W.C. (eds) *Shock Compression of Condensed Matter 1995*, American Institute of Physics, Woodbury, NY, pp. 1221–1224.
(74) Labaste, J.L., Doucet, M. and Joubert, P. (1996) Shocks induced by laser driven flyer plates. 1: Experiments, in Schmidt, S.C. and Tao, W.C. (eds) *Shock Compression of Condensed Matter 1995*, American Institute of Physics, Woodbury, NY, pp. 1213–1215.
(75) Cazalis, B., Boissière, C. and Sibille, G. (1996) Shocks induced by laser driven flyer plates. 2: Numerical simulations, in Schmidt, S.C. and Tao, W.C. (eds) *Shock Compression of Condensed Matter 1995*, American Institute of Physics, Woodbury, NY, pp. 1217–1220.
(76) Davison, L. and Graham, R.A. (1979) Shock compression of solids. *Physics Reports*, **55**, 255–379.
(77) Bourne, N.K., Rosenberg, Z., Johnson, D.J. *et al.* (1995) Design and construction of the UK plate impact facility. *Measurement Science and Technology*, **6**, 1462–1470.
(78) Paisley, D.L. (1990) Laser-driven miniature flyer plates for shock initiation of secondary explosives, in Schmidt, S.*C., Johnson, J.N.* and Davidson, L.W. (eds.) *Shock Compression of Condensed Matter - 1989*, Elsevier, Amsterdam, pp. 733–736.
(79) Trott, W.M. (1996) High-speed optical studies of the driving plasma in laser acceleration of flyer plates, in Schmidt, S.C. and Tao, W.C. (eds) *Shock Compression of Condensed Matter 1995*, American Institute of Physics, Woodbury, NY, pp. 921–924.
(80) McDaniel, O., Moore, C. and Tindol, S. (1990) Laser ordnance initiation, in Schmidt, S.C., Johnson, J.N. and Davidson, L.W. (eds.) *Shock Compression of Condensed Matter - 1989*, Elsevier, Amsterdam, pp. 787–790.
(81) Hafenrichter, E.S., Marshall, B.W. and Fleming, K.J. Fast laser diode initiation of confined BNCP, AIAA Paper 245 (2003).
(82) Chernai, A.V. and Sobel, V.V. *et al.* (2003) Laser Initiation of explosive formulations on the base of di-(3-hydazine-4-amine-123-triazole-copper (II) perchlorate. *Fizika Goreniya i Vzryva*, **39**, 127.
(83) Wsng, Y., Sheng, D., Zgu, Y. and Chen, L. (2008) Study of laser sensitive coordination compound primary explosive. *Initiators and Pyrotechnics*, **2**, 30–33.
(84) Yang, L.C. and Menichelli, V.J. (1971) Detonation of insensitive high explosives by a Q-switched ruby laser rent. *Applied Physics Letters*, **19**, 473.

3

Lasers and Their Characteristics

3.1 Definition of Laser

Laser is an acronym for **L**ight **A**mplification by **S**timulated **E**mission of **R**adiation. The acronym gives a clue to the process by which this amazing light is generated, but not what it is. It may, however, be called a device which produces an amazing 'beam' of light with some unique properties. The term 'laser' is now interchangeably used to mean the source and the beam. In this chapter, we shall deal primarily with the properties relevant to its use for the ignition initiation in energetic materials. In general, however, a laser's unique properties include high coherence, low beam divergence and wavelength tunability, and all these with output power density much higher than that available from any conventional bright light source, such as a search light. These properties allow lasers to deliver very high power density within a very narrow wavelength band onto targets – even on ones located at a considerable distance.

The quantities which characterize light are first defined, this is followed by a comparison between these characteristics for conventional light generated from a very bright thermal source – for example, a flash lamp, search light and so on – and that provided by commonly used lasers. This comparison allows an understanding of the uniqueness of lasers as a versatile optical energy source for a very wide range of applications, including those in the field of ignition initiation in energetic materials to affect events such as deflagration, gas generation, decomposition, detonation and so on. Since a laser is essentially a light source, albeit a special one, it is necessary to have an understanding of the concept of the nature of light, so that the mechanism of laser interaction with energetic materials can be analysed and understood, and this source of optical energy can be effectively used for laser ignition of energetic materials.

Laser Ignition of Energetic Materials, First Edition. S Rafi Ahmad and Michael Cartwright.

3.2 Concept of Light

In 1677, the Danish scientist Ole Römer demonstrated that light is 'something' that travels at a finite speed. The enquiry into what that 'something' could be was first made on a scientific basis by the Dutch physicist, Christiaan Huygens, in a small treatise on optics published in 1678. His vision of light was that it is a form of ripple emanating from every point in a source. The Swiss scientist Leonard Euler was the first to propose, in 1768, that, just as the wavelength of sound waves determines its pitch, the wavelength of the light ripples determines the colour of that light. However, the concept of light as a propagating ripple or some kind of wave did not stand up against the theory that light is a stream of minute particles (corpuscular theory) propounded by the British scientist Isaac Newton in 1704. Nearly a century on, in 1801, a British medical doctor, Thomas Young, by his famous 'light and two-slit' experiment, firmly established the wave nature of light from the observation of fringes – that is, dark and light patterns on a screen.

Newton imposed the doctrine that the theory of light must finally be based on the 'particular' nature of light and Young contradicted this by conclusive and easily reproducible evidence that light is a 'kind of wave'. A wave and a beam of particles are very different things. A particle can exist only at one point at an instant of time, whereas a wave pervades a volume of space at any instant of time. It was proved later that the 'wave' and 'particle' need not be taken as mutually exclusive, but that sort of duality concept did not exist in the nineteenth century.

A completely different approach was needed to achieve this understanding, and it came out of intense research on electricity and magnetism. It was discovered that electricity and magnetism were inexorably interrelated – one can be created from the other. In 1845, the English physicist, Michael Faraday, after successfully demonstrating the relationship between the electricity and magnetism, also succeeded in demonstrating the relationship between light and magnetism. Twenty years later, a Scottish scientist, James Clark Maxwell, transformed Faraday's dream into reality by connecting the electric field, the magnetic field and the light through a grand mathematical theory known as 'the electromagnetic theory of light'. Later, it transpired that electromagnetic radiation encompasses a wide range of spectrum, from the high-energy gamma radiation down to low-energy infrared radiation, all characterized by the frequency of the electric field vectors. Light is simply the visible part of the entire electromagnetic spectrum, within the wavelength band of approximately 400–700 nm, as is shown schematically in Figure 3.1.

In the year 1900, by ingenious trial and error, the German scientist Max Planck put together an empirically based formula which correctly related the observed colour (quantified by wavelength, λ) and the corresponding intensity of a glowing object, such as heated metal (known as a blackbody). The intensity of radiation, $I(\lambda, T)$ emitted by a unit surface area into a fixed direction (solid angle) from a blackbody at a fixed temperature, as a function of wavelength, is given as Planck's Radiation Law:

$$I(\lambda, T) = (2hc^2)/\lambda^5\{\exp(hc/kT\lambda) - 1\}^{-1} \tag{3.1}$$

where:

λ (lambda) is the (peak) wavelength, associated with the light,
ν (neue) is the corresponding frequency ($= c / \lambda$) of the light,

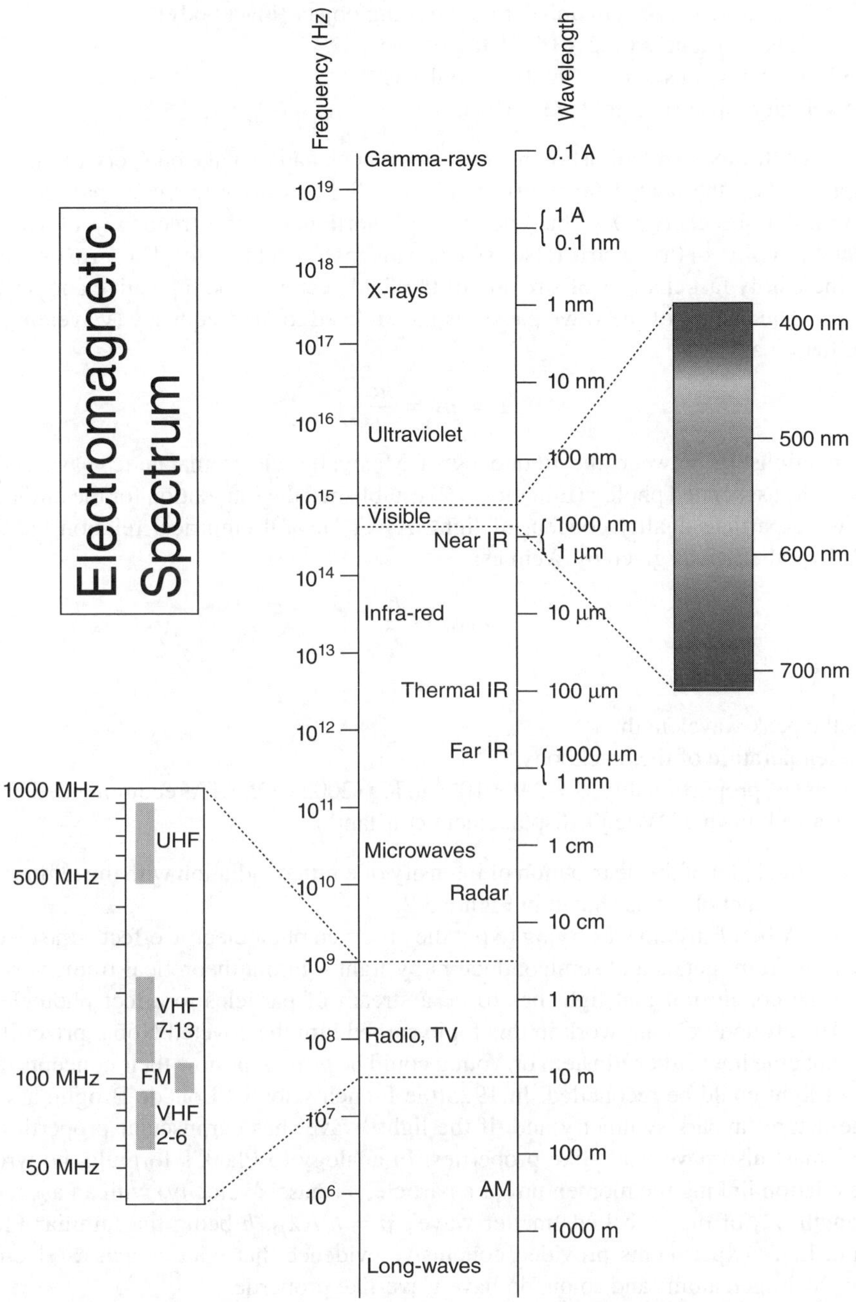

Figure 3.1 *Electromagnetic spectrum extending from radio wave up to gamma radiation. Reproduced from http://upload.wikimedia.org/wikipedia/commons/8/8a/Electromagnetic-Spectrum.png.*

T is the temperature (in Kelvin unit) of the emitting object (blackbody),
h is the Planck constant, $\approx 6.62 \times 10^{-34}$ Joules-sec,
k is the Boltzmann constant $\approx 1.38\ 10^{-22}$ Joules K^{-1},
c is the velocity (speed) of light $\approx 3 \times 10^8$ m sec^{-1}

To deduce the above relationship theoretically Planck had to make one very fundamental assumption, that, the heated (excited) atoms emit light as discrete 'wave packets', each carrying a definite energy, E. This energy is proportional to the frequency of vibration (oscillation), ν(*nu*), of the electric (also associate magnetic) field vector. The term oscillation means incredibly fast changes of strength of the field vectors in both positive and negative directions. The energy of the wave packet is then related to the frequency (wavelength) of the oscillation as:

$$E = h\nu = \frac{hc}{\lambda} \tag{3.2}$$

The relationship between the frequency of Maxwell's electromagnetic wave and the energy of the associated photon (Equation 3.2) established the foundation for the unification of the wave-particle duality concept of light. A very useful empirical relation based on Plank's hypothesis was given by Wein as:

$$\lambda_{\max} = \frac{b}{T} \tag{3.3}$$

where:

$\lambda_{\max}$ is the peak wavelength,
T is the temperature of the blackbody,
the constant of proportionality, $b \approx 2.9 \times 10^{-3}$ m K s (2002 CODATA recommended value); this is also known as 'Wien's displacement constant'.

A theoretical plot of the distribution of intensity of emitted radiation with the temperature of an incandescent object is shown in Figure 3.2.

In 1905, Albert Einstein was trying to put the observed photoelectric effect – the ejection of electrons from metals and semiconductors by light – into a theoretical framework. He came to the conclusion that light has to be a stream of particles to affect photoelectric events. His groundbreaking work in this topic earned him the coveted Noble prize. It was hard to imagine how either Einstein or Young could be wrong, or how their concepts of the nature of light could be reconciled. In 1925 the French scientist Lois de Broglie asserted that the nature favours symmetry and, if the light 'wave' has corpuscular properties, the 'matter' must also have wave-like properties. In analogy to Plank's formula, he wrote a simple relation linking the momentum of a particle, p (mass $\times$ velocity) with an associated wavelength, λ_D of the so called 'matter wave', $p = h / \lambda_D$, h being the familiar Planck constant. Later experiments provided conclusive evidence that matter such as electrons, protons, hydrogen atoms and so on, do have wave-like properties.

To bring discord to the solution, one of the great scientists of the early twentieth century, Schrödinger, provided a satisfactory description of the behaviours of particles, including light particles, by a mathematical formalism defining the particle by a phantom wave function, $\psi\ (x,\ t)$. The only significance of this representation is that the square of the modulus of this wave function (in the space-time continuum) gives the probability of finding the light particle (photon) within a specified volume of space at a certain instant

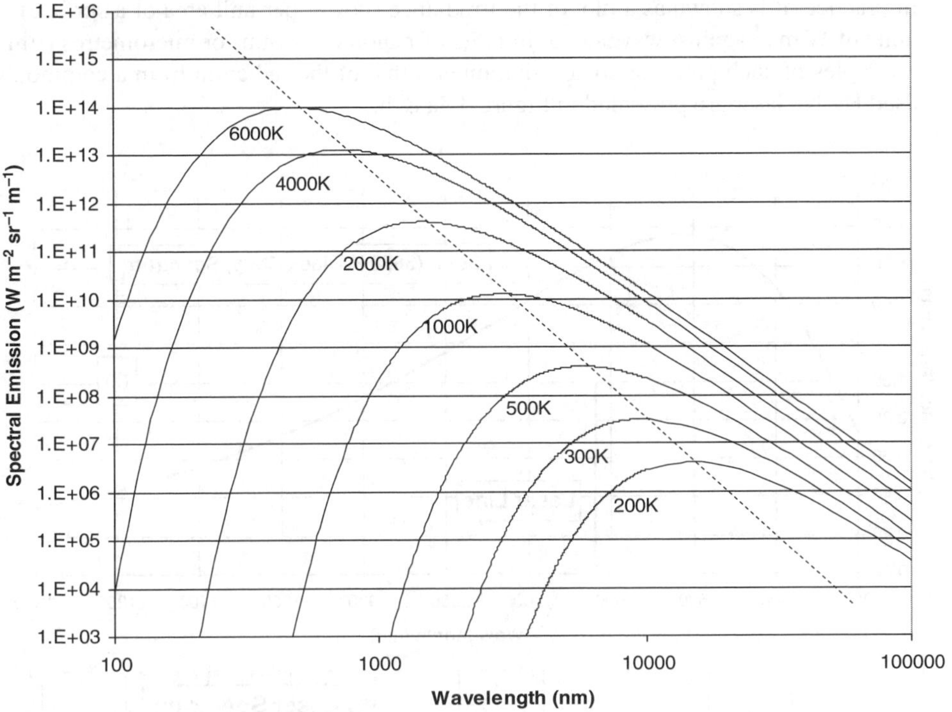

Figure 3.2 *Theoretical spectra of radiation emitted from an ideal 'black body' at different temperatures (based on Plank Radiation Law).*

of time. The famous Schrödinger's equation was formally accepted as the only available interpretation of the behaviour of light, albeit without knowing what light exactly was, as was commented on by Einstein. This accord is known as the Copenhagen agreement.

This gave rise to a mathematical formalism known as 'quantum electrodynamics' which allows us to solve problems associated with the interaction of light with atoms and molecules. However, the application of such formalism is limited to well-specified atomic systems and not real-world samples containing atoms and molecules, as in most complex configurations. In order to understand the mechanism of the interaction of light with energetic materials, the established laws governing the meso-scale absorption by real-world systems, along with the dynamics of conversion of optical energy into the thermal form and its subsequent propagation in a medium, need to be considered.

3.3 Parameters Characterizing Light Sources

Thermal light in general, and laser light in particular, are characterized by five major parameters. These are governed by the way the light is generated and used, and are defined as follows:

i. *Line width* defines the purity or the uniqueness of the colour of the light, and it parameterizes the quality of a beam or a source of light known as 'monochromacity'.

In practice, it is shown as a plot of the irradiance (power per unit area of a surface) in units of W m^{-2} against wavelength in units of nanometres (nm) or micrometres (μm). Examples of such plots for solar radiation and that of the radiation from a commonly used He-Ne laser are presented in Figure 3.3a & b.

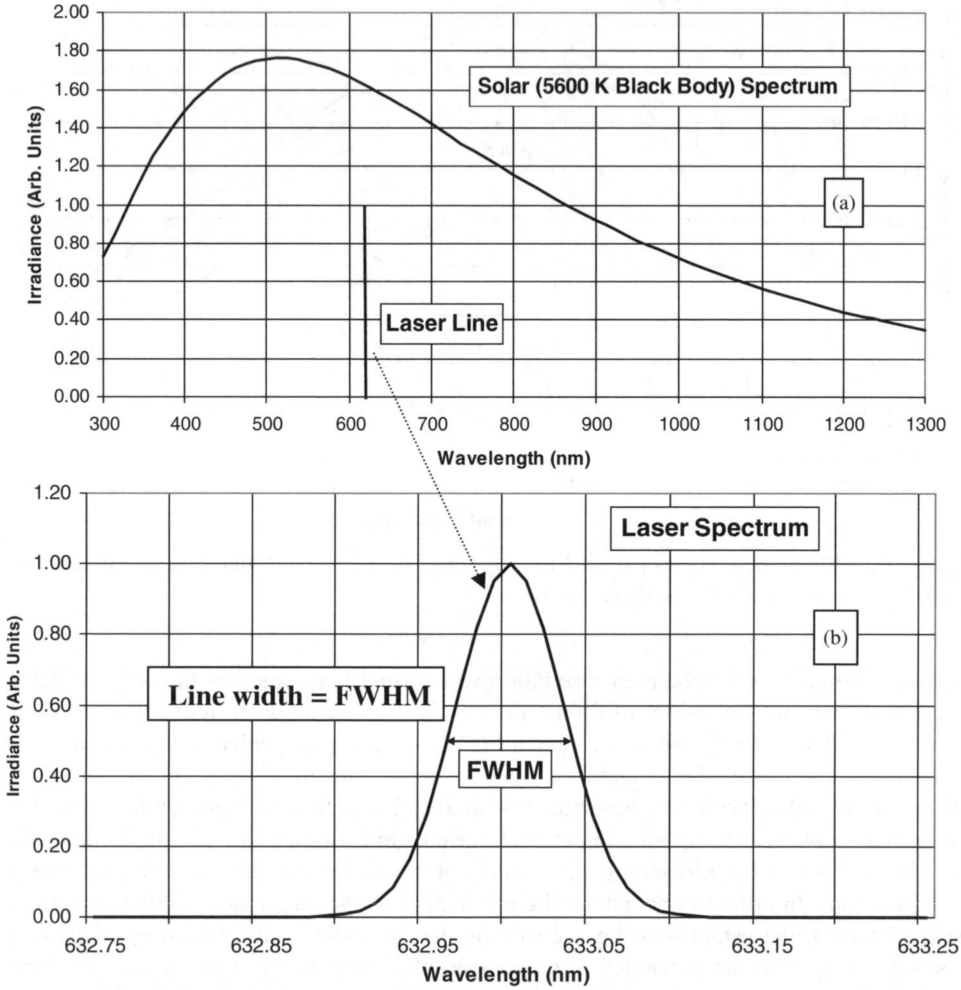

Figure 3.3 *Comparison between: (a) broad band spectrum of solar radiation and, (b) narrow band spectrum of a laser beam (concept of laser line and line width).*

In comparison with any man-made light source, lasers can provide output that has line width many orders of magnitude smaller. Therefore, for most practical purposes, laser sources are considered to be monochromatic ones. Unless the spectral recording is carried out using a high-resolution spectrometer, it will appear as a vertical line – hence the term 'laser line' or a 'spectral line'.

ii. *Pulse width* defines the temporal extent, i.e. the duration of the beam emanating from the source, and applies when the beam is either a single or a repetitively pulsed output from a source. A 'pulsed' beam can be created from a continuous beam of light, either mechanically, such as using a chopper or a shutter, or electronically, such as by switching the power 'on' and 'off'. Laser beams can also be pulsed using a pulsed flash lamp as the excitation source. Additionally, lasers can be made to provide output in a single pulse of an extremely short duration. The width of a laser pulse is conventionally measured as the full width at the half maximum (FWHM) position of a plot of irradiance versus time, as shown in Figure 3.4 for a Q-switched output of a typical solid state laser.

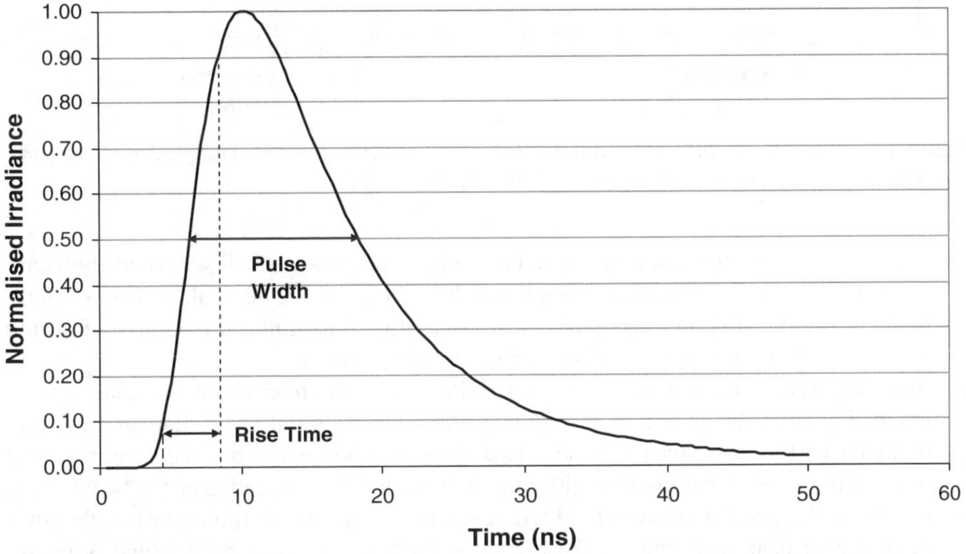

Figure 3.4 *Example of temporal history of a solid-state pulsed (Q-switched) laser.*

Conventionally, the pulse width of such output is defined by the duration of the beam at the half maximum point – again, that is, by the FWHM. Detected pulse rise time (10–90% of peak value) will be modulated by the rise time or the response time of the photo detector and the associated electronic circuitry. Therefore, for extremely short pulses, a detection device with high band width (e.g. 500 MHz for nanosecond pulses) is required.

iii. *Beam profile* is the intensity distribution along the cross-section of a laser beam. The profile is governed by the way the laser beam is produced, and it may consist of longitudinal and transverse mode structures, giving a non-circular and inhomogeneous intensity distribution with some mode structure. A single-mode laser beam will usually have a Gaussian intensity profile, as shown in Figure 3.5a. The colour within the cross-sectional area of the beam represents the intensity, red being the highest. The profile (i.e. the plot of intensity against the cross-sectional diameter with the origin at the centre of the circular beam shape) is an ideal beam profile for diffraction limited focusing with a lens. The beam profile shown in Figure 3.5b is that of a multimode laser. A multimode

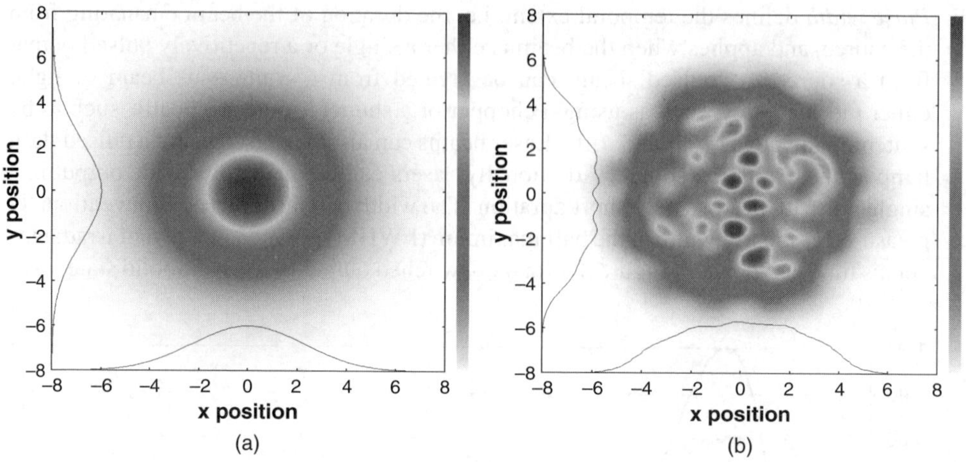

Figure 3.5 *Intensity profiles of a Gaussian beam (a) and a multimode laser beam (b). Reproduced with permission from Rüdiger Paschotta. © 2013.*

laser may assume a top-hat profile (sharp edges like a square pulse) when the beam is homogenized during transmission through a fibre optic cable. Unless it is specially made or the beam is homogenized, most lasers do not exhibit a true Gaussian profile and contain hot spots (points of unusually high intensities).

iv. *Beam divergence* defines the extent of parallelism or unidirectionality of a laser beam emanating from the source. A laser beam will have an initial beam diameter and that diameter will increase over distance. This increase is known as beam divergence. The beam divergence is parameterized by the half angle (θ_d) subtended at the beam waist point (i.e. the point at the near-field where the beam diameter is minimum) by the lines of divergent beam extremities. For practical reasons, it is defined by a quality factor, M^2, as follows:

$$M^2 = \frac{\pi(\theta_d \times w)}{\lambda} \tag{3.4}$$

where w is the radius of the beam at the beam waist. For all intents and purposes, this is measured at the exit window of the laser beam. The product, $(\theta_d \times w)$ in the numerator is defined as the beam parameter product (BPP) and (λ/π) represents that for a Gaussian beam. A laser beam with $M^2 \approx 1$ represents a Gaussian beam profile.

The beam divergence angle varies widely depending upon the type of laser and the way these are produced and operated. For example, a well-designed (TEM mode) He-Ne laser (≈630 nm) may have a half beam divergence angle of less than 1 m rad ($< 10^{-3}$ radian), whereas medium-power (a few watts) diode lasers, which have a prominent role in ignition initiation applications, usually have beam divergence three to four orders of magnitudes higher than that of a good quality He-Ne laser. This property of diode lasers, however, does not allow the beam to be tightly focused for achieving high power density at the target. It can be concluded, from the above equation, that lasers with shorter wavelengths and lenses with shorter focal lengths will allow tighter focusing. Because of the importance of the diode lasers in the present context, characteristics of different types of diode laser including those of some high radiance light emitting diodes (LD), are presented in the next section. It is noted that the output from diode lasers usually has a rectangular beam profile (spatial intensity distribution) and the beam divergence in the two directions is different. Spatial profiles of diode lasers can easily be manipulated to provide a near-Gaussian circular beam profile if necessary. Such a beam will render itself to be focused by a aberration-free lens to a diffraction limited spot size, with a radius, r_f, given as:

$$r_f = \frac{1.7 \times \pi f \lambda}{r_b} \tag{3.5}$$

where:

f = the focal length of the lens,
r_b = beam radius at the exit. Ideally, this should be the radius of the focusing lens when the laser beam just fills the area of the lens.

It is noted from the above equation that for shorter laser wavelengths (λ) and for lenses with shorter focal lengths (f), the dimension of the focused spot and, consequently, the values of irradiance, will increase as the square of the radius of the spot.

For example, a He-Ne laser with 1 mm beam diameter, operating at 633 nm (6.33×10^{-7} m) wavelength and 1 mW output power (a common laboratory laser used for alignment purposes) will have a beam radius of $\approx 4 \times 10^{-6}$ m at the focal point of a 50 mm focal length lens, giving a power density at the target of $\approx 2 \times 10^7$ W m^{-2} (i.e. $\approx 2 \times 10^3$ W cm^{-2}). Since the typical line width of a He-Ne laser is $\sim 10^{-4}$ nm, the spectral power density (spectral brightness) at the focal spot will be $\approx 2 \times 10^7$ W cm^{-2} nm^{-1}. Note that, although a highly monochromatic and parallel light beam can be produced by modifying the output from a strong thermal source using many spectral filters and collimating lenses, the intensity available from such an arrangement will be many orders of magnitude lower than that available from a small laser (such as a pointer).

v. *Pulse repetition rate* defines how quickly the pulses of light beam are emitted from the source. For a continuous wave beam (CW), the repetition rate can be controlled mechanically by a chopper. The rate achievable by this means is limited by the speed at which a chopper can be rotated without unacceptable jitter. The repetition rate of pulsed lasers is controlled by electronic circuitry which provides the high voltage to the flash lamp or other discharge mechanism for laser excitation. The complexity of rapidly switching high voltage normally limits the repetition rates of most high-power solid state lasers to 200 Hz, but this is achieved at the cost of reduced pulse energy. Lasers having a pulse repetition rate in excess of a few KHz are commercially available. The high pulse rate available from a laser has to be traded off by the reduction of output power.

vi. *Strength of an optical beam/source* defines the power available from a source. It may be defined either as the brightness (irradiative strength) of the source or the brightness viewed on an illuminated area. These may be expressed in terms of unit area of either the source or the illuminated area, per unit of line width and per unit of solid angle. For low divergent lasers incident on a target in the near field region, the commonly used quantity is the irradiance (W m^{-2}). However, other parameters may be used for different applications, as is given in Table 3.1 for a continuous laser output.

Table 3.1 *Definition of Radiometric quantities and their dimensions.*

Quantity	Symbol	SI unit	Notes
Radiant energy	Q	Joule (J)	Energy
Radiant flux	Φ	Watt (W)	Radiant energy per unit time, also called *radiant power*
Radiant intensity	I	Watt per steradian (W sr^{-1})	Power per unit solid angle
Radiance	L	Watt per steradian per square metre (W sr^{-1} m^{-2})	Power per unit solid angle per unit *projected* source area. Called *intensity* in some other fields of study.
Irradiance	E, I	Watt per square metre (W m^{-2})	Power incident on a surface. Sometimes confusingly called 'intensity'.
Radiant exitance/ Radiant emittance	M	Watt per square metre (W m^{-2})	Power emitted from a surface.

For a pulsed laser beam, the strength of the beam on a target is quantified by 'peak power or power density' or simply (peak) irradiance (W m^{-2}). For more precise quantitative assessment of optical strength of a pulsed laser beam at the target the term, 'pulse energy' is introduced. This may be defined as the integrated area under the laser pulse (see Figure 3.4) within the 10% pedestal (for reproducibility of values). For most practical considerations the energy in a laser pulse may be estimated by assuming the laser pulse to be a triangular pulse so that the area within the pulse, that is, the integrated pulse energy is given approximately as $E_p = 1/2P_{\text{pk}} \times \tau_p$, (Joules), where, P_{pk} is the peak value of the power (watts) and τ_p (seconds) is the laser pulse width (duration). As with the power density, the energy density is also expressed as per unit area of the irradiated target. For repetitively pulsed lasers, the strength of the source or the beam is quantified by the parameter, 'average power', $P_{\text{av}} =$ Energy in the pulse × pulse repetition rate ($\text{J} \times \text{s}^{-1}$).

It is noted that the optical power is defined as the rate of energy deposition on a target, that is, joules per second. According to Planck's radiation law, (light being considered as a stream of photons never at rest) each of the photons in the stream (beam) has an energy related to its wavelength according to:

$$\varepsilon_p = \frac{hc}{\lambda} \text{(Joules)}$$

where:

h (6.625×10^{-34} J sec) is Planck's constant,
c (2.99×10^{8} m s^{-1}) is the velocity of light.

If the number of photons incident per second (over an area A on the target) is n, the power (J sec^{-1}) is defined in terms of photon rate as:

$$P = n \times hc/\lambda \text{ (watts)} \tag{3.6}$$

The concept of laser power has to be distinguished between lasers operating in a continuous mode (CW), such as the He-Ne, Ar-ion, diode lasers and so on, and those operating in pulsed mode, such as ruby laser, Nd-YAG laser, and so on. In the pulsed operation, if the effect to be analysed relates to a single pulse, then the peak power or energy within the pulse, as defined before, is considered. If the effect of repetitive pulses with a pulse repetition rate, z (hertz, Hz) is considered, the concept of an average power is considered, so that the average power is now represented by the following equation:

$$P_{\text{av}} = z \times E_p \text{ (watts)} \tag{3.7}$$

3.4 Basic Principle of Lasers

'LASER', as mentioned at the beginning of the chapter, stands for Light Amplification by Stimulated Emission of Radiation. To understand the basic principle of laser action, it is necessary to understand the two terms defining it – stimulated emission and its amplification. It is known that 'energy, in any form, cannot be created or destroyed'; however, energy in one form can be converted to another form. Light is emitted when a fuel such as, coal, wood, gasoline and so on is ignited to burn, thereby converting the internal energy of the fuel into light and heat energies. Light is also emitted when electric current is transmitted through a thin wire in an electric bulb, where the electric energy is converted into heat energy in the thin metal wire inside the bulb, which then glows to give light. In nature the sun and stars emit light through thermonuclear reactions in the He and H atoms at extremely high temperatures. It is now obvious that the conversion process must go through a material medium, such as energetic molecules of fuel, atoms in metal wire, atomic gases in solar system and so on. To understand how light quanta are generated, one needs to consider the atomic structure of materials and how atoms interact with energy.

For the present analysis, it is sufficient to accept that atoms are the fundamental building block of all matter. Different kinds of matter have atoms with different structures, combining into different types of molecules. However, the absorption and emission of energy are governed by the negatively charged electrons whirling around the nucleus of an atom, which consists of an equal number of positively charged protons and a similar (but not necessarily equal) number of neutral particles called neutrons. The atom, as a whole is neutral. There are many possible orbitals (i.e. energy states) that electrons can occupy – the higher the orbital, the lower the potential energy of the electrons for moving to a higher orbital or ejection from the atomic nucleus altogether. An example of the visualization of the structure of a carbon atom is shown in Figure 3.6.

Atoms are excited when energized by external stimuli such as heat or light energies. The excited atoms (or molecules) retain the energy (excited state) briefly (a few nanoseconds at a meta-stable state) before returning to the stable state. In this process, radiation is emitted in quanta spontaneously, with no phase difference between emission events – a process taking place in all natural and man-made light sources, and known as 'spontaneous emission'. When the excitation process is carried out by a beam of light, there will be no correlation between the phases and the wavelengths of the incident photon (quanta) and the spontaneously emitted ones.

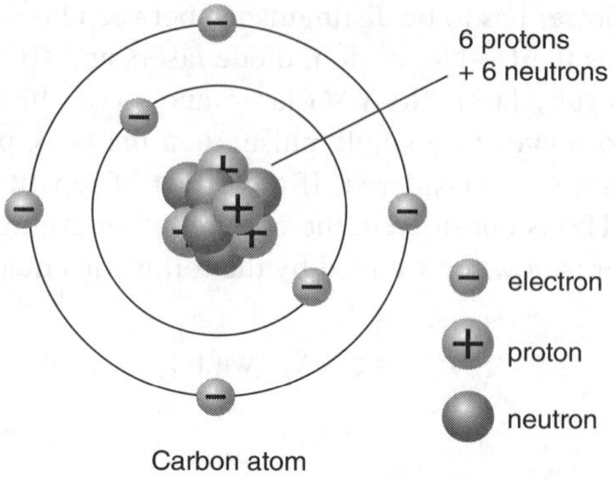

Figure 3.6 *Schematic for the visualization of the cross-sectional structure of electronic orbitals in a carbon atom. Reproduced from Atom Structure by Jerry Coffey on February 22, 2010, http://www.universetoday.com/56747/atom-structure/.*

While trying to theoretically validate Planck's blackbody radiation law, Einstein, in 1917, postulated in a paper entitled 'On the quantum theory of radiation' that the energy balance in that law required the existence of an emission process, albeit with very small probability, which is initiated by a stimulus from another photon (because of its proximity to the excited atom or molecule) having an energy value (wavelength) corresponding to the energy difference between the ground and the excited state of the atom/molecule, as shown in Figure 3.7.

Such stimulated emission quanta (wave packets), both being in the same phase, will be constructively (amplitude wise) added during their propagation along a medium. In a medium where a large number of excited atoms/molecules are sustained, the probability

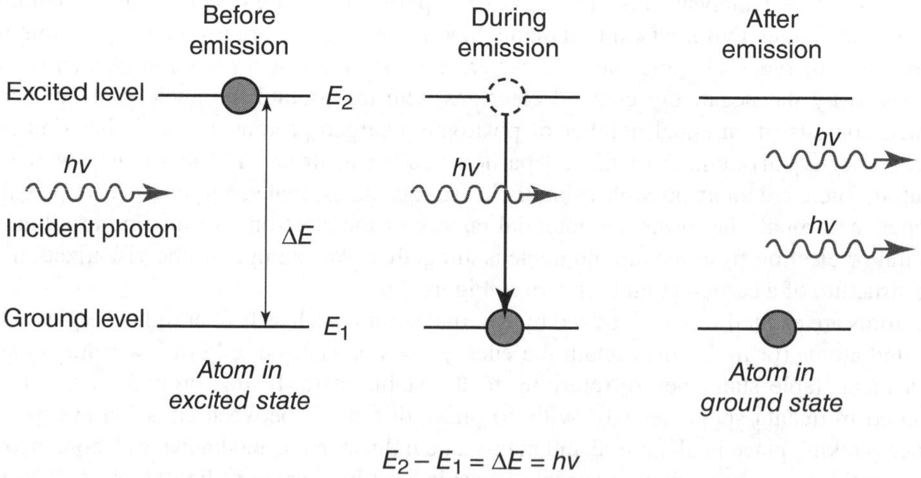

Figure 3.7 *Schematic of atomic transitions illustrating stimulated emission process. Reproduced from http://en.wikipedia.org/wiki/File:Stimulated˙Emission.svg.*

of such emission and addition (amplification) events will be high. This stimulated emission process inherently provides some amplification of radiation during its short passage along the medium containing the target atoms/molecules. But this amplification is far too insignificant to make an emergent (from the medium) beam to be a usable laser source. For a practical laser, technological developments for further amplification of this small stream of stimulated photons (quanta) were necessary.

3.5 Basic Technology of Lasers

A century later, Einstein's hypothesis of 'stimulated emission' was practically demonstrated by further amplifying the stimulated radiation trains using long media and an external cavity to reflect the stimulated radiation to and fro along the length of the medium containing excited atoms/molecules, as illustrated in Figure 3.8.

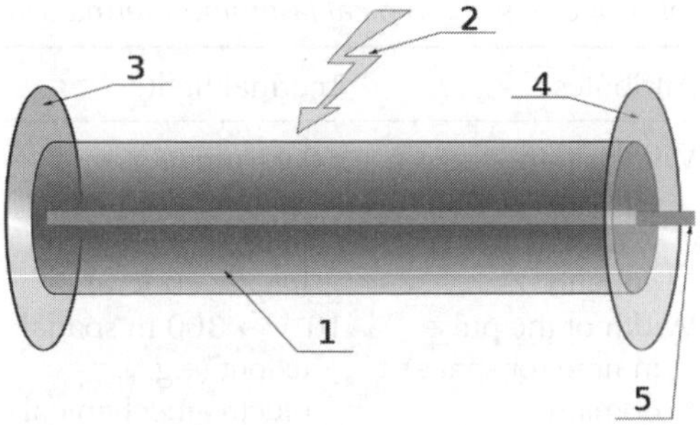

Figure 3.8 An artist's sketch illustrating the components of a practicable laser. Principal components: 1. Gain medium; 2. Laser pumping energy; 3. High reflector; 4. Output couple (low reflector); 5. Laser beam. Reproduced from http://en.wikipedia.org/wiki/Laser.

During the continuation of the amplification process, a fraction of the amplified radiation, 'amount surplus to requirements for sustaining the amplification during the lifetime excitation event (pulsed or continuous)' is siphoned off through one end of a two-mirror cavity with partially reflecting coating. A portion of that beam is the practical and usable laser beam. While there had been a spate of activities to demonstrate the stimulated emission process in the microwave and infrared wavelength regions, the first working laser, operating in the visible wavelength (694 nm) and in the pulsed mode was demonstrated in a ruby crystal by Theodore H. Maiman in 1960.

Later in the same year, a laser beam from a gaseous medium (He-Ne gas mixture) and operating at a visible wavelength (633 nm) and in a continuous mode was first demonstrated by Ali Javan *et al.* Since the above discoveries, lasers have been produced in hundreds of different types of molecules embedded in a plethora of host solid matrices and confined in gaseous media, providing output at many wavelength bands (lines) and many different modes. Output wavelength bands (lines) of some popular laser sources are shown in Figure 3.9. Note that laser output at approximately 9.2–11 μm wavelength in the IR band (CO_2 laser) was tested for laser ignition of energetic materials at the early stages of development of its concept. The output wavelength of this laser is most suitable for heating effects, and it is extensively used for industrial cutting, drilling and welding of metals. However, the heavy weight, large size and high cost of such lasers rendered the prospect of finding applications in ignition initiation in energetic materials poor.

3.6 Comparison between Laser and Thermal Sources

The parameters defined and described in Section 3.3 provide the basis for quantifying the differences between a thermal source and a laser source. In this section, typical values of these parameters for a laser source and those of common thermal sources are compared in Table 3.2. From this table it is clear that the laser can deliver optical power within a very narrow wavelength band (monochromatic) and with very low beam divergence, thereby

Table 3.2 *Comparison of properties of a typical laser and thermal light sources.*

Properties	Attributes	Thermal light	Laser light
Line width (nm) (coherence, i.e. purity of colour)	Width of an emission line in wavelength domain	≈0.01 nm (e.g. Sodium discharge lamp, coherence length ≈0.5 m)	<0.0001 nm (e.g. He-Ne laser; coherence length ≈50 m)
Pulse width (sec) (Duration)	Width of the pulse in time (or space) domain	10^{-6} → 300 m spatial extent (e.g. electro-mechanical switch)	$<10^{-8}$ → 3 m spatial extent. (≈10^{-12} → 3 mm spatial extent for mode locked lasers)
Beam divergence (radians) (unidirectional)	Angle over which light beam diverges from its source	10^{-2} (e.g. searchlight)	≈10^{-9} (specially collimated beam)
Pulse repetition rate (pulses/sec or Hertz, Hz).	The rate at which a pulse of beam is radiated from the source.	≈10^3 – limited either by the speed of mechanical chopper or electrical properties of the lamp circuitry.	≈10^5 – using electro optical modulator.
Radiance (watts m^{-2} sr^{-1}) (Source strength)		≈2×10^7 (sun radiating over 10^{15} Hz bandwidth)	3×10^7 (e.g. 1 mw common He-Ne laser radiating over 10^6 Hz)
Spectral radiance watts m^{-2} sr^{-1} Hz^{-1}		2×10^{-8} (sun radiating per wavelength unit)	3×10^1 (a small He-Ne laser radiating per wavelength unit

providing a very high spectral radiance (i.e. power per unit area and per unit wavelength unit). Such sources of light, operating at many discrete wavelengths within the UV-Vis-IR bands, are now commercially available (see Figure 3.9).

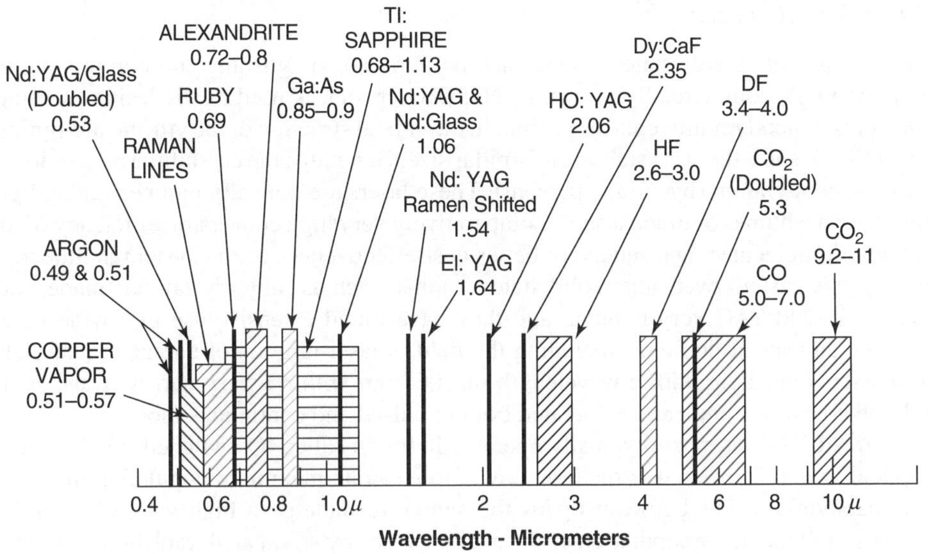

Figure 3.9 Wavelength range of laser outputs from some molecular/atomic species. Reproduced from http://holoinfo.no-ip.biz/wiki/index.php/Laser.

3.7 Suitable Laser Sources for Ignition Applications

For laser ignition of energetic materials, the choice of laser will be dictated by the requirement of a specific application. Energetic materials encompass all types of pyrotechnic, propellants and explosives. These are used for different purposes. For example, in some cases, explosives need to be ignited to lead to detonation, while the propellants are ignited for gas generation in missiles or airbag applications. Also, detonation in explosives may be initiated by two different processes (vide afro) and these will require two different types of lasers. Ultimately, the practical application of laser for ignition initiation in energetic materials will, primarily, be governed by two factors: portability and cost-effectiveness. From these considerations, two types of lasers are now most widely used for research and development. These are:

1. Nd:YAG lasers
2. Diode lasers (including LEDs).

Diode lasers and LEDs allow portability, ruggedness and fulfil the cost-effective criterion, albeit mainly for initiating a burning process in pyrotechnics or propellants, whereas the former type, being capable of providing high peak power within a relatively short-duration pulse, is more suitable for initiating detonation in explosives, albeit these being comparatively bulky and costly. Since both of these types of lasers are important for the present topic, they are briefly described below (details can be found in the bibliography at the end of this chapter).

3.7.1 Nd:YAG Laser

In this type of solid state lasers, neodymium-doped yttrium aluminium garnet ($Nd{:}Y_3Al_5O_{12}$), commercially known as Nd:YAG crystal, is used as the lasing medium. Triply ionized neodymium replaces yttrium in the crystal structure of the yttrium aluminium garnet (YAG), since these crystals are of similar size. Generally, the crystalline host is doped with ≈1% neodymium (by atomic percent). These lasers are optically pumped, using light from either flashtubes or diode lasers. Comparatively very high conversion efficiency of the neodymium molecules, and relatively better cost-effectiveness of the system architecture based on this system over other solid-state systems, such as ruby crystal, has made laser sources with Nd:YAG very popular, and they have found extensive use in a wide range of applied and scientific fields, including the field of laser ignition of energetic materials. These lasers emit light with a wavelength of 1064 nm within a very narrow wavelength band in the infrared, and can be operated both in pulsed and continuous modes.

Pulsed Nd:YAG lasers are typically operated in the so called 'Q-switched' mode, where an optical switch is inserted in the laser cavity to generate maximum population inversion in the neodymium ions before it opens the switch to undergo a high level of emission through simultaneous depopulation of the excited states by stimulated transition to a lower state within an extremely short time scale. In this mode, output powers in excess of 250 megawatts within durations of 10–25 nanoseconds (pulse width) and a beam diameter of a few mm have been achieved. The high-intensity pulses may be efficiently frequency doubled to generate laser light at 532 nm, or higher harmonics to obtain output at 355 and 266 nm.

3.7.2 Light Emitting Diodes (LEDs)

LEDs are electro-luminescence devices based on conventional electronic semiconductor diodes with p-n junctions. In the operational mode, a reverse bias voltage is applied so that the electrons and the holes (regions with charges opposite to that of the electrons) combine at the junction of the p- and n-type materials and release the energy as quanta of light (photons). The LEDs are generally super radiant devices, and often outputs from many tiny devices are cleverly combined to generate very high power output within relatively broad wavelength bands. The centre wavelength (i.e. the colour of the light) from an LED is governed by the band gap energy of the semiconductor material used for making the LEDs. The impurities used to create the p-n junction can be tailored to create the desired band gap and, hence, output at a wavelength band covering the whole visible wavelength range. There are many different kinds of LEDs, operating in very different regimes of optical output power, wavelength, bandwidth, and other different properties, as shown as examples in Table 3.3. Cheap and cheerful LEDs, providing CW power in excess of 50 W over broad bands in excess of 100 nm, are commercially available.

High beam divergence, practically emitting in all directions, but made somewhat directional by external (integrated) reflectors, low focus ability, very low spatial coherence and

Table 3.3 *Examples of wavelength ranges of some selected semiconductor materials for LEDs and their typical operating voltages.*

Colour	Wavelength band [nm]	Operating Voltage [V]	Semiconductor material
Infrared	$\lambda > 760$	$\Delta V < 1.9$	Gallium arsenide (GaAs) Aluminium gallium arsenide (AlGaAs)
Red	$610 < \lambda < 760$	$1.63 < \Delta V < 2.03$	Aluminium gallium arsenide (AlGaAs) Gallium arsenide phosphide (GaAsP) Aluminium gallium indium phosphide (AlGaInP) Gallium (III) phosphide (GaP)
Orange	$590 < \lambda < 610$	$2.03 < \Delta V < 2.10$	Gallium arsenide phosphide (GaAsP) Aluminium gallium indium phosphide (AlGaInP) Gallium (III) phosphide (GaP)
Yellow	$570 < \lambda < 590$	$2.10 < \Delta V < 2.18$	Gallium arsenide phosphide (GaAsP) Aluminium gallium indium phosphide (AlGaInP) gallium (III) phosphide (GaP)
Green	$500 < \lambda < 570$	$1.9 < \Delta V < 4.0$	Indium gallium nitride (InGaN)/ gallium (III) nitride (GaN) Gallium (III) phosphide (GaP) Aluminium gallium indium phosphide (AlGaInP) Aluminium gallium phosphide (AlGaP)
Blue	$450 < \lambda < 500$	$2.48 < \Delta V < 3.7$	Zinc selenide (ZnSe) Indium gallium nitride (InGaN) Silicon carbide (SiC) as substrate Silicon (Si) as substrate – (under development)
Violet	$400 < \lambda < 450$	$2.76 < \Delta V < 4.0$	Indium gallium nitride (InGaN)
Purple	multiple types	$2.48 < \Delta V < 3.7$	Dual blue/red LEDs, blue with red phosphor, or white with purple plastic
Ultraviolet	$\lambda < 400$	$3.1 < \Delta V < 4.4$	Diamond (235 nm) Boron nitride (215 nm) Aluminium nitride (AlN) (210 nm) Aluminium gallium nitride (AlGaN) Aluminium gallium indium nitride (AlGaInN) – (down to 210 nm)
White	Broad spectrum	$\Delta V = 3.5$	Blue/UV diode with yellow phosphor

relatively low power in a relatively broad wavelength band (≈100 nm), makes the application of such lasers limited primarily to the fibre optics communication industry. In recent times, much development in high-power diode lasers operating in the near-IR wavelength band has taken place. It is envisaged that, with dedicated research and development, such lasers will find widespread applications in the ignition initiation of energetic materials, both in the defence industries and also in the civilian sectors.

3.7.3 Diode Lasers

Diode lasers, sometime referred to as 'laser diodes' (LD), like the LEDs, are also based on semiconductor based technology, where a reverse bias voltage is applied between the ends of a p-n junction device, as shown schematically in Figure 3.10.

In diode lasers, the amplification is obtained by fashioning an integrated cavity like that of any other conventional lasers. Diode lasers are often coupled together by ingenious technology for achieving high output power. Many different types of diode lasers, operating at wavelengths mainly in the near-IR wavelength band, approximately between 800 and 930 nm, and having a wide range of output power, are readily available commercially. These are briefly discussed below:

- *Edge emitting diode lasers* are characterized by the resonator formed by end facets (cleaved edges) of the semiconductor wafer. These normally have a relatively low threshold pump power and high conversion efficiency. These very small-sized diode lasers can

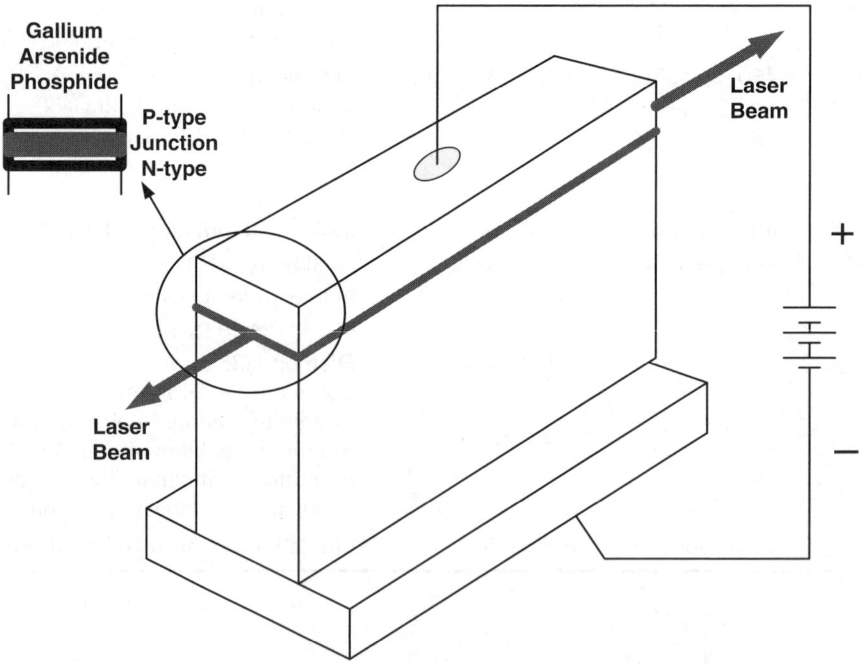

Figure 3.10 *Schematic of the structure and operational principle of a diode laser.*

generate up to half a watt of output power in a beam with high beam quality ($M^2 \approx 1$). The output may be coupled into a single-mode optical fibre, allowing beam homogenization and a near-Gaussian, but still rather highly divergent, beam output.

- *Surface emitting diode lasers* are a different kind of semiconductor laser. The emission direction is perpendicular to the wafer surface. In this category, the high brightness type has wide stripes, typically 300–400 μm wide and 100–150 μm long, capable of generating up to a few watts of output power. The beam quality is significantly poorer than that of lower-power LDs, but better than that of diode bars (see below). Tapered broad-area lasers can exhibit an improved beam quality and brightness.
- *Small LDs* are made as distributed feedback lasers (DFB lasers) or distributed Bragg reflector lasers (DBR lasers) with short resonators. These may achieve single-frequency output and can operate with wavelength tunability.
- *External cavity diode lasers* contain a laser diode as the gain medium of a longer laser resonator, with additional optical elements such as laser mirrors or a diffraction grating. They are often wavelength-tuneable and exhibit a relatively small emission line width.
- *Broad-area diode lasers* (also often called broad stripe laser diodes, wide stripe lasers, or high brightness diode lasers) generate up to a few watts of output power. The beam quality is significantly poorer than that of lower-power LDs, but better than that of diode bars (see below). Tapered broad-area lasers can exhibit an improved beam quality and brightness.
- *Slab-coupled optical waveguide lasers (SCOWLs)*, contain a multi-quantum well gain region in a relatively large waveguide. These can generate a watt-level output in a diffraction-limited beam with a nearly circular spatial profile.
- *High-power diode bars* contain an array of broad-area emitters, generating tens of watts with poor beam quality. Despite the higher power, the brightness is lower than that of a broad-area LD.
- *High-power stacked diode bars* (i.e. *diode stacks*) are stacks of multiple diode bars for the generation of extremely high powers of hundreds or thousands of watts.
- *Monolithic surface-emitting semiconductor lasers (VCSELs)* typically generate a few milliwatts with high beam quality. There are also external-cavity versions of such lasers (VECS). Laser diodes may emit a beam into free space, but many LDs are also available in fibre-coupled form. The latter makes it particularly convenient to use them, for example, as pump sources for fibre lasers and fibre amplifiers which, in turn, can generate much higher powers with still excellent beam quality.

3.8 Beam Delivery Methods for Laser Ignition

3.8.1 Free Space Delivery

For efficient laser ignition, the laser beam needs to be delivered to the target over a very small focused area. The tighter the beam is focused, the higher will be the power density, albeit with a local thermodynamic process restricting the lower limit for the dimension of the focused beam spot. This limit is dictated by the inherent wave nature of light, allowing it to cause diffraction at the edges of optical components (e.g. lenses). For a highly parallel beam and an ideal lens (negligible aberrations – spherical or chromatic), the diffraction limited spot size is given by Equation 3.5, cited earlier in this chapter. Free space beam delivery systems utilize lenses and mirrors to accomplish this purpose. Usually an up-collimator (lens for expanding beam) is used to increase the size of the beam so that, when a down-collimator (lens for collimating beam) is used, the beam divergence is reduced. A mirror may be used to direct the beam towards the target material and an objective lens is used to focus the beam onto the sample.

The difficulties with the free space beam delivery system arise from the inherent divergence of the laser beam, albeit very small for some lasers. For a long-distance beam delivery situation, the beam can become large, requiring corresponding increases in the diameters of the required optical elements. In the case of the objective lens, increasing the diameter limits the minimum focal length and causes spherical aberration. This has the consequence of reducing the minimum focused spot size. In addition to this problem, conventional beam delivery systems are rather inflexible. Changing the relative positions of any of the elements can cause misalignment problems. As the distance between the laser and the objective lens changes, the focused spot size also changes. Unless the optics are fixed with respect to the target, any small variation in the distance between the objective lens and the target will cause a large difference in the spot size and will thereby compromise reproducibility of any laser interaction event. For applications in laser ignition of energetic materials, delivery of the laser radiation through a flexible optical fibre system is highly desirable for a variety of reasons. For example, this will provide flexibility (position and orientation) in positioning the focused spot, afford safety through laser beam confinement within the fibre optic cable, allow a constant beam diameter over a range of distances, and so on.

For most practical purposes, the output of a very low beam divergent beam (a few milliradians) from, for example, an Nd:YAG laser or an Ar-Ion laser, may be focused to a 'diffraction limited' spot size, given by the formula in Equation 3.5, and this is illustrated using beam tracing approach, as shown in Figure 3.11.

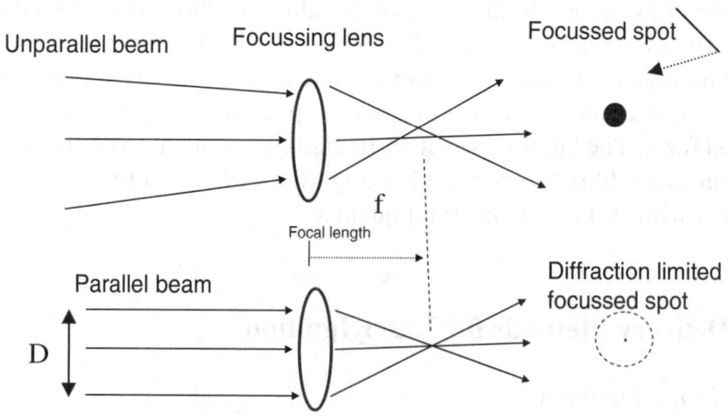

Figure 3.11 *Diffraction-limited spot size at the focus of a lens for a parallel laser beam.*

3.8.2 Fibre Optics Beam Delivery

The output of diode lasers and light emitting diodes are usually highly divergent. It is, therefore, not possible to focus such a beam to a small enough diameter to obtain high

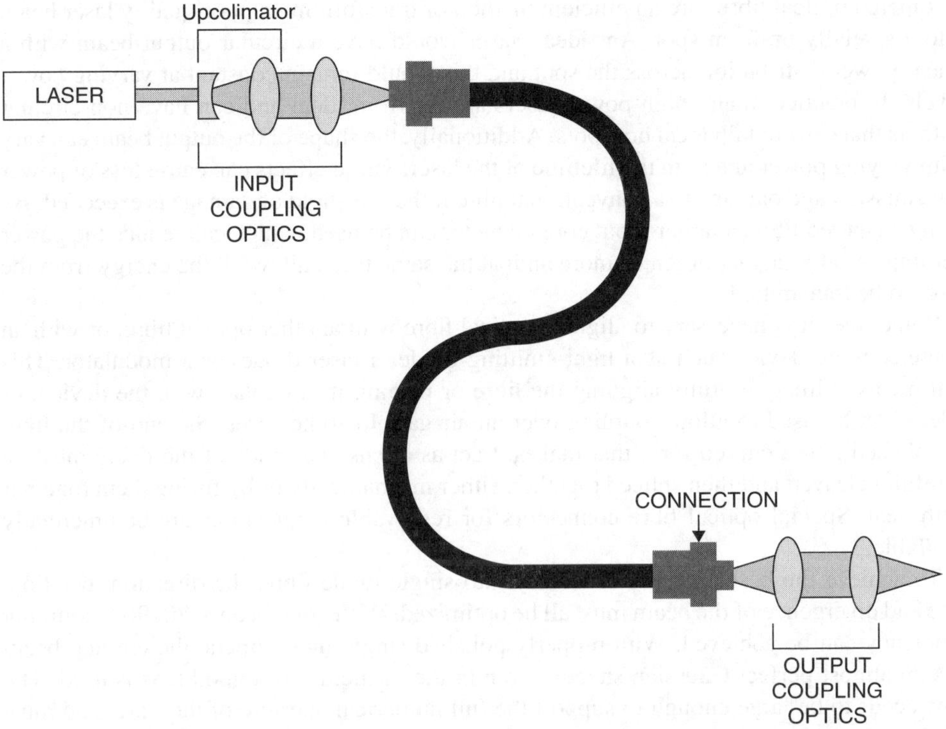

Figure 3.12 *Schematic of a typical fibre optics laser beam coupling system.*

power density. The beam from these sources are often coupled to other devices or delivered to targets through optical fibres. For most efficient coupling such fibres, having highly polished end faces, are directly and closely coupled to the active element of the laser diode. A scheme of such a coupling utilizing focusing lens systems is illustrated in Figure 3.12.

Optical fibres have a diameter between ≈0.1 mm and ≈0.6 mm, are made of very pure glass (silica) and are usually flexible enough to withstand some limited bending. Typically, an optical fibre cable consists of a transparent core surrounded by a transparent cladding material with a lower index of refraction. This is again coated with high-performance plastic material such as Kevlar for mechanical stability and protection from dust and humidity. The light propagates within the core by total internal reflection, thus sustaining minimum loss and making it act as a waveguide. Fibres that support many propagation paths or transverse modes are called multi-mode fibres (MMF), while those which only support a single mode are called single-mode fibres (SMF). Multi-mode fibres generally have a larger core diameter and are used for short-distance communication links and for applications where high power must be transmitted. Single-mode fibres are used for most long-distance communication links. For high-power, short-range applications, such as those for application in laser ignition of energetic materials, the fibres need to have a surrounding core with stabilizing flexible metal fibres and further plastic covering.

Tapered optical fibres are an efficient method of transforming a poor-quality laser beam into a spatially uniform spot. An 'ideal' laser would have a circular output beam with a linear power distribution across the spot and this would remain constant at varying power levels. In practice, many high-power lasers are far from ideal and can have non-circular outputs that can contain local hot spots. Additionally, the shape of the output beam can vary with varying power and with the lifetime of the laser. These effects can cause loss of power transmission and damage to a conventional fibre if the threshold of damage is exceeded. By using a tapered fibre, a larger input core diameter can be used, which can reduce the power per unit area by factors of ten or more and, at the same time, allow all the energy from the laser to be transmitted.

Sometimes it is necessary to align an optical fibre with another optical fibre, or with an optoelectronic device such as a light-emitting diode, a laser diode, or a modulator. This can involve either carefully aligning the fibre or placing it in contact with the device, or a lens can be used to allow coupling over an air gap. In some cases, the end of the fibre is polished into a curved form that makes it act as a lens. The ends of the fibers must be carefully cleaved and then spliced together, either mechanically or by fusing them together with heat. Special optical fibre connectors for removable connections are commercially available.

To achieve the best injection efficiency into single-mode fibre, the direction, position, size and divergence of the beam must all be optimized. With good beams, 70–90% coupling efficiency can be achieved. With properly polished single-mode fibers, the emitted beam has an almost perfect Gaussian shape – even in the far field – if a good lens is used. The lens needs to be large enough to support the full numerical aperture of the fibre, and must not introduce aberrations in the beam. Aspherical lenses are typically used.

There are important limitations to selecting the appropriate (smallest) fibre size. Clearly, the overall mechanical stability of the laser system is important. The laser beam-focused spot size has to be smaller than the fibre core to avoid heating effects and to allow for the mechanical tolerance of the fibre-optic connectors used for easy interchange of fibres without realignment. However, the most fundamental limitation is imposed by the combination of the laser's beam quality and the numerical aperture (NA) of the fibre. Together, these prevent arbitrary usage of any focal length lenses in the launch optics. Working from the equation for the focused spot size for an M^2-times diffraction-limited beam, only two assumptions are necessary to determine the minimum fibre size.

The beam quality of a laser is defined as the ratio of the beam waist and the beam divergence of the laser under consideration to that expected from the one with Gaussian beam profile at the same wavelength. The beam quality of the fibre output depends on the distribution of optical power over the fibre modes. This, again, depends on the launching conditions. For a step-index multimode fibre, the beam quality M^2 factor can be roughly estimated, with the assumption that the power is well distributed over the modes so that the numerical aperture represents a reasonable (perhaps slightly too high) estimate for the actual beam divergence. This leads to an equation as follows:

$$M^2 \approx (\pi\alpha) \times \lambda^{-1} \times \mathrm{NA} \tag{3.8}$$

where α is the fibre core radius (i.e. half the core diameter). Best beam quality is achieved when $M^2 \approx 1$ (i.e. the diffraction limited case).

3.9 Laser Safety

The laser is an effective tool for a plethora of applications. Examples include as a surgical scalpel tool in biomedical applications; as cutting, drilling and burning tools in a variety of industrial and military applications; and so on. It is accepted that an effective tool must have a sharp edge, and the laser is no exception. It is, therefore, mandatory to understand the mechanism of laser interaction with biological tissues, so that adequate precautions can be taken for safe use of lasers, particularly for application in research and development on laser ignition of energetic materials. The topic is an important and expanding branch of 'laser interaction with materials', and details may be found in many text books and review papers. Following is a brief analysis of the interaction processes for various types of lasers and tissues, the damaging effects due to such interactions and the precautions to be undertaken.

3.9.1 Laser Interaction with Biological Tissues

Two types of biological tissues are of concerned in connection with laser safety, and these are the eyes and skin. For low to moderate power/energy densities, such as those expected from diode lasers or Nd:YAG lasers, the effect will be severe for tissues within the eye. Therefore, this section is primarily devoted to the effects of laser irradiation on the eye.

The pupil of the eye is a perfect lens, i.e. free from any spherical or chromatic aberrations, and thus allowing diffraction-limited focusing of a near parallel laser beam. For a continuous wave (CW) laser such as the Helium-Neon laser (632.8 nm) or a pulsed laser such as the Nd-YAG laser (1643 nm), the power density at the focal point of this lens could be five to six orders of magnitudes higher than that at the laser exit end. For highly divergent diode lasers, the enhancement may not be as drastic, yet it may be high enough to cause permanent damage to sensitive retinal tissue cells. Lasers operating over the entire visible to near-IR wavelength range (approximately 400–780 nm), at low to moderate energy densities, will penetrate the pupil (eyeball) to cause retinal damage, primarily through a photochemical process. Laser output in the near-IR wavelength band (approximately 780–1400 nm) may cause both cataract (clouding of the lens) and retinal burn. However, lasers operating in the very short wavelength range (180–315 nm) and over the IR band (1400–3000 nm) will not penetrate the region adjacent to the pupil, and will be absorbed within the cornea and give rise to photokeratitis (inflammation of the cornea), cataract and corneal burn, primarily through a photo-thermal interaction mechanism.

3.9.2 Precaution against Ocular Hazards

For the safe use of lasers, both in the laboratory and field applications, appropriate protective goggles need to be used. For this, it is necessary to know the minimum laser energy density that the eye can tolerate without suffering any physiological damage when the beam, either directly (intrabeam viewing), or scattered or reflected, is exposed to the eye (cornea). Laser goggles are designed with wavelength-selective light-absorbing coating on the surface of the viewing glasses. The transparency factor needed for safety during laser operation for a specific laser depends on the 'minimum permissible exposure (MEP)' factor. However, as stated earlier, the interaction processes are different for lasers with different wavelengths and different modes of operation (e.g. single pulse, repetitively pulsed or continuous wave), in addition to their different levels of output power/energy densities.

In order to choose appropriate protective goggles, the MPE factor (W cm^{-2}/J cm^{-2}) must be judiciously established for the particular laser type under consideration. The appropriate goggles will then provide protection to the eye when the beam is exposed to the cornea of the human eye. The factor that defines the probability of damage to the eye to be very low is roughly parameterized as 10% of the dose that has 50% chance of creating damage under a worst-case scenario. Note that for visible radiation, the MPE is usually very low. For example, for a CW He-Ne laser, it is as low as 10^{-2} W m^{-2} (corresponding to $\approx 10^{-8}$ W for a beam of 1 mm^2 beam dimension). For pulsed visible lasers, the MPE can be an order of magnitude lower than that for the long-viewing condition for CW lasers. It is generally accepted that most lasers, including some laser pointers, deliver output which exceeds the recommended MPE for direct viewing.

Lasers are classified in accordance with their ability to cause ocular damage. This classification provides a guide to the use of appropriate safety goggles to limit the exposure of the eye (either within the blink response time or pulsed duration) to the beam (either directly or diffusely reflected). These categories are as follows:

- *Class 1*: lasers belonging to this class are, for all intent and purposes, safe. Some laser pointers and high-divergent diode lasers belong to this category.
- *Class 1M*: lasers belonging to this class are safe for all conditions of use, except when passed through magnifying optics such as microscopes and telescopes.
- *Class 2:* lasers of this class are generally safe, due to the blink response, which will limit the exposure to less than ≈0.25 seconds. This only applies to visible-light lasers (400–700 nm). Class 2 lasers are limited to 1 mW continuous wave, or more if the emission time is less than 0.25 seconds, or if the light is not spatially coherent. Some laser pointers and measuring instruments are put into class 2.
- *Class 2M*: lasers of this class are safe, as those of class 2, but unsafe when viewed through an optical instrument such as a microscope.
- *Class 3*: lasers in this class are considered to be safe if handled with restricted beam viewing. MPE for visible continuous lasers in this class are limited to 5 mW. For other wavelengths and for pulsed lasers, other limits apply. Direct exposure must be avoided for these lasers. However, diffused reflections, such as that from matt white paper, may safely be viewed with naked eye.
- *Class 4*: lasers belonging to this class are highly dangerous and will cause ocular damage even when viewed from a diffused reflection source. By definition, a class 4 laser can burn the skin, or cause devastating and permanent eye damage as a result of direct, diffuse or indirect beam viewing. These lasers may ignite combustible materials, and thus may represent a fire risk. Use of such lasers must be controlled by mechanical means (e.g. shutter), administrative precautions (follow safety protocol) and, of course, by adequate laser safety goggles.

The administrative protocol for the safe use of lasers is provided in the European directive, EN207, which also provides specifications for appropriate laser safety goggles. All commercial laser-eye protection goggles must be certified and levelled with the CE mark. The US norm (ANSI Z 136) for laser safety is comparatively less stringent than that of the EU one. An EN 207 specification might read *IR 315–532 L6*. Here, the letters *IR* indicate the laser working mode, in this case a *pulsed* mode. The range *315–532* indicates the wavelength range in nanometres. Finally, the scale number *L6* indicates a lower limit for the optical density, i.e. the transmittance within this wavelength range is less than 10^{-6}.

Bibliography

Xinju, L. (2010) *Laser Technology*, 2nd edn, CRC press, ISBN: 142009081X.

Hecht, J. (2008) *Understanding Lasers*, 3rd edn, IEEE press, ISBN: 978-0-470-08890-6.

Hitz, B., Ewing, J.J. and Hecht, J. (1998) *Introduction to Laser Technology*, 3rd edn, IEEE press, ISBN: 0 7803 3440 X.

Csele, M. (2004) *Fundamentals of Light Sources and Lasers*, Wiley, ISBN 0-471-47660-9.

Siegman, A.E. (1986) *Lasers*, University Science Books, ISBN 0-935702-11-3.

Silfvast, W.T. (1996) *Laser Fundamentals*, Cambridge University Press, ISBN 0-521-55617-1.

Crisp, J. and Elliott, B. (2005) *Introduction to Fiber Optics*, 3rd edn, Newnes/Elsivier (pub.), ISBN 0750667567.

Schroder, K. (ed.) (2000) *Handbook on Industrial Laser Safety*, Technical University of Vienna (pub.).

British Standard BS EN 6085: 1992 Radiation safety of laser products, Equipment Classification, requirements and users guide, BSI Linford Wood, Milton Keynes, MK14 6LE.

Paschotta, Rudiger (2007), 'Beam Profiler', http://www.rp-photonics.com/beam_profilers.html (Last accessed 3rd December 2013).

Coffey, Jerry (2010), Atomic Structure, http://www.universetoday.com/56747/atom-structure/ (Last accessed 3rd December 2013).

Vladislav, I (2008), 'Description of Stimulated Emission', http://en.wikipedia.org/wiki/File: Stimulated_Emission.svg (Last accessed 3rd December 2013).

Wikipedia (2013), Laser – Design, http://en.wikipedia.org/wiki/Laser (Last accessed 3rd December 2013).

Wikipedia (2013) Laser-holowiki, http://holoinfo.no-ip.biz/wiki/index.php/Laser (Last accessed 3rd December 2013).

4
General Characteristics of Energetic Materials

4.1 Introduction

Before considering the characteristics of energetic materials, particularly explosives, it is important to appreciate that the simple definition of an energetic material as a source of chemical energy covers such a wide range of materials that it would almost be easier to name non-energetic materials. To be useful to engineers, scientist and military personnel, the energetic material has to be capable of providing this useful energy rapidly in the form of work, without the intervention of any other external material.

There is an important caveat that the material should be safe to make, store and use. It should not function accidentally, but should readily perform to specification on demand. The range of explosive materials is very extensive and ranges through all physical states of matter and from simple compounds to complex mixtures. The simple statement that an explosive is a material capable of causing an explosion assumes knowledge of what an explosion is, and that everything capable of causing an explosion is suitable for use as an energetic material (explosive) which is definitely not the case. A brief discussion of what constitutes an explosion will help the understanding of energetic materials.

4.2 The Nature of Explosions

The phenomenon called an explosion is easier to define than the term explosive: an explosion is a violent expansion of gas at high pressure. This expansion of the gas can be harnessed to do useful work. Explosions are common in nature, ranging over a wide scale of magnitude and location from the cosmic to the terrestrial. On the cosmic scale, the origin of the universe is described as The 'Big Bang', and conditions were such that no living organisms could have survived the event. Island volcanoes, such as Krakatoa, can and have exploded

Laser Ignition of Energetic Materials, First Edition. S Rafi Ahmad and Michael Cartwright.

with great violence, causing extensive damage. At the lower end of the scale, a discharge of lightning can explosively damage objects in the way of its path to earth, such as a tree trunk or buildings not protected by a lightning conductor. These two terrestrial examples are due to the vaporization of water without any chemical change occurring.

Other physical explosions can occur from man-made causes, such as the bursting of a high-pressure boiler. Pressurized gas cylinders can explode if over-pressurized, such as can occur when they are caught in a fire. None of these examples require the existence of what is normally called an explosive substance. Nevertheless, physical explosions are inherently destructive processes, capable of causing damage by air blast and by the propulsion of debris at high velocity. Many workplace hazards arise from physical explosions.

A second class of explosion is the nuclear type. These explosions are the result of the processes of fission or fusion of atomic nuclei. During both processes, some mass is lost and this lost mass is converted into enormous amounts of energy, which is rapidly released as a complete spectrum of electromagnetic (EM) radiation, from short-wavelength gamma radiation, through visible light, to the IR and heat end of the spectrum. The actual expanding gas, in this case, is mostly that of the air surrounding the device, since the temperature is increased from 293 °K to $\approx 2 \times 10^6$ °K in microseconds, creating an intense local pressure rise. Most terrestrial materials in contact with this temperature will be vaporized, adding to the gas pressure. Primary material damage is due to air blast and heat, with additional effects due to radiation. The sun is an enormous fusion explosion system which continually radiates EM radiation in the direction of earth but, because there is no operating medium between earth and sun, none of the blast effects occurring from the nuclear explosion are felt on earth. Solar flares produce significant radiation-induced changes in the earth's atmosphere.

The radioactive elements which produce nuclear explosions are not normally referred to as explosives, since they do not respond to conventional initiation techniques. Incidentally, it is necessary to use an arrangement of chemical explosive charges to trigger a nuclear explosion by forcing sub-critical masses together at high pressure and elevated temperature.

The third class of explosion, the chemical type, is caused by the rapid decomposition of a chemical system, with the evolution of gas and, usually, heat (but note that this latter product is not strictly necessary). In general, the decomposition takes the form of combustion – that is, an exothermic oxidation reaction which produces some visible manifestation such as flame or smoke (exceptions to this are certain compounds that decompose explosively without the formation of oxidation products).

There are numerous substances known to chemists that are so unstable that they may explode at room temperature, even when little or no stimulus is applied. Examples of these systems are the ammoniate of nitrogen tri-iodide, produced by the reaction of concentrated ammonia with iodine and, as discussed later, some of the metallic complexes being investigated as alternative primer systems.

Therefore, if we simply define an explosive as a substance capable of causing an explosion, we shall include many which are of no practical value and which, as a class, constitute little more than a dangerous inconvenience in laboratory work. In order to be of use, an explosive substance must possess properties such that it will explode only when it is required to do so. In practice, this implies that it must be chemically inert to any other substance with which it may commonly come into contact (including air and moisture), and that it must be thermally stable at normal ambient temperatures. At the same time, its ignition temperature must be low enough to allow initiation by some convenient means. It is a characteristic of practical explosives that the minimum energy required for initiation is invariably small by comparison with the subsequent release of energy by the charge.

These various properties collectively form the basis of two essential requirements of a practical explosive – namely, safety and reliability. The material must be safe to manufacture, store and use, but should be readily set off when required and have a high resistance to accidental initiation. The application of these two criteria rules out many explosive substances as useful candidates. One of the principal reasons for examining laser initiation is the potential enhanced safety margins that the method could offer over conventional initiation techniques.

There is a further basic criterion to be considered, namely effectiveness. All chemical explosives generate heat, but some produce little or no gas. This can be utilized in pyrotechnics, where the object of the operation is to produce a special effect other than work (i.e. smoke or light). An adequate work capacity in an explosive demands the production of both heat and gas. Compared with the heat released by a nuclear reaction, that produced by a chemical explosion is many orders of magnitude lower. It is therefore impossible for a sufficiently high temperature to be attained by the surrounding atmospheric gases to utilize them as a work medium, or working fluid, as happens in the case of a nuclear explosion. The chemical explosive must produce its own work medium in the form of evolved gas or gases, and the molar total of such gases must be quite high in relation to the mass of explosive generating it.

We may therefore define a practical explosive as 'a substance which, when suitably initiated, decomposes explosively with the evolution of heat and gas'. We can also compare a chemical explosion with a physical or nuclear one in regard to damage capacity: a chemical explosion can cause damage by shock pressure, air blast, propulsion of material and sometimes by incendiary effect. Generally their effects are intermediate between physical explosions and the devastation of a nuclear explosion.

4.3 Physical and Chemical Characteristics of Explosives

Chemical explosive systems exist in a wide range of physical forms. They can be either single compounds or mixtures. If they are mixtures, their constituents may be all compounds, or compounds with elements or, conceivably, elements only. The physical state of explosives can be solid, liquid or gaseous, and explosive mixtures can be, and frequently are, two-phase systems typified by modern ammonium nitrate (AN)-based emulsion and gels widely used in the quarrying and civil engineering industries. Also important is the mixture of **a**mmonium **n**itrate with various liquid hydrocarbon **f**uel **o**ils, to give ANFOs.

Most practical explosives are solids or liquids, i.e. 'condensed' substances. Gases have such a low density that the change in volume following explosion is much less than for solids and liquids, hence the destructive power of these explosions is less; the velocity of the shockwave can be as low as 2 km s^{-1}, in contrast to the typical 8 km s^{-1} from solids. Nevertheless, gases can represent a significant explosion hazard in the workplace, such as underground mines, where methane gas is often found in coal seams. Although the effect is reduced in comparison with close to the site of a solid explosive, the effect is constant throughout the dimensions of the gas cloud and, hence, can be spread over a wide area. The shock pressure from gaseous systems will still exceed the strength of the human body, and fatalities will result.

Dispersing a combustible fuel in air and then initiating the mixture can produce both a viable weapon, FAX, and a significant explosive hazard, hence the restrictions on smoking or the use of open electrical switch systems at fuel filling stations. Ethyne (acetylene) is a typical example of a gaseous system capable of violent explosion when suitably initiated, and there is an additional hazard with the gas. It will react with some metals, such as copper and silver, to form metal acetylides, which are extremely sensitive to impact, friction and electric spark. As a result, the use of the gas is very strictly regulated, and the metal pipework has to be carefully selected – usually stainless steel or nickel lined.

4.4 Fuel and Oxidizer Concept

The chemical reaction which occurs in most explosion processes is the oxidation of one or more fuel elements. This reaction between a fuel and an oxidizer is critical to all energetic materials. If the desired reaction is to propel a projectile, then useful work has to be performed and the products must be gaseous. On the other hand, pyrotechnics are designed to produce special effects and, as such, they may not be required to produce gas. This opens up the choices for both fuel and oxidizer. Lists of some suitable materials for these two components are given in Table 4.1 below. This is only an indicative table, and many other materials can act as both fuel and oxidizer.

When examining the periodic table of the elements, the only gaseous oxides are those of a few non-metallic elements. Of these, only carbon monoxide (CO), carbon dioxide (CO_2) and hydrogen oxide (H_2O: water), have the correct properties to be suitable gaseous products of an explosion. Oxides of the halogens are gaseous, but are unstable and are sometimes spontaneously explosive, producing oxygen. The oxides of nitrogen are usually a source of oxygen, since the bond between nitrogen atoms is stronger than any of the bonds between nitrogen and oxygen. However, some nitrogen oxides are produced during a chemical explosion as an undesirable by-product, creating acrid fumes.

Carbon and/or hydrogen are usually the fuel elements. The oxidizing element is normally a source of oxygen but, in the case of pyrotechnics, other oxidizers are possible. Oxygen itself, being a gas, has low density and, hence, is not an efficient oxidizer. However, when liquefied, it plays a major role in rocket motors for space exploration (the most obvious alternative, fluorine, is found to be unsuitable for the synthesis of explosive compounds and on toxicity grounds). Normally, the source of oxygen is an anion which liberates oxygen on heating, i.e. nitrate (NO_3^-), chlorate (ClO_3^-) or perchlorate (ClO_4^-). The nitro group ($-NO_2$), when chemically bound to another atom in a covalent bond, is also used (see later). It is a characteristic of most explosives that the oxygen necessary for its combustion is contained within the chemical system in some form. Thus, the heat supplied by the initiatory stimulus and the subsequent onset of decomposition rapidly liberates all the oxygen atoms to combine with other elements.

The fact that explosives are (more or less) self-sufficient in oxygen gives rise to an important distinction between them and other flammable substances, which is that the combustion of explosives can take place equally well either in the presence or the absence of air. The propellant used in a gun burns in the absence of air, and the flash observed is when the hot combustion products meet the air and produces further burning. By contrast, ordinary solid combustibles such as paper, wood and fabrics are not capable of rapid,

Table 4.1 *Typical fuel and oxidizer compounds.*

Fuels Elemental			
Metals		**Oxidizers**	
Aluminium	Al	Oxygen gas	
Iron	Fe	Nitrogen oxides	N_2O_4
Magnesium	Mg	Liquid oxygen	
Manganese	Mn	Fuming nitric acid	
Titanium	Ti	Hydrogen peroxide	H_2O_2
Tungsten	W	**Solids**	
Zirconium	Zr	Hypochlorite	$Ca(ClO)_2$
Lead	Pb	Chlorates	$KClO_3$
Non-Metals		Perchlorates	$KClO_4$
Hydrogen	H	Iodates	KIO_3
Boron	B	Nitrates	$NaNO_3$
Carbon	C	Chromates	$BaCrO_4$
Silicon	Si	Dichromates	$K_2Cr_2O_7$
Sulfur	S	Oxides	Fe_3O_4
Phosphorous	P		Pb_3O_4
Combustible C & H Compounds			ZnO
Charcoal			CuO
Sawdust		Peroxides	BaO_2 (Ba +IV)
Paraffin wax		**Peroxy componds**	
Paraffin oil		Perborates	$Na_2H_2B_2O_8$
Diesel oil		Percarbonate	$K_2C_2O_6$
Napthalene		Persulfates	$K_2S_2O_8$
Glycerine		**Halocarbons**	
Sulfur		Hexachloroethane	C_2Cl_6
Sugar		PTFE polymer powder	
Hexamine (urotropine)			
Hydrazine N_2H_4			
(ii) **Semi–explosive**			
Nitrobenzene			
Nitrotoluene			
Dinitrotoluene			
Ethylene oxide			
(iii) **Explosive**			
Acetylene (ethyne)			

accelerating burning unless brought into good contact with a plentiful supply of air. A chip pan fire can be put out very effectively by covering with a wet cloth, which effectively excludes the air. The combustion of fuel in an engine requires a plentiful supply of air.

The penalty which explosives incur, however, by containing their own oxygen, is that the oxygen atoms are bound to constitute a considerable proportion by mass of the explosive. When combustion takes place, therefore, the total energy released per unit mass is less than

Table 4.2 Comparing energy released for fuels and energetic materials.

Fuel	Molar mass	Enthalpy of combustion	
		kJ Mole^{-1}	kJ g^{-1}
Hydrogen	2	242	121.0
Carbon	12	394	32.8
Methane	16	890	55.6
Ethane	30	1492	52.1
Propane	44	2223	50.5
Butane	58	2882	49.7
Octane (petrol/gasoline)	114	5513	48.4
Explosives			
Ethylene glycol dinitrate EGDN	152	1022	6.73
RDX	222	1139	5.13
Trinitrotoluene TNT	227	926	4.08

that produced by an ordinary combustible material burning in the atmosphere. For instance, a given mass of petrol burning in air generates about eight times more heat than the same mass of high explosive (see Table 4.2), although it has to be supplied with a large volume of air in order to do so. Notice that as the ratio of hydrogen to carbon decreases, so does the heat output.

An explosive is a secondary energy source, and its usefulness derives more from the speed at which it releases its energy, in sometimes confined conditions, rather than from the quantity of energy released (see Table 4.3). Again, data in the table show the importance of speed. Also, an explosive can function in a totally enclosed system, e.g. in a gun cartridge case or underwater.

The burning of the candle takes typically 24 hours, even with multiple wicks required to mix the liquid hydrocarbon with sufficient air to support combustion. The explosive has completely reacted in 22 μs after initiation has generated the shock wave. Notice also that, for the brief period of time that the detonation takes, the power output is an order of magnitude greater than that of a typical carbon fuelled electrical power generating station.

4.4.1 Explosive Mixtures

As we have seen, explosives exist in a number of forms and the simplest way to classify them is into mixtures and single compounds. Taking mixtures first, consider a mixture of gaseous hydrogen and oxygen, in the ratio of 2 : 1 by volume respectively. Complete

Table 4.3 Comparing energy released and power developed by a fuel burning with that of the best-performing explosive.

100 g of material	Energy content (kJ)	Power developed (W)
Candle burning (≈ 1 d)	4360	50
Detonating explosive (≈ 22 μs)	513	2.3×10^{10} (23 GW)
Typical electrical power station		3×10^{9} (3 GW)

mixing occurs naturally, and the system can be easily exploded by a spark or small flame. The reaction is given by the equation:

$$H_2\,(g) + {}^1\!/_2O_2\,(g) \rightarrow H_2O\,(g)$$

This yields no less than 13 260 J per gram of mixture, and the violence caused by release on this scale can be simply demonstrated by exploding the mixture in a toy balloon. The violence of this reaction when the stoichiometric mixture is initiated can destroy the container completely by burning to detonation. However, the practical application of the phenomenon in explosives technology is much more difficult because, in the gaseous state, the gases are too bulky for convenience; they can not be liquefied by the application of pressure alone under normal atmospheric conditions. If they are cooled sufficiently, then the liquid state is obtained, but they still have comparatively low density and, in the liquid state, they are dangerously cold; critical temperature for hydrogen is 15 K and it boils at 22 K (−253 °C). The corresponding data for oxygen is 80 K and 90 K (−183 °C). These liquids are thus highly volatile and could easily cause physical explosions on their own.

Attempts to use liquid oxygen as a blasting explosive were given up many years ago for these and other reasons. Perhaps the best-known application of the reaction today is in the field of space rocketry, due to the high specific impulse delivered by hydrogen against oxygen and in outer space, where the ambient temperature is $\approx$ 5 K and the liquids are readily stored. The original moon landing used hydrogen/oxygen rocket motors and they form the main engines of both the original Atlas rockets and the American space shuttle system.

For practical, instantaneous explosions, however, the reacting material is required to be in a compact, 'condensed' form. Therefore, as a second example of an explosive mixture, consider a solid one – and, in fact, the oldest explosive of all, known as black powder or gunpowder. Its three ingredients have remained unchanged to this day, although their proportions and the method of manufacture have been steadily improved over the centuries.

Roger Bacon (c. 1214–1292), believed to have been the first European to write a description of black powder in one of his ecclesiastical writings, gave its composition as equal parts of sulfur, charcoal and saltpetre (converted here into approximate percentages) as:

- Potassium nitrate: 37.5
- Charcoal: 31.25
- Sulfur: 31.25

With the development of modern science, it was realized that the proportion of potassium nitrate (the oxidant) had to be increased, while that of the fuel elements charcoal (carbon) and sulfur decreased, so that the composition of modern black powder is approximately:

- Potassium nitrate: 75
- Charcoal: 15
- Sulfur: 10

The sulfur, as well as being a fuel element, reduces the porosity of the grains and also reduces the ignition temperature of the mixture. The method of preparation involves rather more than simple mixing, in order to bring the components together as intimately as possible. Modern 'corned' (granulated) black powder burns much more rapidly than the pulverized type of older times. The individual ingredients are ground to dust and mixed together with a little water to form a paste, which is then pressed to form a solid pellet. After the pellet has been dried, it is then broken up into easily handled small particles. There is no dusting problem, but the burning characteristics are extremely good because of the intimate nature of the mixture. Gunpowder has been used over the centuries for all kinds of applications, including pyrotechnics, demolitions, as a propellant in guns and rockets, as a filling for explosive projectiles and for rock-blasting. It fulfils some of these roles better than others, but even at its best it has serious limitations as an explosive. These can be attributed to two main facts:

1. It is a mixture of non-explosive ingredients, which take an appreciable time to react together, and the explosion is therefore relatively slow. It is impossible for gunpowder to burn to detonation but it can be shocked to detonation.
2. It contains 36% of the metallic element potassium, so therefore an even greater percentage of the products of combustion are necessarily solid rather than gaseous.

Black powder was invented long before the advent of modern chemistry and, even now, it is not possible to describe its combustion adequately in a simple chemical equation. An over-simplified attempt, such as may be found in a school text-book, is as follows:

$$4KNO_3(s) + 7C(s) + S(s) \rightarrow 3CO_2(g) + 3CO(g) + 2N_2(g) + K_2CO_3(s) + K_2S(s)$$

From this, it can be seen that only three of the five main products are gases; in fact, experiment shows that only 43% by mass of a typical black powder is converted into gases. Expressed in terms of volume, this is only about a quarter of the gas produced by an equal mass of an efficient modern explosive. Moreover, because the quantity of the heat evolved is largely governed by the amounts of CO_2 and CO formed, black powder produces only half the heat that a modern explosive generates. Compared on a same-mass basis with the hydrogen-oxygen mixture mentioned earlier, it produces only a quarter of the heat.

Currently, black powder is not used in large quantities for any application, and even in its use as an intermediate in the ignition of propellants, it is being slowly replaced. Modern explosive mixtures, which have largely replaced it in commercial blasting operations, are more effective in most respects. The most common are those based on ammonium nitrate (AN). Ammonium nitrate is explosive in its own right; witness the major accident which occurred when a fire started in the hold of a cargo ship carrying a cargo of AN fertilizer tied up in a Florida harbour. Closing the hatches to exclude air led to a pressure rise which accelerated the rate of burning until it transitioned to detonation, with disastrous consequences for both ship and harbour. Pure AN is not easy to initiate, and requires a minimum diameter of over 1 m before it can sustain detonation – hence, it is easy to forget it is an explosive. Mixing AN with a solid fuel such as aluminium (ALAN) or sugar (ANS) gives better performance than gunpowder. Using AN solid with hydrocarbon fuels, such as diesel, we obtain ANFO. This is often the choice for improvised explosive devices, because of the ready availability of the ingredients and ease of manufacture. Also, the critical diameter to sustain detonation in these mixtures is reduced to centimetres rather than the metres of the pure AN.

Dissolving AN in hot water and forming an emulsion with the hydrocarbon fuel produces very safe 'emulsion' explosives, which are almost impossible to set off accidentally. The added advantage is that they are made from cheap, readily available materials, which can be mixed on site immediately prior to use. Emulsion explosives are widely used in the mining and quarrying industries because of their safety. They are not normally used without a booster between detonator and emulsion charge. It would be difficult to laser initiate emulsion explosives without some form of sensitizer which would absorb the laser radiation and then exothermically decompose.

However, even these mixtures are inferior to single explosive compounds, in which the oxygen and fuel elements are combined as atoms within each molecule, and it is to these materials that we now turn. Quality control of these single compounds is much easier than with mixtures of often disparate (in terms of particle size and density) solids. Ammonium nitrate and sugar have different densities so, on standing for any period of time, they can be vibrated into separating from the mixture, degrading the performance considerably. Even emulsion preparations can have dispersed phase size and density problems, depending on the manufacturing process.

Other mixtures of fuels and oxidizers involve liquids. Mixtures of hydrazine and hydrogen peroxide are used as liquid rocket propellants. Nitromethane, when mixed with amines, becomes a detonator-sensitive explosive with a comparable performance to TNT. Some of these mixtures can spontaneously ignite when mixed, and are classed as 'hypergolic'. Their use is mainly in rocket propellants. Two other areas where mixtures of fuel and oxidizer are very much to the fore are pyrotechnics and solid rocket propellants, some brief details of which are given below. More extensive discussion on both of these topics will be undertaken later

4.4.2 Pyrotechnics

One area of energetic materials that still relies heavily on mixtures of different compounds is pyrotechnics. Here, the chemical reaction produces a special effect. These are summarized in Table 4.4, with an indication of their uses.

In the table, only in the cases of sound and gas generation is there a requirement to produce gas but, in a number of reactions, gas is a by-product.

Pyrotechnics are often the mechanism whereby the initiatory stimulus (e.g. stabbing, impact, etc.) are converted to ignition, and this topic is covered in Chapter 7 on initiation. The rate of burning of a pyrotechnic is dependent on the following parameters of the materials used:

- Composition of mixture
- Particle size of ingredients
- Dimensions of burning surface
- Configurational heat loses (dependent on container material, thickness and geometry)
- Ambient pressure.

Table 4.4 *Pyrotechnic special effects.*

Effect	Uses
Heat	Ignition source, priming, incendiary, heater composition, delay
Light	Illumination, signalling, tracking, infrared emission, strobes
Smoke	Screening (visual and infrared), anti-personnel
Sound	Maroon, railway warning devices, battlefield simulators, whistles
Gas	Pyromechanisms, propulsion, inflation
Chemicals	Production of ceramics, alloys

Incendiary pyrotechnic systems fall into two types: liquid mixtures and solid mixtures. Simple hydrocarbon fuels are liquids which are easily dispersed over an area and are easily ignited, but not generally by laser. Mixing liquid aluminium alkyls with the fuel can cause spontaneous ignition on contact with air, and this represents a serious storage hazard. Their heat output is sufficient to combust soft targets such as wood and fabrics. The solid mixtures are usually more difficult to ignite, and one of the best contains aluminium fuel and ferric oxide oxidizers, identified as 'Thermite' according to the equation:

$$8\text{Al} + 3\text{Fe}_3\text{O}_4 = 4\text{Al}_2\text{O}_3 + 9\text{Fe} \quad q = -3.7\ \text{kJ g}^{-1}$$

Although the heat output is not particularly high, the temperatures achieved are very high and the iron metal produced is molten and sinks through the less dense aluminium oxide slag, which is a good insulator. Thus, the thermite reaction is ideal thermal reaction. This mixture is not easily initiated, requiring a strong primer such as silicon and lead oxide to start the reaction. This starter material has a lower thermal output than the thermite mixture above, but produces hot molten slag with only a small quantity of gas, due to the vapour pressure of molten lead at the high temperatures produced. The silicon-lead oxide mixture is very spark- and friction-sensitive, requiring careful handling. The thermite mixture is also used for the *in situ* welding processes necessary to produce continuous welded railway lines necessary for high-speed rail travel, with kilometres between the plated joints, and in conditions where normal gas or electric welding is impossible.

Delay elements burn to deliver an initiatory level quantity of heat/flame to the main energetic filling in a device after a predictable and specific time interval. Delays must be contained in a non-combustible tube and burn in a cigarette mode. The simplest device containing a delay element is the hand grenade, which allows the thrower to initiate the device before throwing, but only delivers the output after several seconds. The delay time is controlled by a number of parameters: the exact composition of the formulation; the particle sizes of the components; the confinement; and the length of the tube containing the delay. The effect of the composition parameter on the burn time of a common delay composition are given in Table 4.5 below.

Note that these compositions were all made from the same batch of chemicals, and extensive procedures were followed to avoid any separation of particle sizes in the samples and also to achieve perfectly homogeneous mixing. Compositions were pressed to the same pressure, and five trials at each composition were tested and results averaged. Safety fuze is an example of a delay element and is used in plain detonators (see Chapter 7 on initiation).

Table 4.5 *Pyrotechnic delay burn time as a function of composition.*

Ingredients	Composition %			
Tungsten	27	33	49	80
$BaCrO_4$	58	52	41	12
$KClO_4$	10	10	5	5
Superfloss	5	5	5	5
Burn rate s cm^{-1}	**16**	**11**	**4**	**0.6**

The light output from pyrotechnics depends critically on the purpose to which it is applied. Flash Illumination suitable for high speed photography requires high intensity light sources, typically 200×10^6 cd, but is only required for $\approx$ 10 ms. On the other hand, illuminating an area for observation purposes requires a lower intensity – typically 6×10^6 cd – but this output is required for $\approx$ 3 minutes. The most efficient light emitters are hot solids so, the more hot solids in the products, the higher the light output. The optimum light output is derived by reacting metals such as magnesium and aluminium, with oxidizers typified by perchlorates and nitrates. Control of the burning rate by combinations of particle size of the ingredients and the operating geometry are crucial. High surface area solid materials, ignited over the whole surface, produce the flash output, whereas larger particle size materials, burnt in a cigarette configuration, produce an illuminating output lasting for several minutes if necessary.

Perchlorates are the preferred oxidizer, since the presence of chloride species enhances the proportion of energy emitted as light. Control of the light colour is based on the counter-ion used. If white light is required, then the counter-ion is usually ammonium but, if coloured light is required, then metal counter-ions such as alkali or alkaline earth metals from Group I and Group II in the periodic table are used. Copper is also used, and here the colour of the light is either blue or green, depending on the exact conditions in the flame. One interesting light system is the stroboscopic light source. An oscillating chemical reaction, controlled by the composition and the method of burning, produces an oscillating light output. As the burning material is lifted off the surface, the reaction rate falls and so does the light output. The burning material then falls back on to the surface, is heated, and the rate of reaction and the light intensity increases. One unusual oxidizer is polytetrafluoroethylene (PTFE), which is used in mixtures with magnesium to produce the intense light output for decoy flares. The hot radiating solid is magnesium fluoride, produced by the reaction. The energy output of these devices, ≈ 10.2 kJ g^{-1}, is greater than most explosive formulations, giving a distracting source for heat-seeking missiles, and high intensity light is also a bonus. The maximum emission power is ≈ 20 kW sr^{-1} and this is achieved in less than one second after initiation. In the correct burning configuration, this level of output is maintained for about five seconds.

Smokes are used for three purposes: screening, signalling, and anti-personnel in public disorder situations. Smokes are suspended particles in the air which interfere with the transmission of radiation. Although water makes fog, which is a good screening material, solids are the best screens, particularly if they cover a complete range of particle sizes. White screening smokes are easily generated by chemical reactions, either by the production of volatile metal halides such as zinc chloride, or by the combustion of phosphorous (preferably the red isomer, since the white isomer spontaneously combusts on exposure to air or oxidizers and is thus difficult to store). A less toxic white smoke is produced by the sublimation of some organic materials, such as benzoic acid.

Coloured signal smokes are more difficult to produce and often rely on the use of clouds of highly coloured organic dyes. The dyes are vaporized by the heat from a pyrotechnic thermal generator (usually potassium perchlorate and lactose), and condense back into solid particles in the cold air. Producing the correct conditions for the sublimation is important because, if the mixture burns too strongly, the dye will start to burn rather than vaporize and the smoke will become black rather than coloured. These coloured clouds can be excellent absorbers/scatterers for laser radiations of the correct wavelengths. Coloured dyes can be added to energetic materials to increase their laser sensitivity. There are several books available on the chemistry and physics of pyrotechnics, and these are detailed in the bibliography at the end of this chapter.

Production of sound and gas usually go together. The only difference is in the speed and pressure at which the gas is produced. For a 'bang' sound, the rapid production of gas in a sealed container is required, and the pressure at which the container ruptures determines the sound level. The stun grenade uses a 'bang' sound to paralyse the auditory senses for a short time to disorientate humans. Whistling sounds are produced by oscillating burning gas producing pyrotechnics in a tube which is closed at one end. The chemical reaction's burning surface oscillates between being in contact with the bulk material and being lifted away from the bulk by the gas produced, reducing the burn rate, and thus allowing the gas to escape in shorts bursts (i.e. oscillations), and this is the whistle recipe. Frequency is controlled by the particle size of the ingredients and their method of presentation.

Propulsion systems require similar burn times, but without the confinement and pressure rise. Two areas of pyrotechnics dependent on gas production are the deployment of safety features such as airbags in motor vehicles, and the deployment of passenger escape chutes in aircraft. The original mixtures in these systems contained sodium azide, but this has been replaced by nitrocellulose-based propellant systems because of the toxicity of undecomposed azide. Aircraft ejector seats are gas-producing pyrotechnics, but they are better described as rocket propellants, which will be discussed in the following section.

Pyrotechnic mixtures can be used to produce high-temperature ceramics which are difficult to produce by conventional chemistry. An example of this is the production of mixed metal oxides, which can be difficult to produce by fusion of the relevant oxides but which can be produced by reaction of a metal with an appropriate oxidizer as the source of the second metallic element. A ceramic such as aluminium silicate requires temperatures of >1500 °C to produce a smooth, continuous, homogeneous sample, but this can be readily achieved using an aluminium-based pyrotechnic.

There are a number of other ingredients added to pyrotechnic mixtures to improve their properties, as summarized under the headings below. Some of these may have a role to play in laser initiation systems, to:

- provide the required physical properties to the mixture;
- improve the safety and handling characteristics;
- modify the burning rate;
- provide or enhance the special effect.

Mixtures of solid powders can separate on storage and thus require materials – binders – to hold the ingredients in intimate contact. The binders can also provide strength to the device, reduce its sensitivity and enhance the special effect. Natural binders include paraffin wax, beeswax, boiled linseed oil, shellac and lithographic varnish.

Man-made binders include bakelite resin, polyester resin, epoxy resin and Thiokol resin. These binders tend to have greater mechanical strength than natural binders. The can also reduce the burning rate, but often a burn rate modifier such as calcium oxalate is added to the mixture to provide a more accurate control of the burn rate. These binders can also be chosen for their absorption sensitivity to laser irradiation, increasing the initiation options for the device. PVC can meet most of the requirements, but it has a strong disadvantage regarding its toxicity and also has a tendency to produce hydrogen chloride under some conditions, which could react with the metals in the mixture, destabilizing the product and increasing the hazard.

There are several books available on the chemistry and physics of pyrotechnics, and these are detailed in the bibliography at the end of the chapter.

4.4.3 Rocket Propellants

Again, mixtures of fuels and oxidizers are widely used in solid rocket propellants as alternatives to the conventional energetic compound systems. Two types of rocket propellants are used: liquid and solid.

4.4.3.1 Liquid Propellants

Liquid systems are typified by hydrazine fuels, either anhydrous or as the hydrate, which is easier to store and handle. Oxidizers take the form of hydrogen peroxide, nitrogen dioxide or dinitrogen tetroxide. Inhibited red fuming nitric acid (IRFNA) is a readily storable mixture of concentrated nitric acid containing dissolved dinitrogen tetroxide and $\approx 0.5\%$ hydrofluoric acid to inhibit attack on the stainless steel storage containers. These systems are hypergolic, and they spontaneously ignite when the fuel and oxidizer come into contact in the rocket motor. The thrust or specific impulse of these systems is very comparable to the best performing system of hydrogen and oxygen.

Liquid-propelled rocket systems are easily controlled by switching of the pumps delivering the liquid to the motor, but they involve complex plumbing and are very hazardous to the user, should leaks develop. A famous series of incidents occurred with the ME 262 rocket plane, which relied on hydrogen peroxide against methanol/hydrazine fuels. Bullets through the peroxide holding tank caused the contents to leak out onto the parachute packs of the pilots, which then spontaneously ignited, leaving the pilot with no emergency escape route. None of these systems are particularly suitable for laser initiation.

4.4.3.2 Solid Rocket Propellants

The most common solid rocket propellant mixture is the gel formed by mixing nitrocellulose and nitroglycerine, two energetic materials which are safer to handle when mixed than individually. They can be easily fabricated into complex shapes, to give a variety of burning surface characteristics, and this controls the specific impulse of the motor. They have high performance, but are very vulnerable to accidental initiation and have long-term stability problems. Materials which have to be added to double-base rocket systems in order to counteract these problems can enhance the laser sensitivity but still require considerable laser power, as will be discussed in later chapters. Improvements in performance of these double-base systems can be achieved by incorporating aluminium and an oxidizer to give composition-modified double-base propellants.

Other, safer propellants rely on mixtures of hydrocarbon-based polymeric fuels and solid oxidizers. The most widely used oxidants are perchlorates and, in particular, the ammonium salt. This material is thermally stable, non-hygroscopic and fairly insensitive to friction, impact and static accidental initiation. Unfortunately, its combustion with hydrocarbon fuels produces hydrochloric acid, which is environmentally damaging. Ammonium nitrate is less environmentally polluting, but it suffers from ready sublimation and its rate of combustion is inferior to the perchlorate at higher pressures. Mixtures of perchlorate and nitrate have been tried, but with limited success. Ammonium salts are ideal, because they yield all-gaseous products of low atomic mass – ideal for propulsion systems. One drawback of ammonium perchlorate is that, when mixed with fuels, it can burn to detonation. A mixture of Vaseline and AP is used as an improvised plastic explosive, and mixtures of aluminium and AP can also detonate if suitably initiated.

The fuels are often polymeric materials based on a hydrocarbon chain. There are two sorts, thermoplastic and thermosets, identified as rubbery compositions. Typical thermoplastics used are polypropylene, polyisobutene and PVC. Typical thermoset polymers used are polyurethane, polybutadiene, and also both hydroxyl- and carboxy- terminated varieties. Rocket motors can be manufactured in complex shapes by mixing the oxidizer with the pre-polymer and then adding the curing agent to the mixture in a mould, producing the polymer *in situ*. These systems contain compounds of low sensitivity to accidental initiation when compared to the standard energetic compound systems. Unfortunately, some of the cross-linking agents are fairly toxic. Aluminium powder is also added to the mixture to improve performance by increasing the enthalpy of combustion. This will be further discussed in the section on explosive power. Solid products are undesirable in rocket propellants, since they reduce the thrust available, but the abrasive wear on the rocket nozzle is not as important as it is in gun propellants, where extensive barrel erosion results from hot solids being blown over the surface.

The performance of solid mixtures of fuels and oxidizers depends very strongly on two factors; the particle size of the ingredients and the extent of mixing – hence, the use of corning in gunpowder. These factors control the rate of the chemical reaction at the interface between the solids. How much more intimate the ingredients could be if the fuel and oxidizer were contained on the same molecule, where the distances between them are nanometres rather than the micrometres of solid mixtures. Explosive molecules will be the subject of the next section.

4.5 Explosive Compounds

One of the important discoveries made during the rapid growth of organic chemistry in the mid-nineteenth century was that by treating various compounds with a mixture of nitric and sulfuric acids, a vast new family of compounds could be obtained. Each of these contained a higher proportion of the elements nitrogen and oxygen than did its parent compound, and it was deduced that, in each case, one or more 'nitro-groups' ($—NO_2$) had replaced a hydrogen atom or atoms. Such products were called nitro-compounds. Later, with the development of chemistry, molecular explosives based on the nitro group NO_2 were classified into three different classes of 'nitro compound', depending on the linkage of the NO_2 group to the rest of the molecule. One advantage of this classification by chemical linkage is the prediction of sensitivity, and more will be said about this in later sections.

4.5.1 Chemical Classification

4.5.1.1 Nitro Esters

The first group of nitro compounds to be considered are identified correctly as nitrate esters, since they contain nitrate groups ($—ONO_2$). They are the product of the reaction of an alcohol with an acid. In this process, the NO_2 group replaced hydrogen atoms of

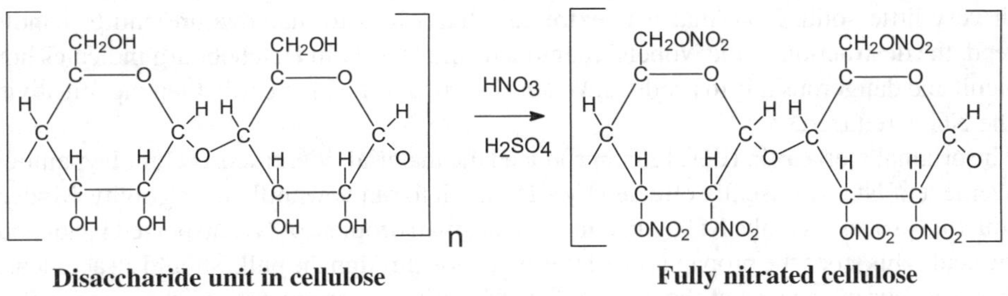

Figure 4.1 Reaction scheme for NC production.

one or more hydroxyl groups (—OH). Thus, the NO_2 is linked to the fuel carbon through an oxygen atom. In general, the new classes of compound were found to be less stable to heat than their parent compounds. They were also more easily ignited and, in some cases, so unstable that a small impact was sufficient to cause them to explode. The earliest example, nitrocellulose or NC, is produced by the action of nitric/sulfuric acid mixtures on the alcoholic OH groups present in cellulose fibres (a polymer formed by the linkage of saccharide, sugar moeties, a polysaccharide), as indicated below with *n* typically >200, and the production reaction is summarized in Figure 4.1.

Note that it is only the free alcoholic oxygens which are nitrated, and not the ring and chain oxygens. Because cellulose is an organic polymer, not all of the OH groups are identical and some can be in sterically hindered sites on the polymer chain, therefore the nitration reaction proceeds slowly and is usually far from complete. The extent of the nitration depends on the mean molecular weight of the polymer, the concentration of the acids and the temperature of the nitration reaction. Because the reaction may not be complete, the product is a mixture of mono-, di- and trinitro ester. The exact composition depends on the extent of nitration and is represented by measuring the percentage of nitrogen in the product. If the material was all dinitro, then the percentage of nitrogen would be 11.1%; if it were all trinitro ester, the percentage of nitrogen would be 14.14%. The extent of the nitration can be approximately estimated by measuring the solubility in a range of solvents. If the cellulose is first hydrolyzed, by boiling with strong alkali to give the individual sugar molecules, then there are six hydroxyl groups on each molecule and all can be used to form nitro esters, but these are best prepared from sugars such as mannose, which yields mannitol hexanitrate.

A most important nitro-ester compound nitroglycerine – NG – is obtained by the treatment of glycerol, a trihydric alcohol, with nitric acid (sometimes the mixture with sulfuric acid are used) at temperatures below ambient (see scheme below). NG is correctly named propane-1,2,3-triol, trinitrate.

$$\begin{array}{l} CH_2OH \\ | \\ CHOH \\ | \\ CH_2OH \end{array} \xrightarrow{HNO_3+H_2SO_4} \begin{array}{l} CH_2ONO_2 \\ | \\ CHONO_2 \\ | \\ CH_2ONO_2 \end{array} + [3\ H_2O]$$

It is a dense, oily liquid, which can explode violently under little provocation. It is clearly a better explosive than gunpowder, since all the ingredients readily form gaseous products with very little solid. Note that it is extremely hazardous to manufacture and to handle. Indeed, the destruction of the Nobels' manufacturing facility in Helenebourg indicates how difficult and dangerous it is to produce. When absorbed on an inert earth, then the sensitivity of the NG is reduced.

The original inert earth, Kieselguhr, produced the materials identified as Guhr Dynamites. Movement of NG was usually effected by allowing it to run downhill under gravity. Modern production processes make NG as an emulsion of small droplets of NG dispersed in aqueous nitric acid. This stops the propagation of the explosion through the bulk, should an accidental initiation occur in one part of the plant. NG is difficult to store and transport, so generally it is used immediately after manufacture. Its principal use is in mixtures with NC as double-base gun propellants and, as such, the manufactured suspension of NG is added to moist NC to form a gel. NG is also used in commercial blasting gelatins, but this has to be stored as an explosive and is subjected to rigorous storage and use regulations.

The molecular structure breaks down on explosion to a momentarily disorganized mass of atoms, which then recombine immediately to give entirely gaseous products, thus:

$$C_3H_5N_3O_9(1) \rightarrow 3C + 5H + 3N + 9O \rightarrow 3CO_2(g) + 1\tfrac{1}{2}N_2(g) + \tfrac{1}{4}O_2(g)$$

This process causes a much more rapid explosion than that of gunpowder, because the reacting atoms, being on the same molecule, are in much more intimate contact with each other. The separation of fuel and oxidizer is nanometres, rather than the millimetres or micrometres of powder mixtures. The heat evolved by the formation of 3 moles of CO_2 and $2\tfrac{1}{2}$ moles of H_2O for every 227 grams of nitroglycerine gives a net heat output of 6275 J per gram of nitroglycerine, compared with 3035 J per gram of gunpowder. The total volume of gas evolved is 740 $cm^3\ g^{-1}$ (as measured at STP), compared with 265 $cm^3\ g^{-1}$ for gunpowder. Clearly, nitroglycerine is a highly effective explosive.

Other important nitrate ester explosives are nitroglycol (EGDN), used as an additive to NG to reduce sensitivity by lowering the melting point, and pentaerythritol tetranitrate (PETN), which is one of the main choices for the booster charge in electric detonators. The formulae for all of these can be found in the appendix at the end of the chapter. These nitroesters are among the best-performing energetic materials, but are also the most sensitive. All of these nitroesters need to be carefully purified from acid residues arising from the manufacturing process in order to prevent decomposition during storage. The formation reaction is reversible, and traces of acid catalyse the reverse reaction with water.

4.5.1.2 *Nitrocompounds*

True nitrocompound explosives contain the group $-NO_2$ linked directly to a carbon atom, the simplest of which is nitromethane, CH_3NO_2. This readily available liquid has been used in a number of explosive formulations because it is cheap and inherently safe, requiring a booster charge or a sensitizer to make it detonator-sensitive. Its explosive performance can almost match that of TNT. The most important nitrocompounds are those based on the nitration of molecules containing the benzene aromatic ring. These include trinitrotoluene (TNT), picric and styphnic acids. These compounds are made by the reaction of mixed concentrated nitric and sulfuric acids on the parent hydrocarbon. Although trinitrobenzene has a higher explosive performance and melting point than TNT, it is easier to nitrate toluene than benzene. A simplified scheme for the nitration products is given below in Figure 4.2.

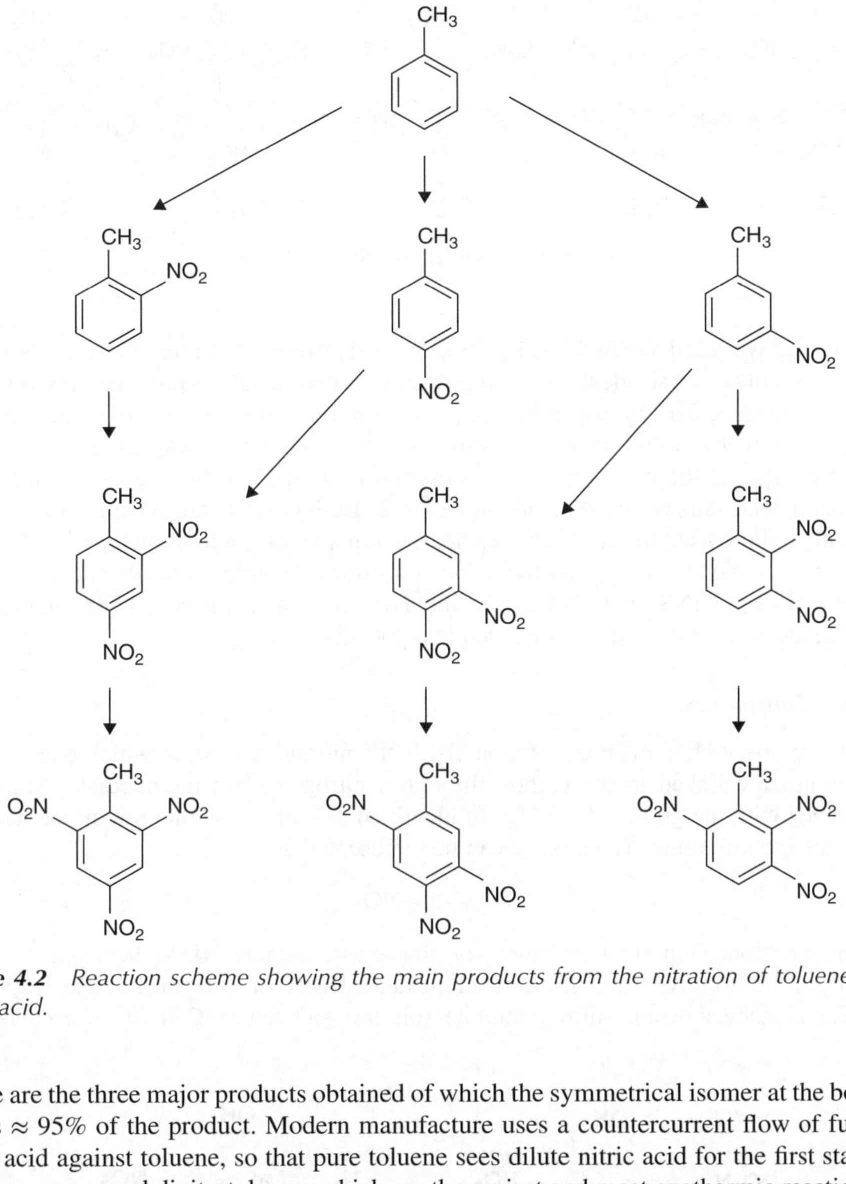

Figure 4.2 Reaction scheme showing the main products from the nitration of toluene with nitric acid.

These are the three major products obtained of which the symmetrical isomer at the bottom left is ≈ 95% of the product. Modern manufacture uses a countercurrent flow of fuming nitric acid against toluene, so that pure toluene sees dilute nitric acid for the first stage to produce mono- and dinitrotoluene, which are the easiest and most exothermic reactions.

Because the substitution of the nitro groups for three hydrogens in the aromatic ring can occur in five different positions occupied by hydrogen, a number of different isomers can be produced. The five shown in Figure 4.3 are unwanted, because their melting points are close to or above 373 K (100 °C), and the sixth isomer is the desired symmetric 2,4,6 isomer. This isomer melts at (82 °C) and ignites at ≈ 270 °C. Thus it can be safely melt-cast using hot water. Picric acid and styphnic acid are produced by the reaction of nitric acid with phenol, hydoxybenzene, and resorcinol, 1,3 dihydroxy benzene respectively, as shown in Figure 4.4.

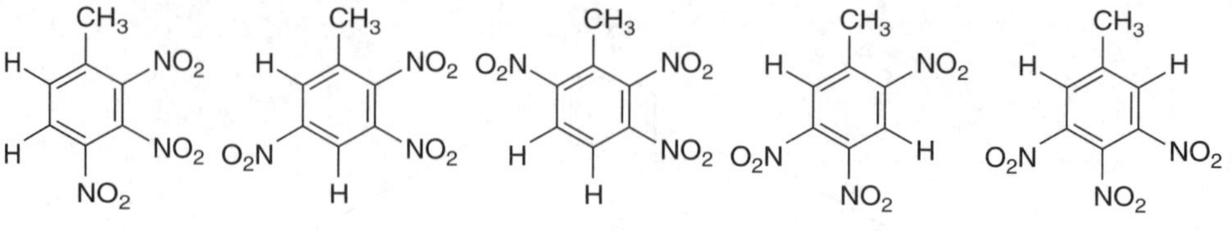

Figure 4.3 Isomers of trinitrotoluene.

Picric acid was used as a high explosive in the early twentieth century but was discontinued after a number of accidents involving reactions with metals to produce very sensitive metal picrate salts. The hydrogen on the phenol groups makes them very acidic, and the reaction with metals to produce salts is used to make initiatory compounds such as lead styphnate. Also, if the nitration of the resorcinol is stopped at the doubly nitrated stage then, again, lead salts are used as initiator LDNR. Both of these are sensitive to a variety of stimuli and are used to set off lead azide in detonators or ignite propellants. Other aromatic nitro compounds are discussed in other sections. Diaminotrinitrobenzene (DATB), hexanitrostilbene (HNS) and TNT can be melted in hot water, whereas TNB can not. The formulae and structures of these are given in Appendix 4.A.

4.5.1.3 Nitramines

A third important class of 'nitrocompounds' is the nitramine type, in which one or more nitro-groups are linked to the carbon through a nitrogen atom intermediate. Since the compounds with the group —C—NH_2 are identified as amines, so the nitroamine name is abbreviated to nitramine. The linkage can be represented as:

$$—C—N—NO_2$$

There are four important members of this class, namely RDX, hexogen or more correctly 1,3,5 trinitro, 1,3,5 triazo cyclohexane, HMX or octagen,1,3,5,7, tetranitro-1,3,5,7,tetraazacyclooctane, nitroguanidine (picrite) and tetryl (CE) (the formulae and

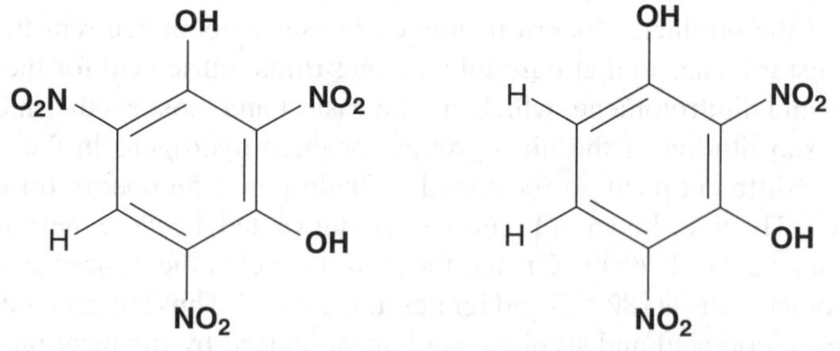

Figure 4.4 Structure of picric acid and dinitroresorcinol.

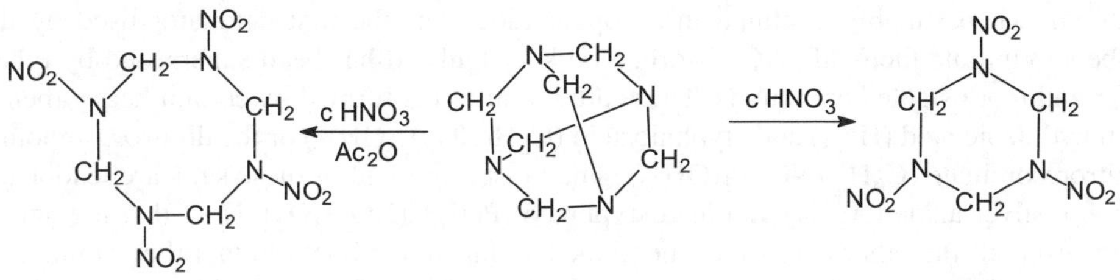

Figure 4.5 *Production of HMX and RDX from hexamine.*

properties of all these also can be found in Index 4.1 at the end of the chapter). Picrite is the easiest to make and only requires water removal from guanidine nitrate. The guanidine nitrate is slowly added, with stirring, to an ice-cold sample of concentrated sulfuric acid, ensuring that the temperature does not exceed 278 K (5 °C). When all of the guanidine nitrate has been added, the completely mixed sample is slowly added to iced water and white picrite is precipitated. It is easily recrystallized from hot water.

RDX and HMX are manufactured by the nitration of hexamethylene tetramine 'hexamine', a bicyclic amine often used in camping burners. The reactions are summarized in Figure 4.5. The conditions determine which is the major product, as indicated in the diagram. The products obtained are always contaminated with varying amounts of the alternative product and require purification.

The yield of RDX can be improved by the addition of ammonium nitrate to the nitric acid. Many of the new high-performance explosives fit into this classification by chemical linkage. One advantage of this classification by chemical linkage is the prediction of sensitivity. The sensitivity to accidental and proscribed initiation decreases as the strength of the chemical bond to nitrogen increases. Nitroester tend to be very unstable, due to the nitrogen-to-oxygen bond being weak, whereas nitro compounds (e.g. TNT) are much more stable due to the strong carbon-to-nitrogen bond. Nitramines (RDX, etc.) are somewhere in between, since the nitrogen-to-nitrogen single bond is of intermediate bond strength. Note that it is possible to add chemicals to nitrocompounds such as TNT to increase their sensitivity. This is by interaction with the nitro group, reducing the effective strength of the carbon-to-nitrogen bond.

An unusual energetic compound is tetrazene. This compound, correctly named as tetrazolyguanyltetrazene, crystallizes as the monohydrate. This water of crystallization is required for stability. While this material is only a weak explosive, its principal use is in sensitizing other materials. Addition of 2% tetrazene to lead azide increases its percussion and stab sensitivity by a factor of ten.

4.5.1.4 Metal Derivative Explosive Compounds

One further class of explosive compounds needs to be mentioned – the metal derivatives. The original compound in this class was mercury fulminate, prepared by treating mercury with nitric acid in the presence of ethanol when the diisocyanate was precipitated. This material, when suitably confined in a copper tube, was the first detonator used by the Nobels to initiate their safe NG absorbed on kieselguhr. It has been superseded by other, safer and more stable compounds. These are mainly salts formed by certain heavy metals with hydrazoic acid (HN_3) and styphnic acid ($C_6H(NO_2)_3(OH)_2$) or the dinitro compound dinitroresorcinate (C_6H_2 $(NO_2)_2(OH)_2$). The most important of these salts are lead azide (PbN_6), silver azide (AgN_3) and lead styphnate ($PbC_6H(NO_2)_3O_2$). Note that the azides differ from all the other explosives mentioned so far in this book, in that they contain no oxygen at all. However, the compound still conforms to the fuel and oxidizer principle, in that the lead ion (Pb^{2+}) oxidizes the azide (N_3^-) ion by removal of the electron to form nitrogen gas. The thermochemical performance of all these metal derivatives in inferior to that of the $C_aH_bN_cO_d$, type as exemplified by RDX, but they all have a low initiation energy and have an essential role to play in the initiation of the latter type, as will be seen later.

Lead azide is prepared by the addition of an aqueous solution of sodium azide to an aqueous solution of lead nitrate **with extreme care**, because the precipitating solid is prone to spontaneous ignition. To overcome this problem, it is either precipitated in the presence of dextrin, which coats the solid and reduces the occurrence of self-ignition, or a mixed azide/hydroxide of lead is precipitated.

The hydrogens on the phenol groups of trinitrodihyroxy benzene (styphnic acid) are, as the name suggests, acidic and, as such, they will react with bases to form salts. Treatment of a solution of styphnic acid with magnesium hydroxide gives magnesium styphnate, which can then be treated with a solution of a lead salt to precipitate lead styphnate, a primary explosive material. Similar techniques using dinitrodiphenol gives lead dinitroresorcinol (LDNR), another important primary explosive.

Other metal derivative salts are starting to be investigated, both as new explosive compounds as well as replacement for existing primary initiating explosives. These will be considered in a later section.

Other important simple explosive compounds based on nitrogen are hydrazinium nitrate ($N_2H_5NO_3$) and urea nitrate. Both of these are made by adding nitric acid to solutions of the respective bases hydrazine and urea. In the case of urea, the nitrate is precipitated by an excess of nitric acid, but with hydrazine, careful control of the acid level is required, since a dinitrate can be formed with excess acid. This dinitrate solid has greater sensitivity to accidental initiation and does not show good long-term stability. Hydrazinium nitrate is particularly good because of its high velocity of detonation, ≈ 8700 m s^{-1}. This compound forms the basis of the Astrolite group of explosive mixtures when mixed with finely powdered aluminium.

4.6 Thermodynamics of Explosions

All chemical explosives are unstable and are slowly decomposing at all temperatures above absolute zero, 0 °K. Fortunately, at normal ambient conditions, the rate of decomposition is extremely slow. Since the decomposition of the explosive into gaseous products is exothermic (i.e. it evolves heat), the enthalpy (heat) change, ΔH, on explosion is by definition negative (the energy is lost to the system). The gaseous products comprise mainly oxides of the fuel elements present in the explosive, usually carbon dioxide (CO_2), carbon monoxide (CO) and gaseous water (H_2O). The composition of the product gases depends on the ratio of fuel to oxidizer (see later) The formation of these highly exothermic compounds accounts for most or all of the total enthalpy change, ΔH, which occurs:

$$C + O_2 \rightarrow CO_2 \qquad \Delta H = -393.3\ \text{kJ mol}^{-1}$$
$$C + {}^{1}\!/_{2} \rightarrow CO \qquad \Delta H = -111\ \text{kJ mol}^{-1}$$
$$H_2 + {}^{1}\!/_{2}O_2 \rightarrow H_2O(g) \qquad \Delta H = -242\ \text{kJ mol}^{-1}$$

If, in the explosive system which provides these elements, they are bonded together in a compound, the compound is bound to decompose in the process, and its own heat of formation (which may be either negative or positive) will be given up. The heat of formation of a compound is defined as the energy change when one mole of the compound is formed from the elements in their standard state. This heat of formation therefore has to be taken into account in calculating the overall heat change which takes place – the 'heat of explosion', ΔH (explosion), the modulus of which is sometimes called Q. Then we can say:

$$\Delta H(\text{explosion}) = \Sigma\Delta H_f(\text{products}) - \Delta H_f(\text{explosive})$$

Invariably, the value of $\Sigma\Delta H_f$ (products) is much greater than that of ΔH_f (explosive) so, regardless of whether ΔH_f (explosive) has a positive or negative value, there is usually heat liberated in the process. For improved performance, if the enthalpy of formation of the explosive is positive, then this will be added to the overall output of the reaction. Also, since the system is an entropic (gas-producing) one, the value of the entropy change ΔS is also large and, in this case, positive. Gases are more random than solids, and hence their entropy is greater than that of solids.

All explosions, whether physical, nuclear or chemical, can be fitted into the broad thermodynamic picture. They all release energy, some of which can be translated into useful work. The free energy function ΔG provides a measure of the work capacity of the system, and for spontaneous reactions ΔG must be negative. For all explosions we can take the value of ΔG to be large and negative.

Now, since:

$$\Delta G = \Delta H - T.\Delta S$$

where T = temperature and ΔS is the change in entropy during the reaction, it follows that for large negative values of ΔG, we require large negative values of ΔH and/or large positive values of ΔS. Both of these conditions are provided by chemical explosives, since not only are the reactions exothermic but there is a dramatic increase in entropy going from solid to gas. We can compare in a simplistic way the thermodynamic differences between the three types of explosion as shown in Table 4.6.

Table 4.6 *Thermodynamic characteristics of explosions.*

Type	ΔG	ΔH	ΔS
Physical	large negative	effectively zero	large positive
Nuclear	large negative	Very large negative	Limited effectively zero
Chemical	large negative	large negative	large positive

4.6.1 Oxygen Balance

In the section above, the combustion of carbon can give either carbon monoxide (CO) or carbon dioxide (CO_2). The product will determine the enthalpy output of the reaction, since the combustion to CO only produces about a quarter of the thermal output when CO_2 is produced. Which of these two products is produced will depend on the amount of oxygen available. All three classes – nitrate esters, nitro-compounds and nitramines – contain only the elements carbon, hydrogen, nitrogen and oxygen. Their composition can therefore be expressed as:

$$C_aH_bN_cO_d$$

where a, b, c and d are usually small numbers, not exceeding 12. The first step, which occurs in the detonation of nitroglycerine, applies to all of them, so we can write:

$$C_aH_bN_cO_d \rightarrow a\,C + b\,H + c\,N + d\,O$$

Precisely what happens in the next step, the recombination into gases, depends on the number of oxygen atoms (*d*) relative to the number of carbon and hydrogen atoms ($a + b$). If *d* is large enough, the products formed will include CO_2, H_2O, CO and N_2. If *d* is smaller, there will be CO, H_2O and N_2. If *d* is markedly deficient, there will be CO, H_2, N_2 and free carbon (visible as black smoke). The proportion of oxygen atoms to those of carbon and hydrogen in the explosive is called its **oxygen balance**, designated Ω and defined as:

> *The percentage by mass of oxygen, positive or negative, remaining after detonation when the products of detonation are carbon dioxide and water.*

Since two atoms of oxygen are required to fully oxidize one atom of carbon, and half an atom of oxygen to oxidize one atom of hydrogen, the value of Ω must be proportional to ($d - 2a - \frac{b}{2}$). If the value of this expression is near to zero, it means that there is approximately the right amount of oxygen present to oxidize all of the fuel elements fully or, in other words, the composition is stoichiometric. The production of heat per unit mass of explosive will therefore be maximum.

Looking again at the formula of nitroglycerine, $C_3H_5N_30_9$, we can evaluate

$$2a = 6, \frac{b}{2} = 2^1\!/\!_2 \text{ and } d = 9.$$

$$\text{Then } \left(d - 2a - \frac{b}{2}\right) = 9 - 6 - 2^1\!/\!_2 = +\,0.5.$$

This represents a surplus half-atom of oxygen (relative mass 8) in a molecule of relative mass 227, a very small but positive oxygen balance:

$$\Omega = 0.5 \times 16 \times 100/227 \cong 5\%.$$

The result is the quarter molecule of free oxygen shown in the equation for the explosion. The near-zero oxygen balance of nitroglycerine is one of the properties which make it such an effective explosive. Most military explosives have a negative oxygen balance. This is partly because the number of nitro- or nitrate groups, which can be introduced into the parent compound is limited, either by the practicalities of manufacture or by the inherent instability of a highly nitrated compound. Attempts have been made to solve the problem of oxygen deficiency. For example, trinitroluene (TNT) $C_7H_5N_3O_6$ which is too oxygen-deficient to be a highly effective explosive, is sometimes mixed with ammonium nitrate,

NH_4NO_3, a source of additional oxygen, to form 'Amatols' to correct this deficiency. The oxygen-balanced mixture is ≈ 75% AN and 25% TNT, which contains a high solid loading which the molten TNT can not wet effectively to yield a reliable melt cast process.

Alfred Nobel (1833–1896) made one of the most fundamental discoveries in explosives chemistry when he found that a mixture of nitroglycerine, which is oxygen-rich, and nitrocellulose, which is oxygen-deficient, forms a gel which is almost oxygen-balanced. This gel, which is much less prone to accidental initiation than is liquid nitroglycerine, can be used as a matrix in which to suspend a whole range of possible oxidants such as inorganic nitrates, and fuels such as woodmeal. Nobel proceeded to develop a whole series of explosives based on this discovery, with a wide range of properties and the advantage of water-resistance (another advance on black powder). The versatile gel was also used by Nobel to form the first double-base gun propellant, ballistite – the forerunner of many present-day conventional propellants.

Appendix 4.A

A.1 Data for Some Explosives

Key to Data:		
	Q	the heat output for one gram of explosive.
	Gas Vol:	The gas output for one gram of explosive
	T_d:	detonation temperature in Kelvin
	m. pt.	the melting point of the explosive
	V	Detonation velocity

A.1.1 TNT (Trinitrotoluene)

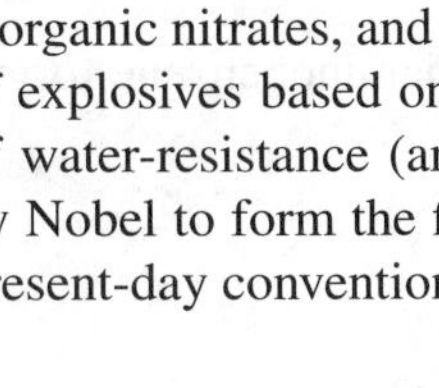

Type: nitrocompound	
Formula:	$C_7H_5N_3O_6$
Q:	4080 Jg^{-1}
Gas Vol.:	790 cm^3g^{-1}
T_d:	2595 K
m. pt.	80.7 °C
V (@1.6 $g\,cm^{-3}$)	6900 ms^{-1}

TNT is nowadays used as an 'energetic binder' in conjunction with higher melting point explosives such as RDX in the production of melt-cast charges.

A.1.2 HNS(Hexanitrostilbene)

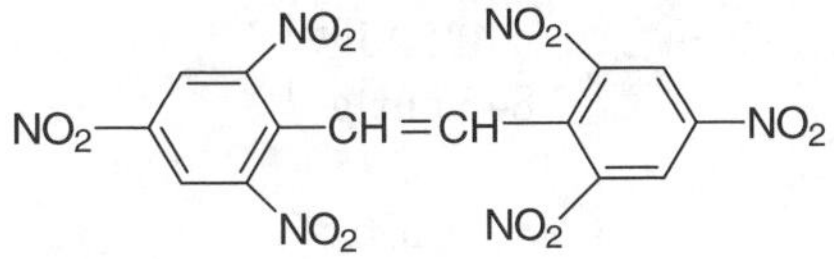

Type: nitrocompound	
Formula:	$C_{14}H_6N_6O_{12}$
Q:	4208 Jg^{-1}
Gas Vol.:	700 cm^3g^{-1}
T_d:	2900 K
m. pt.	318 °C
V (@1.74 $g\,cm^{-3}$)	7100 ms^{-1}

Manufactured by the oxidation of TNT. Used to modify the crystal habit of cast TNT and as a high melting explosive in hot environments.

A.1.3 DATB (1,3,Diamino,2,4,6,trinitrobenzene)

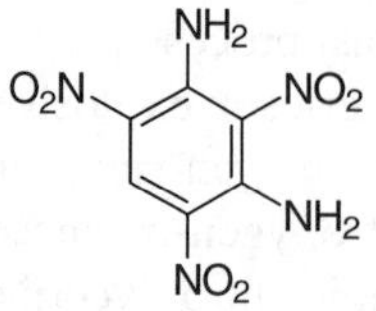

Type: nitrocompound	
Formula:	$C_6H_5N_5O_6$
Q:	3805 Jg^{-1}
Gas Vol.:	1015 cm^3g^{-1}
T_d:	2595 K
m. pt.	286 °C
V (@1.79 $g\,cm^{-3}$)	7520 ms^{-1}

Manufactured by the action of mixed concentrated acids on 1,3 diaminobenzene forms as a bright yellow crystalline solid.

A.1.4 TATB (1,3,5,-Triamino-2,4,6-Trinitrobenzene)

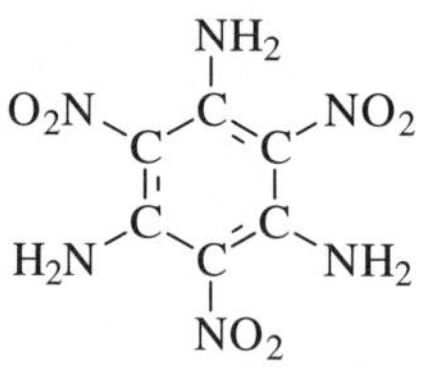

Type: nitrocompound	
Formula:	$C_6H_6N_6O_6$
Q:	3496 Jg^{-1}
Gas Vol.:	781 cm^3g^{-1}
T_d:	2880 K
m. pt.	340 °C
V (@1.8 $g\,cm^{-3}$)	7350 ms^{-1}

Manufactured by the action of ammonia on trichloro trinitro benzene (dry or wet process), TATB is a high melting point low sensitiveness explosive. Its use is likely to increase with the drive for the production of 'insensitive' munitions. Used in some nuclear warheads. It is currently extremely costly to produce.

A.1.5 Picric Acid (2,4,6,trinito- hydroxy benzene)

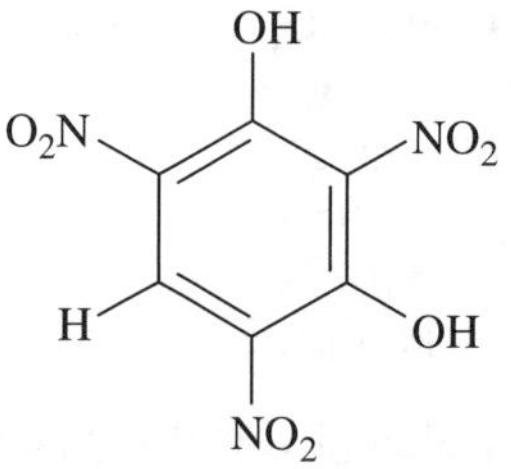

Type: nitrocompound	
Formula:	$C_7H_5N_5O_8$
Q:	4350 Jg^{-1}
Gas Vol.:	845 cm^3g^{-1}
T_d:	3545 K
m. pt.	130 °C
V (@1.71 $g\,cm^{-3}$)	7570 ms^{-1}

A.1.6 Styphnic Acid (2,4,6,trinito-1,3, dihydroxy benzene)

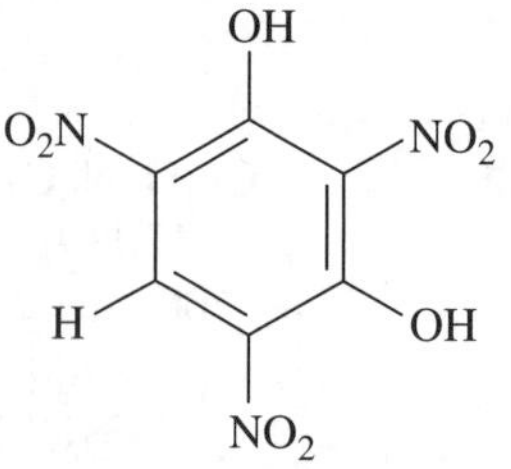

Type: nitrocompound	
Formula:	$C_7H_5N_5O_8$
Q:	4350 Jg^{-1}
Gas Vol.:	845 cm^3g^{-1}
T_d:	3545 K
m. pt.	130 °C
V (@1.71 $g\,cm^{-3}$)	7570 ms^{-1}

Manufactured by mixed nitric/sulfuric acids on resorcinol. Used only as an intermediate for preparation of the lead salts – see below

A.1.7 Tetryl or CE (Composition Exploding)

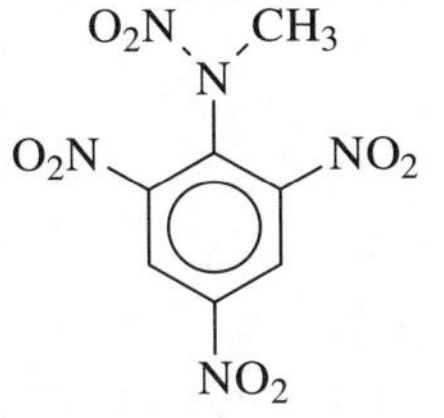

Type: nitrocompound/nitramine	
Formula:	$C_7H_5N_5O_8$
Q:	4350 Jg^{-1}
Gas Vol.:	845 cm^3g^{-1}
T_d:	3545 K
m. pt.	130 °C
V (@1.71 $g\,cm^{-3}$)	7570 ms^{-1}

Tetryl has been widely used in booster compositions. Its toxic hazards are leading to a replacement with RDX-based compositions.

A.1.8 PICRITE (Niroguanidine)

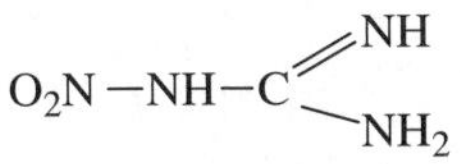

Type: nitramine	
Formula:	$CH_4N_4O_2$
Q:	2680 Jg^{-1}
Gas Vol.:	1077 cm^3g^{-1}
T_d:	2100 K
m. pt.	204 °C
V (@1.76 $g\,cm^{-3}$)	7650 ms^{-1}

Manufactured by the action of sulfuric acid on guanidine nitrate. Used extensively in triple-base propellants due to high gas evolution per gram and cool burning.

A.1.9 RDX (Research Department eXplosive)

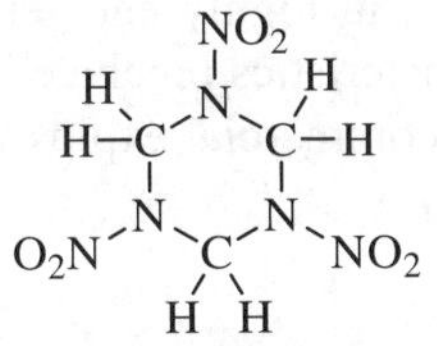

Formula:	$C_3H_6N_6O_6$
Q:	5130 Jg^{-1}
Gas Vol.:	908 cm^3g^{-1}
T_d:	4255 K
m. pt.	204 °C
V (@1.76 $g\,cm^{-3}$)	8750 ms^{-1}

Manufactured by the action of nitric acid on hexamine in aqueous solution. RDX is the main high explosive for military use, in plastic explosives such as PE 4 and SX 2. It is used in the production filling formulations either for melt-cast or press-filled devices.

A.1.10 HMX (High Molecular-weight eXplosive)

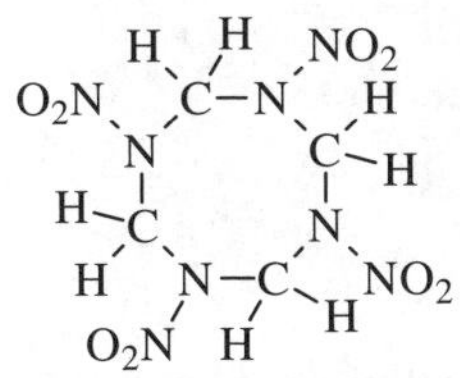

Formula:	$C_4H_8N_8O_8$
Q:	5130 Jg^{-1}
Gas Vol.:	910 cm^3g^{-1}
T_d:	4200 K
m. pt.	278 °C
V (@1.9 $g\,cm^{-3}$)	9100 ms^{-1}

Manufactured by nitric acid on hexamine in non-aqueous media. Also as a by-product of RDX manufacture. HMX is a more costly explosive than RDX but has a better performance due to its higher filling density. There is a drive towards the use of HMX to extend the performance capabilities in, for example, shaped charge explosives formulations.

A.1.11 EGDN (Nitroglycol)

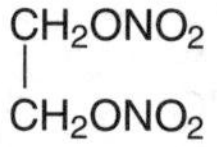

Type: nitroester	
Formula:	$C_2H_4N_2O_6$
Q:	6730 Jg^{-1}
Gas Vol.:	740 cm^3g^{-1}
T_d:	4830 K
m. pt.	−22 °C
V (@1.59 $g\,cm^{-3}$)	8100 ms^{-1}

Manufactured by the action of nitric/sulfuric acids on ethylene glycol at low temperatures. Straw-coloured liquid when pure.

A.1.12 NG (Nitroglycerine)

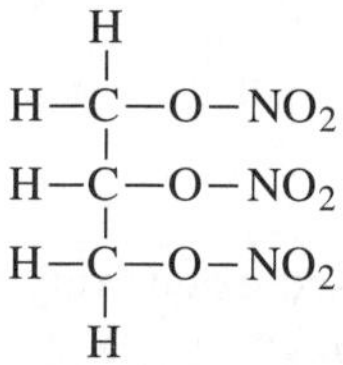

Type: nitroester	
Formula:	$C_3H_5N_3O_9$
Q:	6275 Jg^{-1}
Gas Vol.:	740 cm^3g^{-1}
T_d:	4900 K
m. pt.	13 °C
V (@1.59 $g\,cm^{-3}$)	7600 ms^{-1}

Manufactured by the action of nitric/sulfuric acids on glycerol at low temperatures. Pale yellow oily liquid. NG is a major constituent, along with NC, in highly energetic double-base propellants. Its ease of initiation and poor storage characteristics preclude its use as a military high explosive. It is, however, commonly used in commercial explosive compositions in the gelatinous form with nitrocellulose.

A.1.13 NC (Nitro-Cellulose)

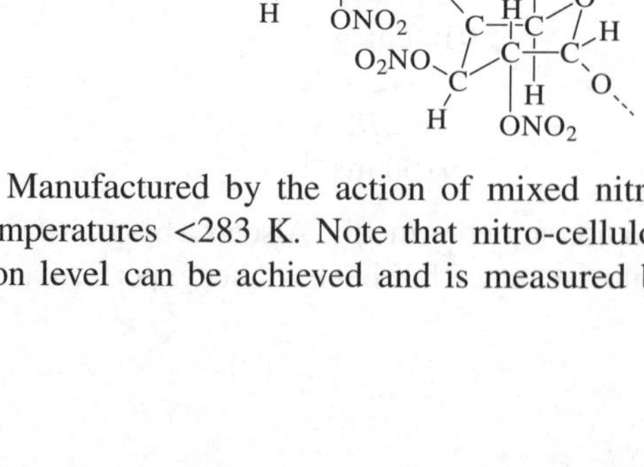

Type: nitroester	
Formula:	$(C_6H_7N_3O_{11})_n$
Q:	3745 Jg^{-1}
Gas Vol.:	880 cm^3g^{-1}
T_d:	3290 K

Manufactured by the action of mixed nitric/sulfuric acids on cellulose fibres at low temperatures <283 K. Note that nitro-cellulose is shown fully nitrated. A lesser nitration level can be achieved and is measured by the nitrogen percentage in the material.

Nitro-cellulose is present in all conventional propellants, acting both as an energetic material and to provide structural strength to the propellant grains.

A.1.14 PETN (Pentaerythritol Tetranitrate)

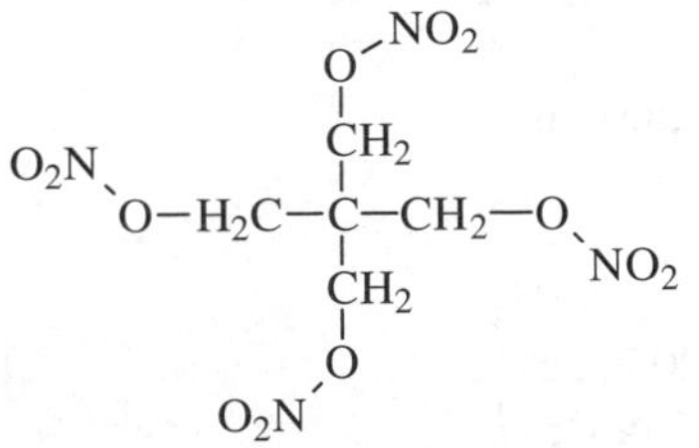

Type: nitroester	
Formula:	$C_5H_8N_4O_{12}$
Q:	5940 Jg^{-1}
Gas Vol.:	790 cm^3g^{-1}
T_d:	4625 K
m. pt.	141 °C
V (@1.7 $g\,cm^{-3}$)	8400 ms^{-1}

PETN is prepared by the action of nitric/sulfuric acid mixture on the parent tetrahydric erythritol. Mainly used in the production of detonating cords and in the base charge in some demolition detonators.

A.1.15 Metal Salts

A.1.15.1 Lead Dinitro Resorcinate (LDNR)

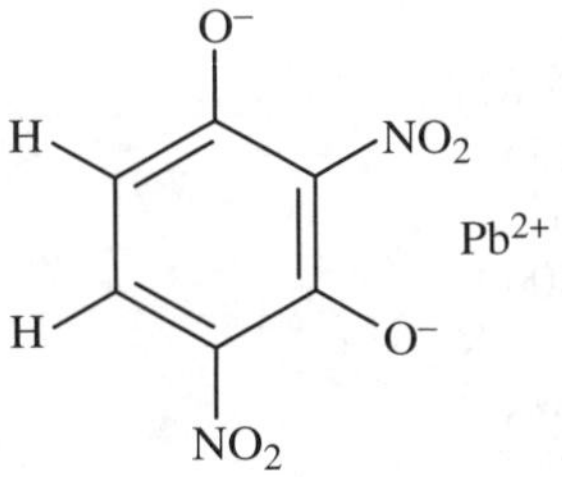

Formula:	$PbC_6H_2N_2O_6$
Q:	4350 Jg^{-1}
Gas Vol.:	845 cm^3g^{-1}
T_d:	3545 K
m. pt.	130 °C
V (@1.71 $g\,cm^{-3}$)	7570 ms^{-1}

Manufactured by reacting lead carbonate with dinitroresorcinol. Used in initiating systems, often in mixtures.

A.1.15.2 Lead Styphnate (lead 2,4,6, trinitroresorcinate)

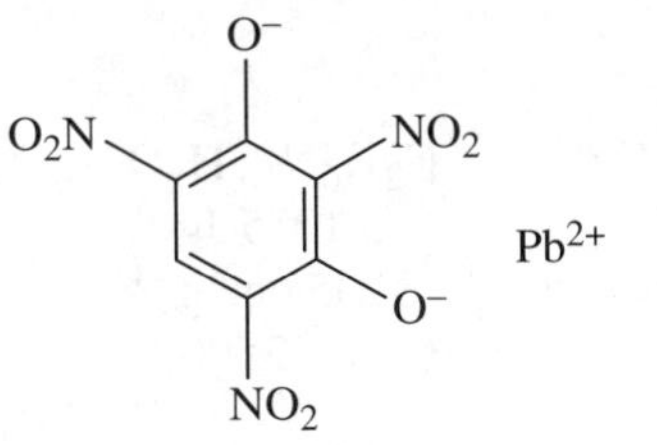

Formula:	$PbC_6H_3N_3O_9$
Q:	1885 Jg^{-1}
Gas Vol.:	325 cm^3g^{-1}
T_d:	4255 K
m. pt.	274 °C(dec)
V (@1.76 $g\,cm^{-3}$)	5200 $m\,s^{-1}$

Manufactured by reaction of lead nitrate with magnesium styphnate, prepared by reacting styphnic acid with magnesium carbonate. Widely used in admixtures with azides, etc. in initiation systems.

A.1.15.3 *Lead Azide*

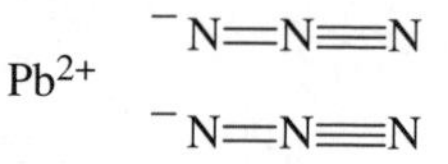

Formula:	$Pb(N_3)_2$
Q:	1610 Jg^{-1}
Gas Vol.:	230 cm^3g^{-1}
T_d:	4625 K
m. pt.	?
V (@1.7 $g\,cm^{-3}$)	4500 ms^{-1}

A.1.15.4 *Silver Azide*

Formula:	AgN_3
Q:	2050 Jg^{-1}
Gas Vol.:	225 cm^3g^{-1}
T_d:	4625 K
m. pt.	X
V (@1.7 $g\,cm^{-3}$)	4400 ms^{-1}

Both lead and silver azides are manufactured by the reaction between the metal nitrate and sodium azide in the presence of a coating agent to prevent initiation on crystallization of the azide. Both of these azides are used as the detonating agent in detonators.

A.1.15.5 *Mercury Fulminate*

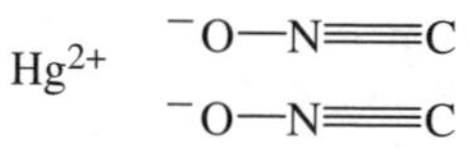

Formula:	$HgC_2N_2O_2$
Q:	1755 Jg^{-1}
Gas Vol.:	235 cm^3g^{-1}
T_d:	2500 K
m. pt.	X
V (@1.7 $g\,cm^{-3}$)	4500 ms^{-1}

Manufactured by the reaction of mercury with nitric acid and the contents poured into 95% ethanol. All of these metal salts are used as initiating explosives, with the lead and silver azides being particularly important as detonating systems.

A.2 Unusual Explosives

A.2.1 *Tetrazene*

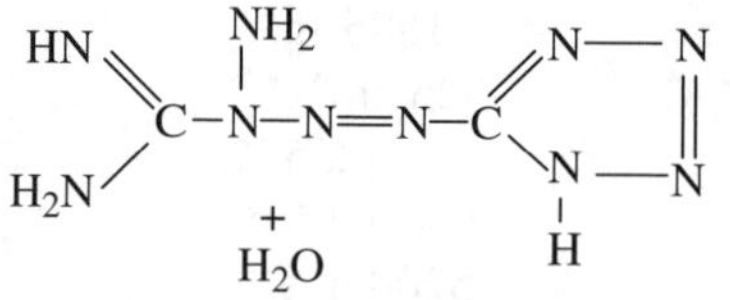

Formula:	$C_2H_6N_{10}.H_2O$
Q:	1755 Jg^{-1}
Gas Vol.:	235 cm^3g^{-1}
T_d:	2500 K
m. pt.	X
V (@1.7 $g\,cm^{-3}$)	4500 ms^{-1}

Manufactured by complex chemical synthesis. There is some discussion as to which is the correct structure – either the structure with a three-nitrogen chain between the carbon atoms or the alternative with the side chain NH_2 being incorporated, to give four nitrogens between the carbon atoms. Notice the absence of oxygen, other than the water molecule of crystallization. Tetrazine is used to sensitize other initiating explosives, such as lead azide, to a stab stimulus, and it is also used in direct impact fuses with lead azide.

Bibliography

Urbanski, T. (1983) *Chemistry and Technology of Explosives*, 1st edn, Pergamon Press, Oxford, ISBN 0-08-026206-6.

US Army (1983) *Encyclopaedia of Explosives and Related Items*, ARRADCOM, Dover NJ.

Kirk-Othmer (1998) *Kirk-Othmer Encyclopaedia of Chemical Technology*, 4th edn, John Wiley & Sons, Chichester.

Myler, R., Kohler, J. and Humbive, A. (2002) *Explosives*, 5th edn, Wiley VCH, Weinheim, ISBN 3-527-30267-0.

Cooper, P.W. and Kerowski, S.R. (1996) *Introduction to the Technology of Explosives*, Wiley VCH, NY, ISBN 0-471-18635-X.

Boddu, V. and Redner, P. (eds) (2011) *Energetic Materials*, CRC press (Taylor and Francis), ISBN 978-1-4398-3513-5.

Akhavan, J.A. *Chemistry of Explosives*, 2nd edn, RSC publish Cambridge.

Teipel, U. (2005) *Energetic Materials Prep & Characterisation*, Wiley VCH, ISBN 3-527-30240-9.

Shidlovsky, A.A. (1964) *Principles of Pyrotechnics*, 3rd edn, American Fireworks News (trans Wright-Patterson Air Base 1977.

Conkling, J.A. (1985) *Chemistry of Pyrotechnics*, Dekker.

Ellern, H. (1968) *Military and Civilian Pyrotechnics*, Chemical Pub. Co. Inc., New York, ISBN 0-08206-032-2.

Russell, M.S. (2009) *The Chemistry of Fireworks*, 2nd edn, Royal Society of Chemistry, Cambridge, 978-0-85404-127-5.

Kosanke, K., Kosanke, B., Von Maltitz, I. *et al.* (2004) *Pyrotechnic Chemistry*, Journal of Pyrotechnics Inc, 978-1-88952-615-7.

5

Recent Developments in Explosives

5.1 Introduction

TNT, Nitroglycerine, RDX and HMX provide high explosives of excellent performance in terms of power and brisance. Extensive research has been undertaken to develop improvements in explosive performance, and this can be discussed under the headings:

1. Improvements in explosive performance (power and brisance).
2. Improvements in storage and life time.
3. Improvements in handling properties.
4. Reduction in vulnerability to accidental initiation.

Some of these aspects may be investigated for specific applications but, generally, explosives of significantly higher performance are unlikely to be discovered. There are two reasons for this, as discussed below.

5.2 Improvements in Explosive Performance

5.2.1 *Heat of Explosion ΔH_c (Q)*

Power and brisance are both determined by the heat of explosion (Q), which is the thermochemical energy in the explosive (in joules g^{-1}) available to drive the detonation wave and the gas velocity. For organic compounds consisting of C, H, N and O, Q is limited by the heats of formation of carbon dioxide and water in the products of detonation. Q cannot exceed the value of ≈ 7000 J g^{-1} produced by the production of CO_2 in perfectly O_2-balanced systems. Some novel explosives with positive enthalpy of formation can store this as an additional energy which will be released on detonation (the limit is typically thought to be ≈ 500 J g^{-1}), but associated with this additional energy is a decrease in stability and an increase in sensitivity. 5-azidotetrazole is such a system, with an enthalpy of formation of 611 kJ mol^{-1}, but its sensitivity clearly places it in the highly sensitive

Laser Ignition of Energetic Materials, First Edition. S Rafi Ahmad and Michael Cartwright.

primary explosive class. Its velocity of detonation is greater than both lead azide and RDX. More details will be given in the high nitrogen compound section.

Other elements, such as boron, aluminium and fluorine, have a greater Q, but boron and aluminium produce solid oxides and little gas, while fluorine has not proved suitable for the synthesis of stable and safe explosives. The magnitude of Q can be increased by the addition of up to 20% aluminium without significantly reducing the volume of the gases produced, provided that little carbon is produced. The addition of aluminium reduces the VOD of the mixture – and, hence, its brisance – but increases the power for blast and heave. Power is proportional to Q, as is the volume of gas liberated, V. The limit for V for CHNO-type explosives is ≈ 1100 cm^3 g^{-1}, a target almost reached by nitroguanidine (picrite) – hence its addition to triple-base propellants to increase gas volumes.

Additional oxidizers such as ammonium nitrate AN and perchlorate AP can enhance the Q and V values. AN is added to TNT to form amatols, improving power by supporting the combustion of the carbon produced in the detonation of pure TNT. Indeed, Al with NH_4ClO_4 has a Q value of ≈ 10 kJ g^{-1} at the stoichiometric composition, which is dramatically above the 7.5 kJ g^{-1} which is the maximum for C, H, O and N explosives (although increased power can be obtained from using pure fuel with O_2 from air, FAX). Octane gives 43 kJ g^{-1}, whereas NG gives 6.8 kJ g^{-1}. These mixtures can produce enhanced blast effects at longer distances from the centre of initiation, particularly if aluminium dust is added to the cloud, but they are not of primary interest for laser initiation and will not be considered further.

5.2.2 *Density of Explosives*

Research is also directed at achieving higher-brisance explosives. Brisance, or the shattering effect of a high explosive, is proportional to detonation pressure, which is a function of density of explosive and VOD. The velocity of detonation is proportional to the square of the density, therefore the detonation pressure is proportional to the cube of the density of the explosive. Lead azide is so effective because its density, ≈ 4.5 g cm^3, is high in comparison with organic explosives, which generally have crystal densities of less than 2.0 g cm^3. HMX occurs in a number of crystal forms, and the higher-density form, β (1.91 g cm^3), produces the higher VOD.

New materials, such as CL20, have a number of polymorphs and, again, the higher-density form (ε) of CL20 has a crystal density of 2.03 g cm^3 and, thus, higher VOD. These are viable systems. This value is unlikely to be exceeded for any organic crystal, as charge densities are necessarily lower than crystal densities. Any achievable increase in crystal and charge densities is valuable, but it seems probable that the limiting value has been almost reached in HMX and CL20. Others still in developmental stages are hexanitrobenzene (HNB) and Octanitrocubane. Both power and brisance of these two materials have been predicted to exceed all known explosives, but stability and manufacture are very problematical. HNB has been prepared and tested but it has a high vapour pressure and, because it is easily hydrolysed in the lungs, it possesses considerable toxicity.

5.3 Areas under Development

Recent developments in high explosives technology have been concerned with the compatibility of explosives with their operating environment. Today both military and civilian high explosives have to withstand increasingly severe operational stresses. Safety, reliability and performance of explosive fillings must not be degraded by the operational environment.

One of the significant recent developments has been in the area of reduced sensitivity explosives. Normally, RDX is desensitized by mixing with TNT and casting, but recent efforts have demonstrated that RDX can have reduced sensitivity to impacting shock wave [1]. The material is not characterized by reduced sensitivity to the normal sensitivity tests, such as impact friction and electrostatic discharge, but only to explosive shock wave. The reasons behind this are attributed to a reduction in the number and distribution of crystal defects and voids within the structure, which may contain trapped solvent or gas. There is also a form of HMX explosive also identified as having reduced shock sensitivity [2].

Since none of the conventional test enabled these reduced sensitivity explosives to be identified, the standard procedure was to cast the explosive in a PBX formulation and to subject this to the Large Scale Gap Test (see sensitivity appendix), in which the explosive is subjected to the shock wave from a donor charge ameliorated by an inert barrier. The presence of IRDX was indicated if thinner barriers prevented the acceptor charge from fully detonating. In order to be statistically significant, a number of these tests must be performed on each explosive manufacturing batch. These barrier thicknesses can be converted to the detonation pressure required for initiation of the receptor IRDX charge, and they indicate that normal RDX has a detonation threshold of $\approx$ 30, whereas IRDX has a detonation threshold of $\approx$ 50.

Since then, a definitive test, first suggested by the author at a NATO meeting in Meppen [3], has been developed [4], which examines the crystal microhardness of the samples. IRDX shows a higher microhardness than normal RDX. Even though a number of samples are required for each manufactured batch, the sample quantity, 1 g, is much reduced from the large scale gap test requirement of 300 g, and no complex preparation of a PBX formulation which can add further uncertainty to the results is required. There is an additional bonus in that the microhardness test does not require an explosive range facility, and no large scale explosions are required.

5.3.1 New Requirements for Explosive Compositions

5.3.1.1 Heat Resistance

Explosive fillings may be exposed to high environmental temperatures for shorter or longer periods of time. This situation may arise in missiles, where aerodynamic and shock heating could raise the temperature of explosive components to 200 °C. Many explosives melt and decompose rapidly at this temperature (see Table 5.1).

Fillings in this state cannot be expected to function in a reliable manner. Materials which have high ignition temperatures are less likely to undergo accidental initiation, either by cook-off in hot environments or by missile skin temperature heating. Also, high melting explosives are less likely to crack or suffer exudation. Materials such as HNS, HMX, Tacot and TATB offer these facilities. TATB is particularly important, because it is difficult to initiate even deliberately, so accidental initiations are less likely to occur. The reduced sensitivity is largely due to its ability to form extended sheets, which are held together by very strong hydrogen bonds between the oxygens in the nitro group on one molecule and the hydrogens in the amino group in a neighbouring molecule. Also, because there are three symmetrical nitro and amino groups on each planar molecule, a two-dimensional network

Table 5.1 *Melting and ignition point of explosives.*

Explosive	M.Pt °C	Ignition temp	Density g cm^3	VoD m s^{-1}
EGDN	–20	217	1.48	7300
NG	13.5		1.26	7600
TNT	80.6	270	1.65	6900
Tetryl	129.5	185	1.73	7570
PETN	141.3	202	1.76	8400
Picrite	232	232	1.71	8200
Higher melting explosives				
RDX	204 (cook off)	213	1.82	8750
HMX	275	280	1.96	9100
DATB	286	300	1.86	8750
HNS	318	318	1.74	
TATB	350	350	1.93	
NTO		273	1.93	8500
CL20*			2.02	10250
FOX 7			1.91	

is generated, as shown in Figure 5.1. Similar conditions exist in the less sensitive FOX 7,1,1′dinitro2,2′diaminoethene, shown schematically in Figure 5.2. Here, the hydrogen bonding produces long-chain species which stack together in parallel to produce the solid. Some other more exotic high melting point/low sensitivity energetic materials are detailed in the appendix.

5.3.1.2 Mechanical Strength and Cohesion

High-explosive fillings carried in high-performance aircraft and missiles are subject to stresses arising during take-off, flight and impact. Similarly, explosives in boreholes are subjected to abnormal stresses. In order to function effectively, such fillings must possess good mechanical strength and cohesion over a wide range of temperatures and pressures/

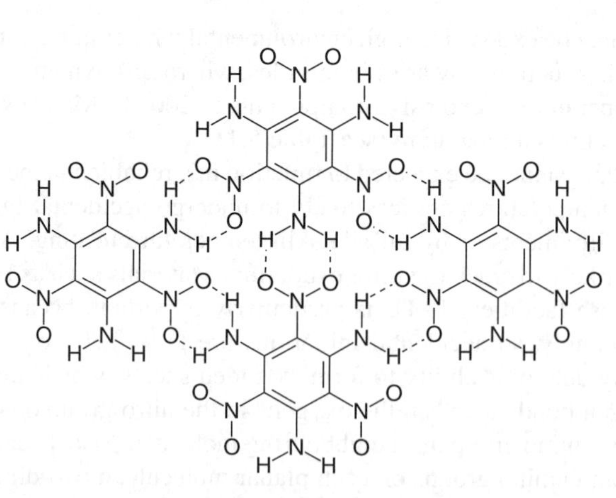

Figure 5.1 *Structure of TATB.*

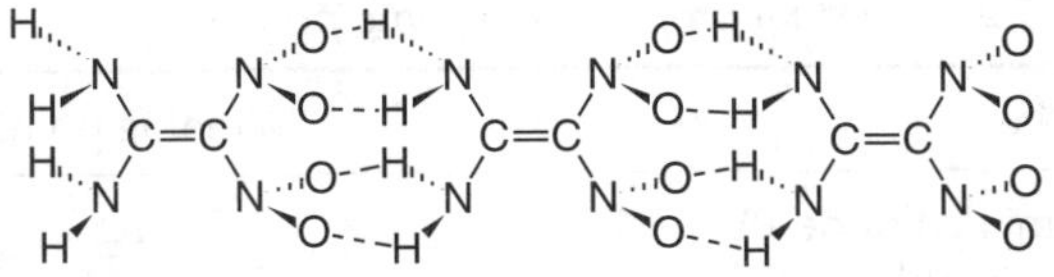

Figure 5.2 *FOX 7,1,1′dinitro2,2′diaminoethene.*

g-forces. This requirement is particularly important in the case of specially formed and shaped charges, the reliability and performance of which depend on a specific charge geometry. Similar considerations apply to caseless charges, which require a high degree of cohesion in the absence of a supporting case. In all instances, high tensile strength and modulus, combined with high impact strength, are desirable. Many explosives cannot be melt-cast but must be powder press-filled. These have poor mechanical strength, and high g-forces could readily deform the charge. Desirable to increase mechanical strength whilst retaining ease of manufacture are:

- *High tensile strength and modulus* are required to prevent the explosive from distortion and viscous flow. Tensile strength and modulus tend to decrease with increasing temperature, and more rapidly as the melting point of the explosive is approached.
- *Impact strength* tends to decrease with decreasing temperature, resulting in brittleness in explosive charges at low temperatures.
- *Workability*: the explosive composition must also be capable of being filled, shaped or moulded without the inclusion of air cavities. The sensitivity of the composition must be such that these operations are safe to carry out in the course of production. One of the ways to increase mechanical durability is to support the explosive in a durable plastic matrix.

5.4 Plastic-Bonded High Explosives

5.4.1 Plastic-Bonded Compositions

The use of long-chain polymers to provide a plastic matrix to bond high explosives offers the possibility of achieving higher melting and softening points, and improved mechanical strength over a wide range of temperatures. Long-chain polymers in the form of commercial plastics are available as thermoplastics and thermosetting materials. Some advantages are gained by using plastic-bonded explosive (PBX). The strength of the PBX is determined by that of the polymer matrix and not that of the powdered HE. The study of PBX materials is a science in its own right. If thermoplastic materials are used, they can easily be melted and extruded. However, when the completed munition is heated, its mechanical strength will decrease and flow may occur. Only high-melting thermoplastics can be used and these demand high melting explosives.

Polymer binders also decrease the sensitivity of the HE by reducing the hot spot production process. Investigations are proceeding to find polymers which are environmentally friendly and offer simple routes for HE recovery and reprocessing from time-expired munitions or non-hazardous disposal routes. The following section investigates some of the current systems.

Table 5.2 *Thermal properties of some polymers.*

Material	Softening point °C
Polyvinyl chloride PVC	82
Polythene (ldp)	≈ 85
Polystyrene	82–105
Polymethylmethacrylate	85–115
Nylon (6 : 6)	220 m.pt.
Terylene	≈ 230
Polytetrafluorethylene PTFE	> 250

5.4.2 *Thermoplastics*

Thermoplastic are stable high molecular weight polymers which soften and melt on heating, and which can be moulded in the plastic state to yield products of varying degrees of hardness and toughness. Typical thermoplastics are listed in Table 5.2. All of these materials have reasonable mechanical strength at ordinary temperatures, but only nylon, terylene and PTFE remain hard and strong above 150 °C. High explosives dispersed in a matrix consisting of nylon or terylene or PTFE would meet the requirement of cohesion and strength at operational temperatures. Special processing techniques would have to be adopted to mould the preparations, in order to avoid thermal decomposition of the employed explosives at the softening point of the thermoplastic.

5.4.3 *Thermosetting Materials*

Thermosetting plastics, which form strong cross-linkages, are extremely strong and cannot be softened by heating. Examples of thermoset materials are phenolics, polyester and epoxy resins, and polyurethanes. Some thermosets become brittle and will not case bond. Epoxy resins also are incompatible, due to amine migration, and amines are well known sensitizers for nitro group explosives. Silicones, in the form of siloxanes, have been used, but the production of solid residues is undesirable. Polyurethanes are particularly good, since there is little swelling on curing. A typical example that has been widely trialled and is close to operational use is hydroxyterminated polybutadiene (HTPB), which is cured by reaction with di-isocyantes, as shown in Figure 5.3. There is difficulty from toxicity, particularly from the curing agent and also, the curing agent reacts with water, liberating gases that can destroy the integrity of the explosive. This reaction is used to produce polyurethane

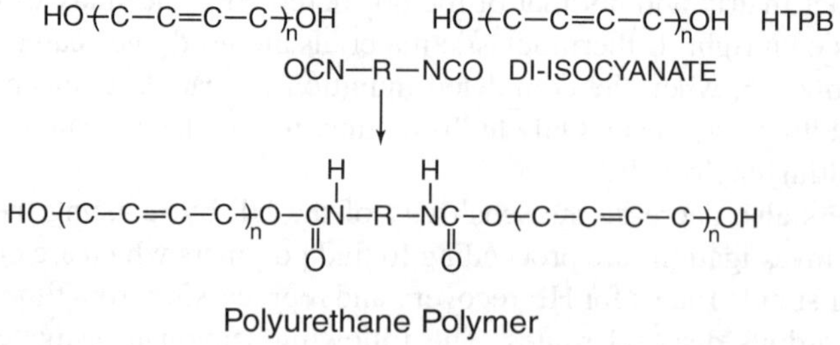

Figure 5.3 *Hydroxyterminated polybutadiene (HTPB).*

foams. The trapped gases in these foams can be used to sensitize liquid explosives to impact initiation [5].

Thermosetting plastics are processed from monomers or partial polymers, in the form of liquid resins or moulding powders. These are polymerized by heat, pressure and catalysts to yield hard plastic materials which cannot be melted or softened by heat. These materials provide varying degrees of hardness and toughness over a wide temperature range, being stable at 200 °C.

Plastic-bonded explosive compositions are conveniently prepared from thermosetting resins or moulding powders, by intimate mixing of the ingredients, followed by polymerization under the influence of gentle heat, pressure and catalysts. Such preparations may be filled into devices and polymerized *in situ*. Alternatively, the operation may consist of moulding the ingredients with the prepolymer, followed by machining the final cured plastic-bonded product. Again, polymerization catalysts can cause toxicity problems, and some work on polyester based PBXs is in progress.

5.5 Choice of High Explosive for Plastic Bonded Compositions

The high explosive selected must have good thermal stability. In practice this is found in explosives of high melting point but more importantly high ignitions temperature. High melting explosives suitable for plastic bonding are shown in Table 5.1. These explosives have the additional advantage of high density.

5.6 High-Energy Plastic Matrices

Development of high-energy plastic matrices, in the form of nitrated polymers might lead to the production of plastic-bonded high explosives of increased performance. Materials such as polyNIMMO and polyGlyn have been trialled but have not reached viable production partly because of stability problems. Their structures and chemical names are given in Table 5.3.

Polymer binders also decrease the sensitivity of the HE by reducing the hot spot production process. Classic examples are the formulations with polymeric waxes – either hydrocarbon or silicone greases. Investigations are proceeding to find polymers that are environmentally friendly and offer simple routes for HE recovery and reprocessing from time-expired munitions or non-hazardous disposal routes. It is very difficult to recover the unused explosive from current polyurethane PBX systems. Supercritical fluids have been tried, but are difficult to operate on the scale required for the quantities of time-expired energetic material currently requiring attention. Some systems are available using polyvinyl alcohols but, since these are water soluble, that also indicates their potential weakness in adverse climates.

Polyphospazenes, polymers derived from phosponitrillic halide systems, are an interesting area being investigated as potential energetic binders. The major investigations involve various substituted nitro ester forms of side chains (R in the structure type shown in Table 5.3). Typical side chains investigated are where R1, R3 and R5 are all the same and are either CH_2ONO_2 or $CH_2CH(ONO_2)CH_2ONO_2$ or combinations of the two. Similar conditions can apply to R2, R4 and R6, but a number of studies have used trifluorethyl

Table 5.3 Energetic polymers trialled.

Polymer	Chemical name	Structure
PolyAMMO	Poly (3azidomethyl-3methyl oxetane)	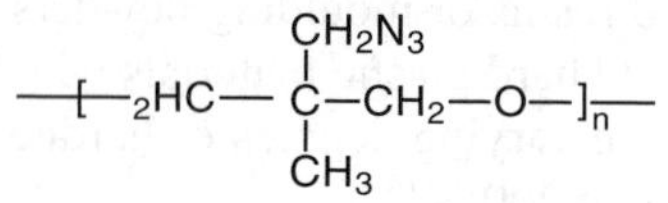
PolyBAMMO	Poly (3,3-bisazidomethyl oxetane)	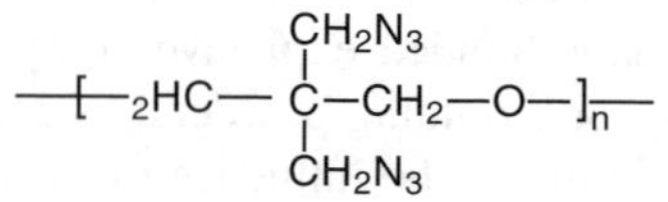
GAP	Poly glycidyl azide	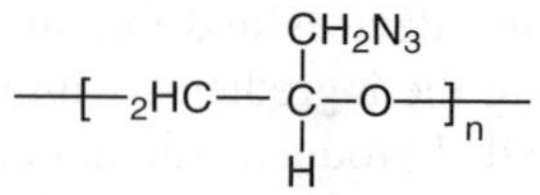
PolyGlyn	Polyglycidyl nitrate	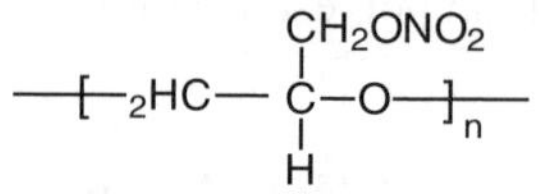
PolyNIMMO	Poly (3-nitromethyl-3-oxetane)	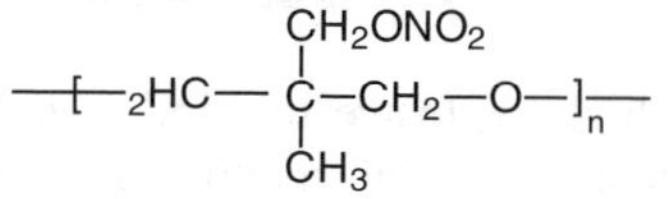
PolyPhosphazene		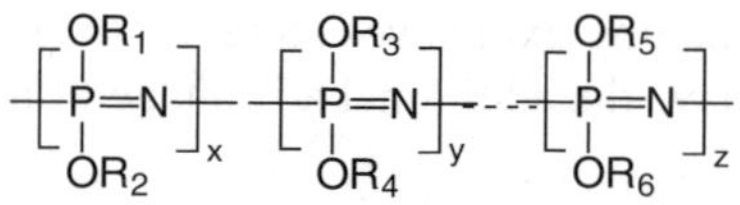

as the even number side chains and chain ends. These have higher thermal stability and mechanical strength, but suffer from reduced performance.

There are a number of different combinations, and the system is sufficiently flexible for x, y and z to have a complete range of integer values. These materials are best synthesized direct from the condensation of trisubstituted phosphite and trimethylsilylazide, followed by an anionic polymerization initiated with a base such as N-methylimidazole, rather than a thermal ring opening of the trimeric and tetrameric phosphonitrilic halides. The various polymer chain lengths can be controlled by choice of polymerization conditions, and the various sub-polymer chains can then be combined for the final product. The enthalpy of combustion increases with the proportion of energetic side chains, such as 3,4 dinitratopropan1-oxy, in the polymer increases. Originally, they were investigated as alternative explosives, but their burn characteristics make them more suitable as binders for PBXs and propellants.

A number of azide-substituted trimer and tetramer phosponitrilic halides were investigated but were found to be very sensitive. However the diaminohexaazide substituted tetramer has been investigated as a primer, and is discussed in the chapter on new primers. Other variants will be discussed later under high-nitrogen compounds in the chapter on new primers (Chapter 8).

High-energy propellants seek to improve ballistic performance by an increase in gas production and a reduction in gun damage. Propellants require the highest efficiency and minimum vulnerability to accidental initiation. Standard double-base propellants can be readily initiated to undesirable detonation. Stray rounds and shaped charges are particular hazards for double-base propellants. RDX in polymer binders has reduced vulnerability and increased explosive performance. The use of energetic binders (e.g. nitrated polymers or polyethylene oxide) offer increased efficiency and can eliminate the metal cartridge cases, which do not contribute to the explosive performance of the propellant. In this situation it is difficult to ensure all the propellant has burnt. The mechanical strength of the round, in the absence of the metal case, is very weak and can lead to major problems when automatic loading systems are less than perfect.

5.7 Reduced Sensitivity Explosives

The increased emphasis on safety and reduction of hazards when explosives are accidentally initiated have lead to a change in emphasis in munitions research away from pure explosive performance. Some of the newer materials have both improved safety and performance.

Triaminotrinitro benzene, TATB, is very strongly hydrogen bonded and as a result forms two dimensional sheets. This gives the compound a high melting point and very low impact sensitivity. It requires a high shock pressure to initiate and, thus, is almost impossible to set off accidentally. For this reason, it has been widely used in missile warheads, where frictional heating of the skin temperature at supersonic speeds could lead to cook-off of conventional high explosive fillings such as RDX and HMX.

Hexanitrostilbene, HNS, has a similar high melting point and low impact sensitivity. Again, it is used in hot environments, and it also is the main explosive in flyer plate-induced shock sensitive detonators or 'slappers'. Two high melting explosives being trialled are TACOT and DIPAM. Other possible high melting explosives are detailed in the appendix. Many of these are of purely scientific interest, since their bulk manufacture is difficult and expensive (see Figure 5.4).

One of the less sensitive explosives being evaluated is 1,1. diamino, 2,2 dinitro ethylene, or FOX 7. This material is comparatively easy to synthesize [6] and has important properties, not the least of which is its insensitivity to impact and friction. It also has a high ignition temperature, and much of this stability arises from having a polymeric structure due to the hydrogen bonding between the nitro and amine groups on adjacent molecules, which is responsible for forming the polymeric structure. Since it is less oxygen-deficient than TNT, it has a higher performance than TNT but is not readily cast, unlike TNT.

Another explosive under evaluation, 5nitro 1,2,4, triazol-3-one NTO has a comparable performance to RDX (NTO VoD 8510 ms^{-1} RDX VoD 8850 m s^{-1}), but is less sensitive to initiation than RDX(Impact FofI NTO 120 RDX FofI 80). It is readily synthesized in high yields from readily available chemicals [7]. One unusual property that NTO possess is a very low toxicity compared to most other high explosives. Whereas RDX has a toxicity level in the 200 mg per kg of body mass range, NTO has a toxicity level of >5 g per kg of body mass. It is very soluble in water, and is readily digestible by bacteria to give harmless products. This is a disadvantage for HE, where keeping the filling dry to avoid exudation and ingress of bacteria can be a problem.

In a typical replacement for RDX TNT compositions, correctly crystallized NTO can be incorporated at >70% level [8], without special casting procedures and, as such, it is still less

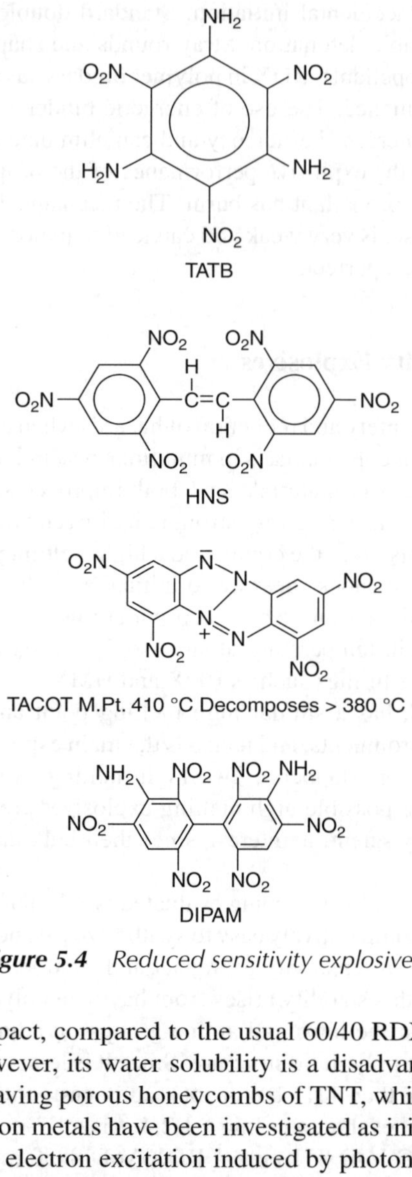

Figure 5.4 *Reduced sensitivity explosives.*

sensitive to accidental impact, compared to the usual 60/40 RDX/TNT formulations used for munitions filling. However, its water solubility is a disadvantage, since water ingress can leach the NTO out, leaving porous honeycombs of TNT, which are sensitive to impact. Salts of NTO with transition metals have been investigated as initiators in explosive trains. These can be activated by electron excitation induced by photon irradiation. Some will be discussed in the section on new primers.

Ammonium dinitramide has explosive properties, but is more widely used in propellants, where it is a source of oxygen for fuel binders and all of its products are gaseous and chlorine-free. It is hygroscopic and susceptible to moisture, but this can be overcome by careful choice of polymer binder.

Tri nitroazetidine TNAZ is more oxygen-balanced than TNT and, as such, would be expected to have a better performance than TNT. It is being evaluated as a replacement for TNT in that, like TNT, it can be safely melted by hot water well below its ignition temperature, and is easily melt-cast. It is less toxic than TNT. The problem is that the synthesis is difficult and not very efficient, so it is comparatively expensive.

Table 5.4 *Explosives under evaluation.*

Common name	Chemical name	Structure
ADN	Ammonium dinitramide	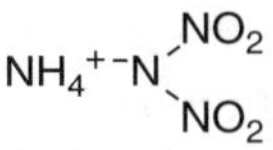
HNIW OR CL20	2,4,6,8,10,12-hexanitro-2,4,6,8,10,12-hexa-aza tetracyclododecane	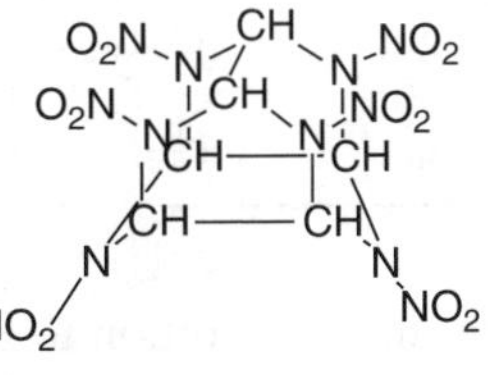
FOX 7	1,1′ diamino, 2,2′ dinitroethene	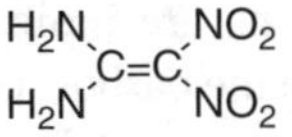
NTO	5-Nitro-1,2,4, triazol-3-one	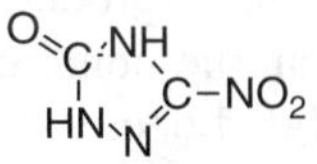
TNAZ	1,3,3-trinitroazetidine	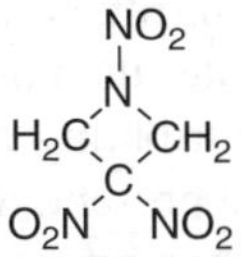

CL20 is a typical strained ring compound and, hence, has a positive enthalpy of formation, due to a combination of the presence of nitrogen in the rings and the ring strain energy. This positive enthalpy of formation adds to the enthalpy of combustion giving improved performance. Also, one of the four isomers (the ε form) has a density of 2.04 g cm^3, giving rise to a VoD of 10 200 m s^{-1}. Again, its synthesis is difficult, and obtaining the correct isomer is very important. For specialist applications, its high performance can make CL20 the explosive of choice. The structure of these explosives under evaluation are given in Table 5.4.

There are a number of other potential materials being evaluated, which can be considered under a number of headings. The aim is to improve performance both detonating and propelling, improve safety, particularly with respect to accidental initiation and reduced toxicity.

Some of the headings are:

- High positive enthalpy of formation molecules.
- High nitrogen-containing materials.
- New heavy metal-free primers (these will be discussed in a later section).

5.8 High Positive Enthalpies of Formation Explosives

Some of these have been discussed in the sections above. Most of these, such as RDX, HMX, NTO and CL20, involve the inclusion of a number of nitrogen atoms in the carbon skeleton of the explosive, either in the straight chain compounds or the aromatic and cyclic skeleton on which the oxidizing species are incorporated. TEX is a nitramine with high

Table 5.5 *Comparison of carbon-to-carbon, carbon-to-nitrogen and nitrogen-to-nitrogen bond strengths.*

Bond type	Carbon	Bond enthalpy kJ mol^{-1}	Carbon to nitrogen	Bond enthalpy kJ mol^{-1}	Nitrogen	Bond enthalpy kJ mol^{-1}
Single	C–C	348	C–N	305	N–N	163
Double	C=C	612	C=N	613	N=N	409
Triple	C≡C	518	C≡N	890	N≡N	945

density, in addition to its strained ring, and thus it demonstrates a high performance [9]. These systems are indicative of the comparative weakness of the carbon-nitrogen bond compared to the carbon-carbon bond, and the increased strength of the nitrogen-to-nitrogen triple bond in the product species. As Table 5.5 shows, the bonding of carbon to carbon is stronger than the bonding of carbon to nitrogen and, whereas the strength of the carbon-to-carbon bond decreases on the formation of double and triple bonds, the bond strength of nitrogen to nitrogen increases as double and triple bonds are formed. The triple bond strength is greater than the sum of the single and double bond strengths

Another contributing factor to the high positive enthalpy of formation for CL20 is the high ring strain present. Some other positive enthalpies of formation compounds are based on strained carbon rings with multi-pendant nitro groups attached. Typical examples are the substituted cubanes, with hexa, hepta and octo nitro substituents [10]. These compounds are a long way from viable systems, but are of scientific interest and could become significant explosive molecules if bulk manufacture could be forthcoming at reasonable cost.

5.8.1 High Nitrogen-Containing Molecules

Nitrogen-rich molecules are potentially "green" energetic materials [11], since the main combustion product is molecular nitrogen gas. They exhibit desirable performance characteristics in high explosives or in propellant formulations, being particularly useful in the latter case by reducing flash and smoke caused by the combustion of carbon monoxide and hydrogen produced by combustion of conventional nitrocellulose-based propellants. Moreover, for gun propellants, high concentrations of nitrogen species in the combustion products of propellants can reduce gun barrel erosion by promoting the formation of iron nitride, rather than iron carbide, on the interior surface of the barrel [12]. The carbide is comparatively easily removed by the passage of the projectile, whereas the nitride formation enhances the steel strength, as demonstrated in high-performance internal combustion engines.

5.8.2 Pure Nitrogen Compounds

Although theoretical studies have shown that a number of polynitrogen molecules in the range from N_4 N_6, N_8, N_{10} and N_{12} are possible, see Figure 5.5, few of these have been observed, probably due to low initial thresholds for decomposition. The simplest member, N_4, would be expected to be tetrahedral, with a structure analogous to that of P_4, the white form of phosphorous used in smoke-generating pyrotechnics. The caged compounds N_6 and N_8 could be more stable than the cyclic forms, analogous to benzene and cyclo-octane.

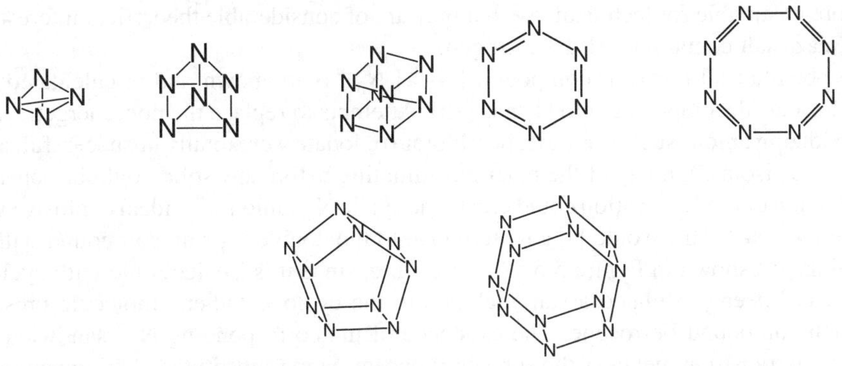

Figure 5.5 *Some suggested structures for pure polynitrogen molecules.*

Valence bond theory predicts that the cyclic N_6 should be stabilized by aromaticity, since it would have alternate single and double bonds. However, lone pair repulsion would indicate a non-planar ring, which would not be aromatic-stabilized.

Typical activation energies for decomposition are calculated for the N_4 species as only <150 kJ mole^{-1} – about the level of primary explosives – and for N_8 it is only ≈ 50 kJ mole^{-1}, with the N_6 showing an activation energy of <40 kJ mole^{-1}, suggesting that it would spontaneously decompose at ambient conditions. The other all-nitrogen molecules, N_{10} and N_{12}, are calculated to have even lower activation energies for decomposition, so may only be stable at low temperatures. The theoretical performance of these three compounds far exceeds that of CL20, with VODs of ≈ 12–14 km s^{-1}. Cubic N_8 has been synthesized in limited quantities under abnormal conditions of 2000 K and 110 GPa in a diamond cell.

Some quantum calculations suggest that the N oxide derivatives of cyclic N_6 should be stable, due to reduction in nitrogen lone pair repulsions, but no molecules have been observed. Chemists are particularly attracted to such nitrogen-rich compounds, since they are on the borderline of stability. Except for molecular nitrogen (N_2) and the cubic nitrogen modification at high temperatures and pressures described by *Eremets et al.* [13], there is no neutral polynitrogen modification known.

The simplest polynitrogen species is the azide ion N_3^-, which has been widely used as the lead or silver salt in detonators for a number of years. Although ammonium azide, with a nitrogen content of 93.3%, is endothermic, (ΔH_f + 85 kJ mol^{-1}) and its explosive properties are quite good, it is not very useful because it is fairly volatile and very oxygen-deficient ($\Omega = -53\%$). A postulated nitrogen compound is the cyclic N_5 ring, but this is unstable, as is the anion formed by electron addition.

Salts containing the N_5^+ cation have been synthesized but initially only with non-energetic, impractical, anions such as AsF_6^- and SbF_6^-. The nitrogen content could be increased by using the SnF_6^{2-} to produce $(N_5^+)_2SnF_6^{2-}$. The structure of this N_5^+ cation is unusual, in that it is not cyclic but a 'V' shape, with an obtuse angle between the two arms. Recently, the salts with the hexaazaphosphide anion, $[N_5^+][P(N_3)_6]$ (N : 91.2%) and the tetraazoboride, $[N_5^+]$ $[B(N_3)_4^-]$ (N : 92%) have been prepared by *Christe et al.* [14]. Unfortunately, these salts are often only stable at very low temperatures, and they spontaneously explode when warmed to ambient. They also require difficult preparations and are therefore unsuitable for technical use, but they are of considerable theoretical interest, with the hexaaza salt containing 91.2% nitrogen.

The potential all-nitrogen compound $[N_5^+]\ N_3^-$ is unknown and is calculated to be unstable, auto-decomposing into N_2 gas [15]. Attempts to replace the non-energetic anions with oxidizing anions such as nitrate, perchlorate or iodate were totally unsuccessful, and no product was isolated, many of the reactions initiating before any solid could be separated.

Reduction of the N_5^+ cation could create the cyclic N_5^- anion. The ideal explosive would then be $N_5^+N_5^-$. This would be an alternative to the double N_5 ring compound with ring cross-linkages shown in Figure 5.5. This N_5^- ring, since it is isoelectronic with cyclopentadiene, has been postulated as an analogue to the cyclopentadiene molecule present in sandwich compound Ferrocene. The existence of the corresponding N_5^- sandwich compound with transition metals is the subject of debate. Some substituted N_5 ring compounds have been reported in mass spectra, but the products have not been isolated.

5.8.3 Other High-Nitrogen Compounds

The first member in the series of compounds containing the highest nitrogen contents is hydrazoic acid HN_3 (N: 97.7%) [4], followed by tetraazidomethane ($C(N_3)_4$) [16], hydrazinium azide ($N_2H_5N_3$, empirical formula N_5H_5) [17], prepared by the neutralization of the unstable hydrazoic acid with the base hydrazine hydrate, ammonium azide (NH_4N_3, empirically tetrazene, N_4H_4) [18] as well as diazene (N_2H_2) [19], all of which have nitrogen contents of 93.3%. Pure hydazoic acid liquid can spontaneously detonate, and hydrazinium azide is very sensitive to accidental initiation. The latter can be desensitized by crystallizing in the presence of excess hydrazine when the hydrazinate ($N_2H_5N_3{\cdot}N_2H_4$) [20] is formed. This has a lower nitrogen content of 91.5%.

More recently a trinitrogen species, the dintramide, $N(NO_2)_2^-$, has been made and characterized as an ammonium salt with a potential use in propellants. This molecule can be considered as a partial oxidation product of the azide ion, in which an oxygen molecule has been added to each of the terminal nitrogen atoms of the azide ion. Unfortunately, the ammonium salt is, like most ammonium salts, fairly hygroscopic, which limits its application in propellants.

Guanidine chemistry has produced a number of interesting compounds used in energetic materials. Nitroguanidine (NQ) [21] has many applications, for example:

1. in pyrotechnic compositions;
2. modified single-based propellants (formulated with nitrocellulose (NC));
3. triple-based propellants (NQ + NC + NG) and in other high performance propellants with RDX and NC.

It is easily obtained by the nitration of guanidinium sulfate or dehydration of guanidinium nitrate. Its aminated analogues aminonitroguanidine (ANG) diaminonitroguanidine (DANQ), are also being evaluated [22] and have some applications in propellants. Also, the ANQ halogenides and the sulfate has been described.

The doubly nitrated dinitroguanidine (DNQ) has intensively studied in the literature in its neutral, as well as its deprotonated, form in salts [23]. Some of these salts have been investigated as potential primers. 5-Amino-1,3-dinitroguanidine has recently been prepared and is currently being assessed.

An unusual polynitrogen anion involves the compound with carbon, which will be discussed later. Other substituted triazines have been synthesized and are being evaluated. Many of these compounds are substituted nitrogen heterocycles, and some of these variations will be examined as part of the nitrogen heterocycle system study.

5.8.4 Nitrogen Heterocycles

Polynitrogen heterocycle compounds are of more potential use. There are a number of different possibilities in terms of size of the ring system and the number of nitrogens in the rings, along with their ring positions when two or more nitrogens are in the ring. Also, there are variations on the number, type and position of substituents present on the ring components. A number of these are very endothermic, but with high activation energies for decomposition. Some of these are detailed in the appendix.

The four-membered ring containing a single nitrogen is typified by the azetidine structure given for 1,3,3 trinitroazetidine. This molecule is particularly important because of its low melting point, meaning that it can act as a replacement for TNT in mixtures that can be prepared using boiling water as the heat source in the casting process. The other four-membered ring compound containing two nitrogens, azine, has substituents other than hydrogen on the carbons for stability; substituted benzene rings are most common, since these can extend the delocalization of the aromatic rings coupled through the azine ring. These materials are of interest because they contain the pseudo-aromatic chromophore C_2N_2, which may indicate a high absorption cross-section for some laser wavelengths, particularly in the infrared and ultraviolet ranges. Compounds such as the dipicryl substituted azine may prove interesting.

5.8.4.1 Five-Membered Rings

Five-membered rings containing one, two or more nitrogens form an important group of compounds currently under evaluation. The five-membered ring systems are characterized by the number and location of the nitrogens in the ring and also the substituents. The variations are given in Figure 5.6.

The thermal stability of these ring compounds is quite high, namely > 473 K (200 °C), and their enthalpies of formation are positive, with their values dependent on the ring positions of the nitrogen atoms. The positive enthalpy of formation assists in their performance. The greater the number of nitrogens in the ring, the fewer the number of nitro groups which are required for oxygen balance, and also increased sensitivity to initiation. The pyrrole ring has a single ring nitrogen, and species containing two and three nitro groups have been trialled as energetic materials. The high-performing tetranitropyrrole is unstable at normal storage temperatures. Some salts of these pyrroles have been suggested as alternative primers, and they will be discussed under that heading in a later chapter.

The di- and tri-carbazoles are again used in their polynitro formations, and some azide-substituted molecules are available. Polynitropyrazole-based energetic compounds have a balanced detonation performance, and the increased generation of nitrogen gas (an environmentally friendly product) gives them an advantage over traditional energetic materials. The number and location of both the ring nitrogens and the nitro groups have an influence on the energetic properties. Figure 5.7 below shows some of the possible variants. Two series are shown in the figure: the first series is with the nitro groups substituted

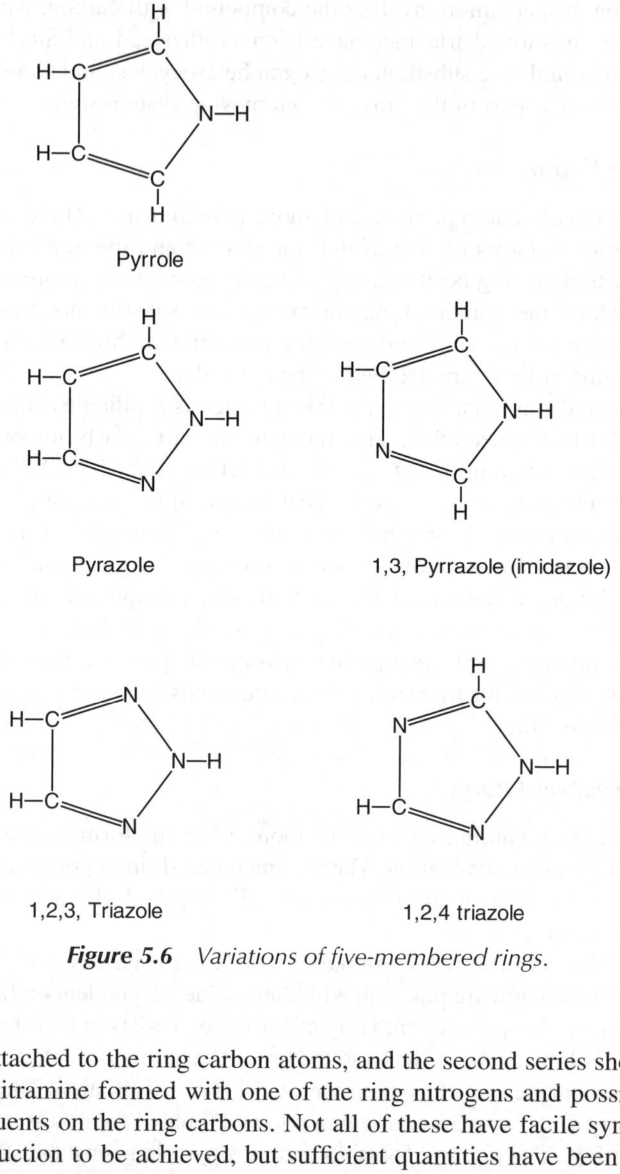

Figure 5.6 *Variations of five-membered rings.*

for hydrogens attached to the ring carbon atoms, and the second series shows compounds containing the nitramine formed with one of the ring nitrogens and possible variants for the nitro substituents on the ring carbons. Not all of these have facile syntheses enabling large-scale production to be achieved, but sufficient quantities have been synthesized for their properties to be investigated.

The high energy content of the polynitro1,2 pyrazoles, when compared to the polynitro 1,3 pyrazoles (or imidazole) stems from the presence of adjacent nitrogen atoms capable of forming nitrogen gas (N_2) and the presence of two or more nitro groups. Many examples of polynitropyrazoles have been synthesized and are the subject of a number of studies [24–33]. Simple examples are 3,5- and 3,4-dinitropyrazoles and 3,4,5-trinitropyrazoles (TNP). TNP possesses an energetic performance (VoD = 8651 m s^{-1}), comparable to RDX (D = 8977 m s^{-1}), but TNT-like impact sensitivity (IS TNP 17 J; TNT 15 J).

Two basic polynitropyrazoles, 3,4-dinitro- and 3,4,5-trinitropyrazoles, have different properties. DNP melts at about 85 °C, which is about 100 °C lower than TNP. Both of

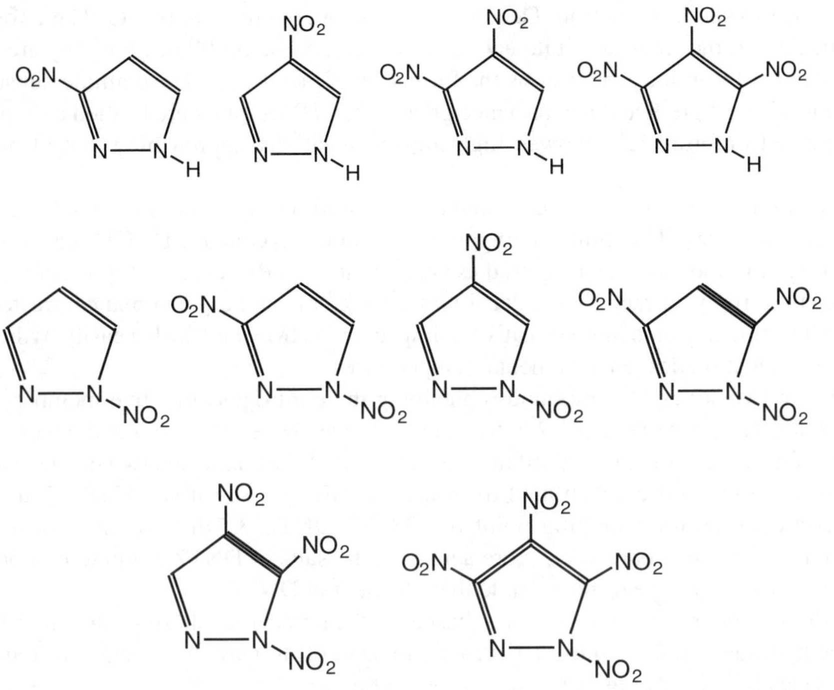

Figure 5.7 *Variations on nitro-1,2-pyrroles, showing amine and nitramine variants.*

these compounds suffer from high volatility, and the sublimation occurs before the onset of decomposition. Kinetics studies gave values for the activation energy barrier of 127 kJ/mol for TNP and 70 kJ/mol for DNP. These values are primary explosive level but, until they have been corrected for evaporation and sublimation, they are in doubt.

Further substitution in the pyrrazole ring can reduce volatility and also assist in stability. 4-amino-3,5-dinitropyrazole (LLM-116) [34, 35], 5-amino-3,4-dinitropyrazole 5-nitramino-3,4-dinitropyrazole [36] have higher melting points and lower vapour pressures. Some of them have high detonation performance in combination with low impact sensitivity, e.g. 4-amino-3,5-dinitropyrazole (VoD = 8490 m s^{-1}, IS > 20 J, 5-amino-3,4- dinitropyrazole (VoD = 8640 m s^{-1}, IS > 50 J).

The hydrogen attached to the ring nitrogen is quite acidic, and a series of salts have been tried to overcome the volatility problems of the DNP and TNP. An alternative method of reducing the volatility is to link two polynitropyrrole rings together – those with two linked pyrazole rings 3,3′,5,5′-tetranitrobipyrazoles-4,4′ and 3,3′,4,4′-tetranitrobipyrazoles-5,5′ [37, 38], 3,6-dinitropyrazolo [4,3-c] pyrazole (DNPP). Thermal stability and thermal decomposition mechanisms depends on the number and mutual arrangement of nitrogroups in the pyrazole core. The tetranitropyrrole in which the acidic hydrogen has been replaced by a further nitro group, although predicted to have a high performance, is not particularly stable and difficult to synthesize in high yields.

One unusual pyrazole compound contains two fused rings. The endothermic (+ 270 kJ mol^{-1}) compound, DNPP (3,6 dinitropyrazolo (4,3 –c) pyrazole) has an explosive performance of ≈ 85% of that of HMX, but its impact sensitivity is half that of HMX and it is not sensitive to spark or friction. Decomposition temperature is >300 °C. The two nitrogens attached to the rings are quite acidic, which can cause problems, but they are easily replaced either by amine groups or by the formation of metal salts. The diamine-substituted compounds have a predicted performance greater than HMX and were trialled as explosive LLM119, but unfortunately showed high impact sensitivity approaching that of primary explosives.

Other fused 1,3, pyrrole rings compounds are based on the glycouracil nucleus, which is derived from urea. The dinitro compound, 1,4, dinitrolycouracil DNGU, has a density close to 2 g cm^3 and has been proposed as an alternative to RDX. The tetranitrocompound, 1,3,4,6,tetranitroglycouracil, has a density of 2.04 g cm^3 and a performance greater than HMX. Unfortunately, it is more sensitive to impact than HMX and is also easily hydrolysed by water, even at modest environmental temperatures.

There are two series of compounds containing three nitrogens in a five membered ring the triazoles. 1,2,3 triazole and 1,2,4, triazole as shown above. The latter is the more stable because of the separation of the third nitrogen atom from the connected two nitrogen atoms. Even so the simple nitro substituted triazoles are still very sensitive. ANTA 5-amino-3-nitro-1,2,4-triazole has a melting point of 238 °C. DNTZ 3,5,dinitro-1,2,4 triazole has improved performance but is also more acidic. Metal salts of DNTZ are potential primary explosives, and they appear more stable than the neutral DNTZ.

A series of important compounds are based on the tetracarbazole ring, shown in Figure 5.8. The hydrogen on the 2 nitrogen is easily ionized off to leave a delocalized ring anion which is capable of forming salts with various cations.

A number of substituents in the X position of the ring are possible. One of particular importance is 5-azidotetrazole, CHN_7 [39, 40]. This compound has the azide N_3 group substituted on to the tetrazole ring, again giving rise to a number of interesting variants. The hydrogen attached to nitrogen is readily removed to give the tetrazolate anion with the formula CN_7^-, which can then be combined with a various cations to give rise to a number of interesting compounds. The lithium and sodium salts are 'relatively' stable towards external stimuli, like the corresponding azide salts, but the potassium and caesium salts spontaneously explode unexpectedly, with the caesium salt being impossible to separate in the dry state. The rubidium salt has been known to detonate even in solution.

Several other metal 5-azidotetrazolates, when completely dried, explode spontaneously without any handling, even in the dark, which makes them of theoretical interest only. While these metal salts are extremely explosive, the nitrogen-rich salts (hydrazinium, ammonium, guanidinium and aminoguanidinium, are all stabilized via hydrogen bonds and represent promising energetic materials, particularly as propellants. In addition, these salts show the highest N-content ever reported for tetrazolate salts – hence their applications in propellants.

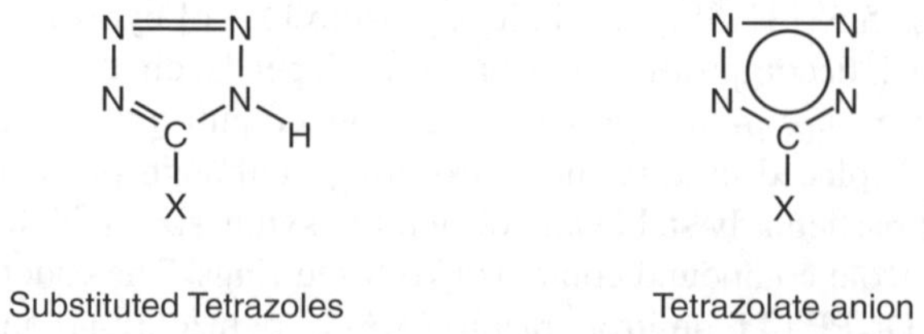

Figure 5.8 Compounds based on the tetracarbazole ring.

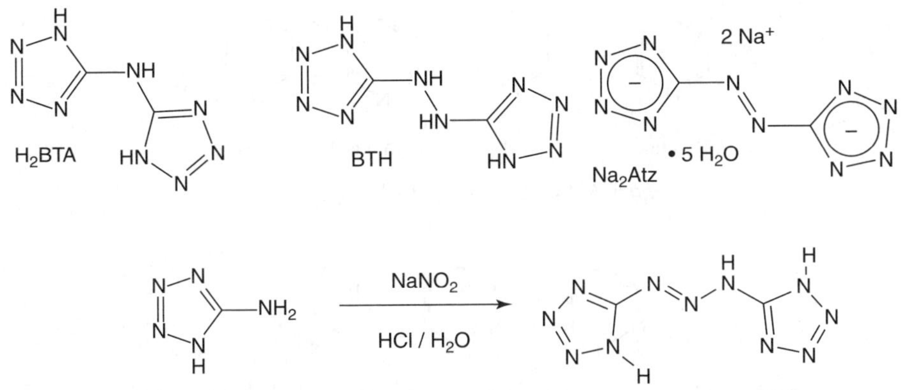

Figure 5.9 *Compounds with multiple tetracarbazole rings linked together either by nitrogen species or carbon-containing moeties.*

Some of these tetrazole compounds have been investigated as primers, both for detonating and burning systems. Some of these will be discussed in the section on new primers.

A number of compounds are available with multiple tetracarbazole rings linked together either by nitrogen species or carbon containing moeties, some examples are shown in Figure 5.9. The simplest linkage is NH to form a secondary amine of which the example below 22′dihyrdo bistetrazolylamine H_2BTA is typical. This material is prepared by reacting aminoterazole with chlorotetrazole, but is also prepared by the reaction of sodium dicyanamide with sodium azide in acidified aqueous ethanol. Systems derived from hydrazine, with two nitrogens forming the linkage between the carbazoles, are possible when amino tetrazole is diazatized, and the triazine with three nitrogens atoms in the linkage between the tetracarbazoles can be thought of as substituted azide ions.

The hydrogens on the heterocyclic rings can easily be nitro substituted to give more energetic compounds which are then oxygen-rich and make viable propellants. Without these nitro substitutions, these compounds require an oxidant such as ammonium perchlorate or nitrate, producing solid rocket propellants. The ring and amine hydrogens can be ionized off yielding anions which can be combined with a number of nitrogen rich cations such as hydrazinium and ammonium. Some of these will also be discussed in the new primers section.

An interesting five-membered nitrogen ring series is the pentazoles. Ideally, the all-nitrogen cyclic pentazolate ion N^{5-} would be combined with the N^{5+} cation to make an all-nitrogen compound. However, the existence of the unsubstituted pentazolate anion is the subject of ongoing debate, and it has only reportedly been observed in solution or in mass spectral decomposition products [41]. Substituted pentazolates with an attached aromatic ring are known, and highly nitro-substituted aromatic ring compounds are under investigation [42]. The presence of an aromatic ring enables the delocalization stabilization of the pentazole to be conjugated with the aromatic delocalized system.

5.8.4.2 Six-Membered Ring Systems

Six-membered ring systems containing different numbers of nitrogen atoms have been investigated. As with the five membered ring systems the number and positions of the

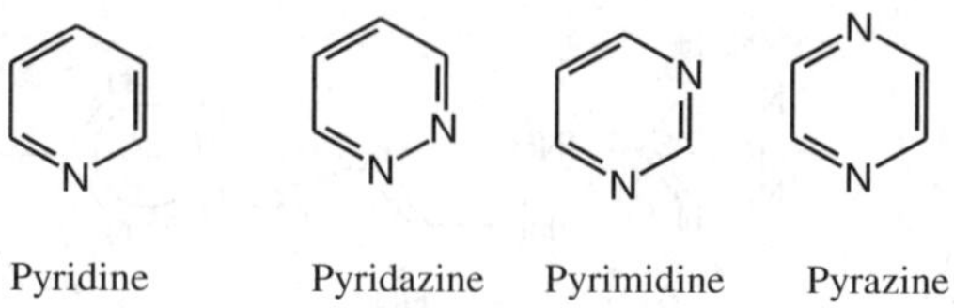

Figure 5.10 *Six-membered ring systems.*

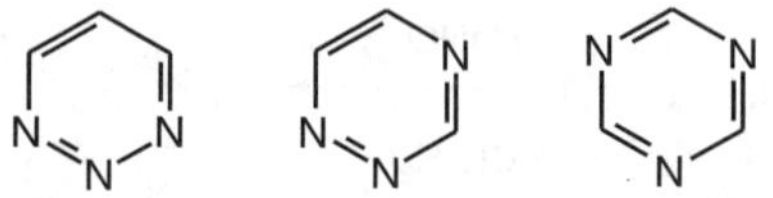

Figure 5.11 *Six-membered ring systems with three nitrogen atoms.*

incorporated nitrogen atoms and the ring substituents have been investigated. The various configurations are shown in the Figures 5.10 and 5.11.

The single-nitrogen compound pyridine and substituted pyridine compounds have been investigated as alternatives to TNT for cast explosives, but they do have a similar toxicity to TNT and do not offer major increases in explosive performance. The two-nitrogen systems are more difficult to synthesize and, as such, will only have specialist applications. The three-nitrogen atom systems are shown in the diagram and there are a large number of possibilities for substituents.

Tetrazines are known, and substituted variants have been used in pyrotechnics and propellants [43]. The most stable of these compounds is with the symmetric arrangement 2,3,5,6 tetra azo benzene. The other unsymmetrical isomers are less stable, and the 1,2,3,4, isomer has primer-like sensitivity [44, 45]. Attempts to prepare the 1,4 dinitrotetrazine by nitration of the 1,4, diamino-tetrazine resulted in a 1,4, di-nitramine (Figure 5.12) [46] which has excellent detonating properties, comparable to RDX, with a VoD of 9350 m s^{-1}. The di-nitrotetrazolyl compound is also a high-performance explosive, but more difficult to synthesize. The dihydrazino compound has been trialled as part of a high performance propellant mixture.

Two phosphazenes, with phosphorous replacing carbon in nitrogen ring compounds, are identified as substituted trimer and tetramer polymeric phosphonitrilic halides. If azide or nitro groups are substituted for the halides, then powerful, extremely sensitive explosive compounds result. See the section on new primers. Replacing the halides with nitro-hydrocarbon chains produces materials with propellant qualities rather than high explosives properties.

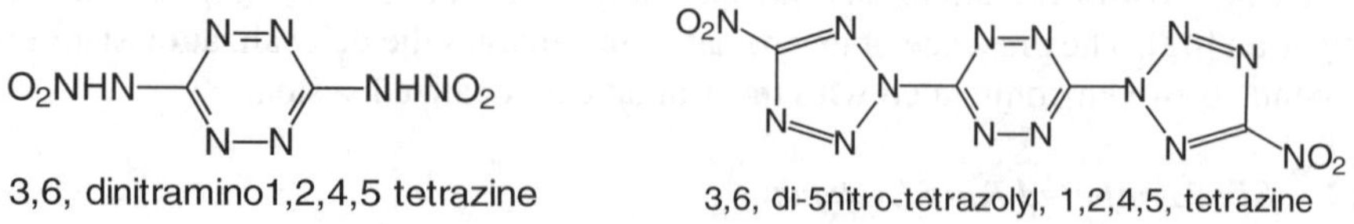

Figure 5.12 *Tetrazines.*

5.8.4.3 Structures of Some Thermally Stable Explosives

ANPZ

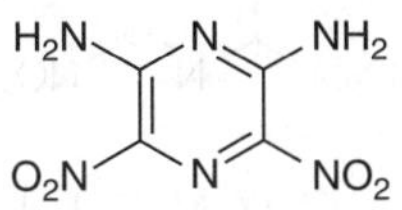

$C_4H_4N_6O_4$

BPABF

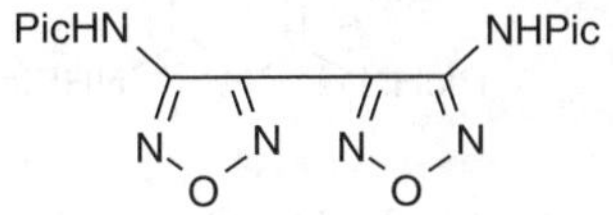

$C_{16}H_6N_{12}O_{14}$

CL-14

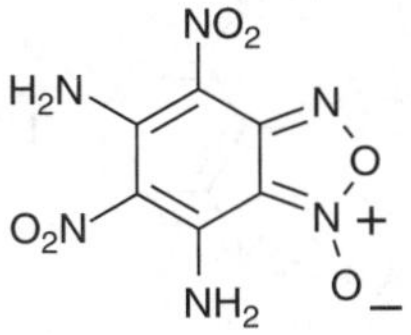

$C_6H_4N_6O_6$

DIPAM

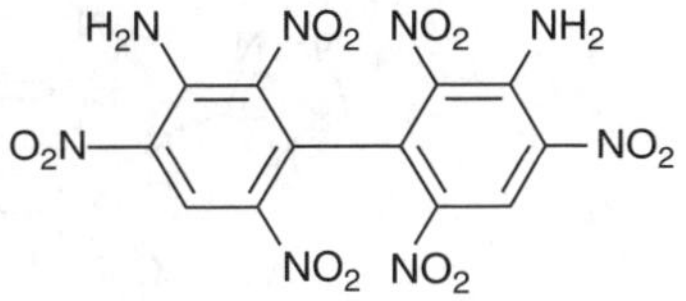

$C_{12}H_6N_8O_{12}$

DNDPF

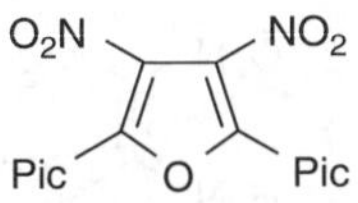

$C_{16}H_4N_8O_{17}$

DNPBT

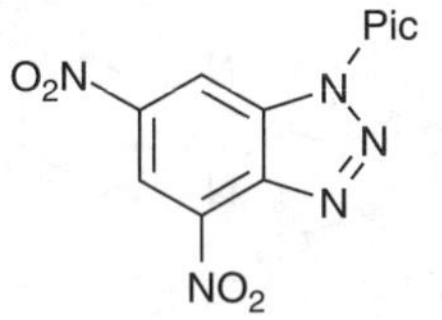

$C_{12}H_4N_8O_{10}$

HNS

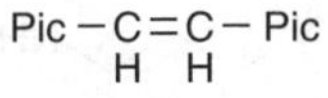

$C_{14}H_6N_6O_{12}$

NDAPDO

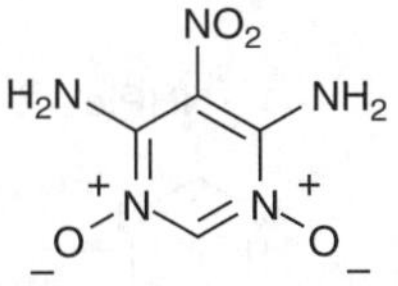

$C_4H_5N_5O_4$

NTAPDO

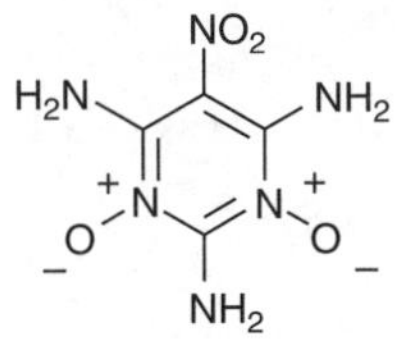

$C_4H_6N_6O_4$

NTPP

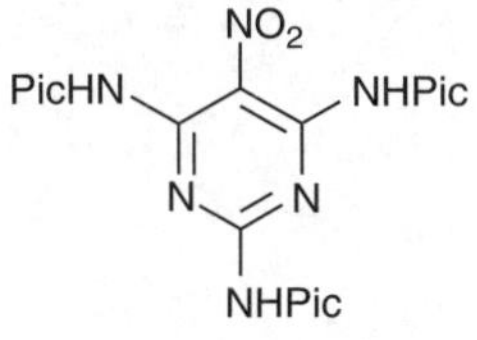

$C_{22}H_9N_{15}O_{20}$

PYX

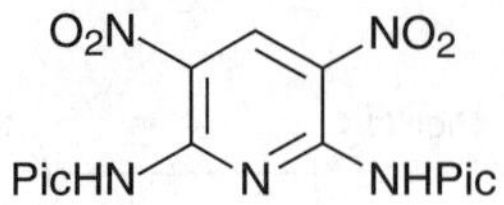

$C_{17}H_7N_{11}O_{16}$

PZO

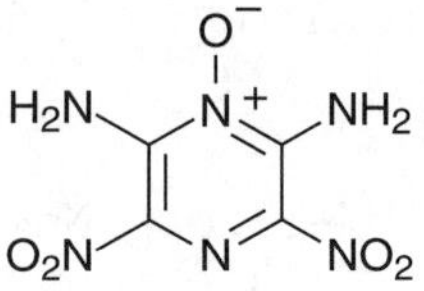

$C_4H_4N_6O_5[-]$

PATO

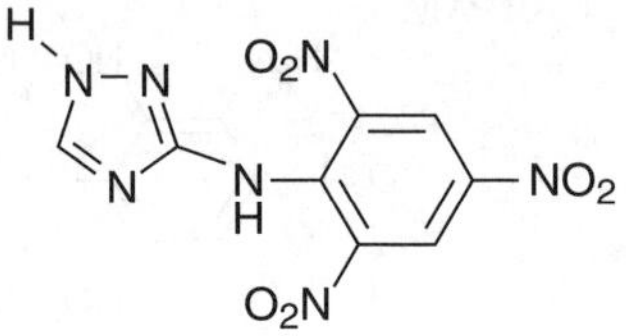

$C_8H_5N_7O_6$

Pic =

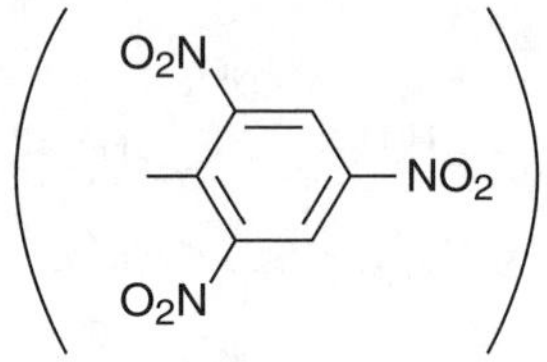

TACOT

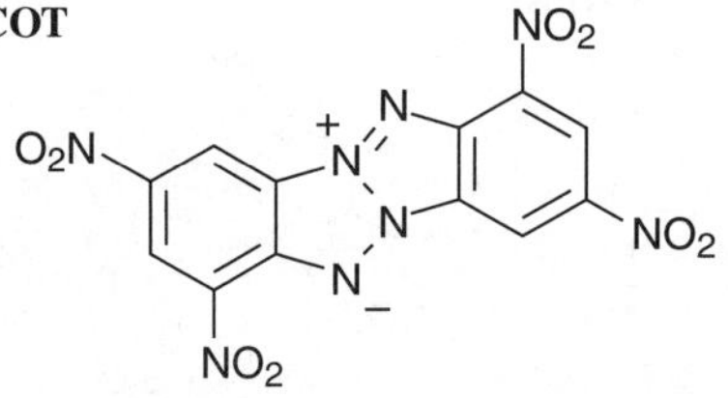

$C_{12}H_4N_8O_8$

TADNP

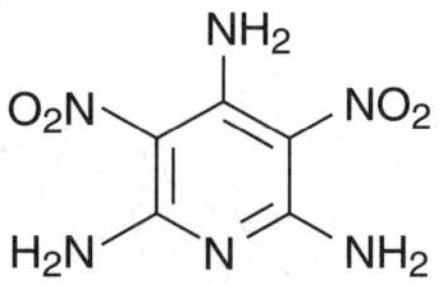

$C_5H_6N_6O_4$

TPM

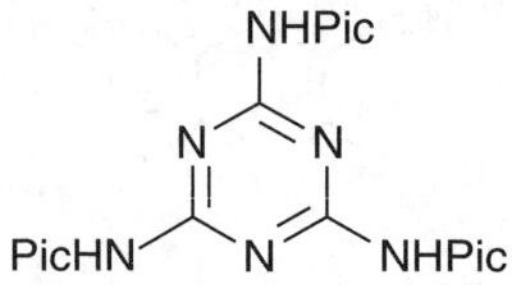

$C_{21}H_9N_{12}O_{18}$

TPP

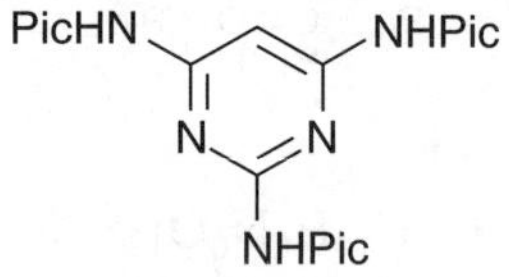

$C_{22}H_{10}N_{14}O_{18}$

Glossary of Chemical Names for High-Melting-Point Explosives

ANPZ	2,6-Di**a**mino-3,5-di**n**itro**p**yra**z**ine
BPABF	4,4′-**B**is(**p**icryl**a**mino)-3,3′-**bif**urazanyl
CL-14	5,7-Diamino-4,6-dinitrobenzofuroxan (**C**hina **L**ake explosive no. 14)
DIPAM	3,3′-Diamino-2,2′,4,4′,6,6′-hexanitrobiphenyl (**Dip**icr**am**ide)
DNDPF	3,4-**Din**itro-2,5-**dip**icryl**f**uran
DNPBT	4,6-**Din**itro- 1-**p**icryl**b**enzo**t**riazole
HNS	2,2′,4,4′,6,6′**H**exa**n**itro**s**tilbene
NDAPDO	5-**N**itro4,6-**dia**mino**p**yrimidine- 1,3- **d**i**o**xide
NTAPDO	5-**N**itro-2,4,6-**tria**mino**p**yrimidine- 1,3- **d**i**o**xide
NTPP	5-**N**itro-2,4,6- **t**ris(**p**icrylamino)**p**yrimidine
PATO	3-(**P**icryl**a**mino)- 1,2,4-**t**riaz**o**le
PYX	2,6-Bis(**p**icrylamino)-3,5-dinitro**py**ridine (e**X**plosive)
PZO	2,6-Diamino-3,5-dinitro**p**yra**z**ine- 1-**o**xide
TACOT	Tetranitrodibenzo- 1,3a,4,6a-tetra-azapentalene (Dibenzo**t**etra-**a**za**c**yclo-**o**ctane, **t**etranitro-; incorrect structure originally assigned)
TADNP	2,4,6-**T**ri**a**mino-3,5-**din**itro**p**yridine
TPM	N,N′,N″-**T**ri**p**icryl**m**elamine
TPP	2,4,6-**T**ris(**p**icrylamino)**p**yrimidine

Bibliography

Klapotke, T.M. (ed.) (2007) *High Energy Density Materials*, Springer, Berlin, ISBN 978-3-540-72201-4.

Marinkas, P.L. (ed.) (1996) *Organic Energetic Compounds*, Nova Science Publ. Inc., N.Y., ISBN 1-56072-201-0.

Agrawal, J.P. and Hodgson, R.D. (2007) *Organic Chemistry of Explosives*, Wiley, Chichester, ISBN 978-0-470-02967-1.

Olah, G.A. and Squire, D.R. (eds) (1991) *Chemistry of Energetic Materials*, Academic Press, NY, ISBN 0-12-525440-7.

Kaplotke, T.M. (2012) *Chemistry of Energetic Materials*, De Gruyter, H., Berlin, ISBN 978-3-11-027358-8.

Boddu, V. and Rednev, P. (eds) (2010) *Energetic Materials*, CRC Press (Taylor and Francis), ISBN 978-1-4398-3513-5.

References

(1) Balliou, F., Dartyge, J.M., Spyckerelle, C.V. and Mala, J. (1993) Proceedings 10th International Symposium on Detonation 816–823.

(2) Kröber, H. and Teipel, U. (2008) Crystallization of Insensitive HMX. *Propellants, Explosives, Pyrotechnics*, **33**, 33–36.

(3) Cartwright, M. (2003) Minutes of Reduced Sensitivity RDX, Proceedings of Nato Technical Meeting WTD91 Meppen, Germany.

(4) Hudson, R.J., Zioupos, P. and Gill, P.P. (2012) Nano indentation as a diagnostic test for IRDX. *Propellants, Explosives, Pyrotechnics*, **37**(2), 191–197.

(5) Cartwright, M., Lloyd Roach, D. and Simpson, P.J. (2007) Non-solid explosives for shaped charges I: Explosive parameters measurements for sensitized liquid explosives. *Journal of Energetic Materials*, **25**(2), 111–127.
(6) Latypov, N.V., Bergman, J., Langlet, U., *et al.* (1998) Synthesis and Reactions of 1,1-Diamino-2,2-dintrothylene. *Tetrahedron*, **54**, 11526–11536.
(7) Chipen, G.L., Bokalder, R.P. and Grinshein, V.Y. (1966) 124traizal 3one ands its nitro and amino derivatives. *Chemistry of Heterocycles Compounds*, **2**(1) 110–116.
(8) Collett, G.C. (1998) EOE thesis Cranfield University.
(9) Karaghiosoff, K., Kaplötke, T.M., Michailowski, A. and Holl, G. (2002) Structure of TEX a nitramine with an exceptionally high density. *Acta Crystallographica*, **C58**, 580.
(10) Zhang, M.-X., Eaton, P.E. and Giliardi, R.D. (2000) Hepta and octanitrocubanes. *Angew. Chemie. Int. Ed.*, **39**, 401.
(11) *Green explosives:* (a) Giles, J. (2004) *Nature*, **427**, 580–581; (b) Carrington, D. (2001) *New Scientist*, 101; (c) Ding, Y.H. and Inagaki, S. (2003) *Chemistry Letters*, **32**, 304. *HEDM general:* Klapötke, T.M. (2007) In *Moderne Anorganische Chemie*, Riedel, E. (Hrsg.), 3rd edn, Walter de Gruyter, Berlin, New York, pp. 99–104. Singh, R.P., Verma, R.D., Meshri, D.T. and Shreeve, J.M. (2006) *Angewandte Chemie*, **118**(22), 3664–3682; *Angewandte Chemie International Edition*, 2006, **45**, 3584–3601. Klapötke, T.M. (2007) In *High Energy Density Materials*; Klapötke, T.M. (Hrsg.), Ed.; Springer, Berlin, Heidelberg; pp. 85–122.
(12) Doherty, R.M. (2003) Novel Energetic Materials for Emerging Needs. 9th-IWCP on Novel Energetic Materials and Applications, Lerici (Pisa), Italy, September 14-18, 2003.
(13) Eremets, M.I., Gavriliuk, A.G. and Trojan, I.A. (2007) Single-crystalline polymeric nitrogen. *Applied Physics Letters*, **90**, 171904/1.
(14) Haiges, R., Schneider, S., Schroer, T. and Christe, K.O. (2004) High-Energy-Density Materials: Synthesis and Characterization of N5 + [P(N3)6]−, N5 + [B(N3)4]−, N5 + [HF2]−n HF, N5 + [BF4]−, N5 + [PF6]−, and N5 + [SO3F]−. *Angewandte Chemie*, **116**, 5027; *Angewandte Chemie International Edition*, 2004, **43**, 4919.
(15) Christe, K.O. (2007) Recent advances in the chemistry of $N5^+$, $N5^-$ and high oxygen compounds. *Propellants Explos & Pyrotech*, **32**, 194.
(16) Banert, K., Joo, Y.-H., Rüffer, T. *et al.* (2007) The exciting chemistry of tetraazidomethane. *Angewandte Chemie*, **119**, 1187; *Angewandte Chemie International Edition*, 2007, **46**, 1168.
(17) Curtius, T. (1891) Neues vom Stickerstoffwasserstoff. *Berichte der Deutschen Chemischen Gesellschaft*, **1891**(24), 3341.
(18) Wiberg, N. (2007) In *Lehrbuch der Anorganischen Chemie /Holleman_Wiberg*, 102nd edn, de Gruyter, Berlin, p. 659 (ammonium azide), pp. 692–693 (tetrazene).
(19) Hünig, S., Müller, H.R. and Thier, W. (1965) The Chemistry of Diimine. *Angewandte Chemie*, **77**; *Angewandte Chemie International Edition in English*, 1965, **4**, 271; (b) Miller, C.E. (1965) Hydrogenation with Diimide. *Journal of Chemical Education*, **42**(5), 254–459.
(20) Chavez, D.E., Hiskey, M.A. and Gilardi, R.D. (2000) 3,3_-Aobis(6-amino-1,2,4,5-tetrazine): a novel high-nitrogen energetic material. *Angewandte Chemie International Edition*, **39**, 1791.
(21) McKay, F. (1952) Nitroguanidines. *Chemical Reviews*, **51**(2), 301–346.
(22) Fischer, N., Klapötke, T.M., Martin, F.A. and Stierstorfer, J. (2010) Energetic Materials Based on 3-amino-1-nitroguanidine, New Trends in Research of Energetic Materials. Proceedings of 13th, Seminar, Pardubice, Czech Republic, **1**, pp. 113–129.
(23) Altenburg, T., Klapötke, T.M., Penger, A. and Stierstorfer, J. (2010) Two Outstanding Explosives Based on 1,2-Dinitroguanidine: Ammonium- dinitroguanidine and 1,7-Diamino-1,7-dinitrimino-2,4,6-trinitro-2,4,6-triazaheptane. *Zeits. für Anorgan. und Allgem. Chemie, (ZAAC?)*, **636**, 463–471.
(24) Janssen, J.W.A.M., Koeners, H.J., Kruse, C.G. and Habraken, C.L. (1973) Pyrazoles. XII. The Preparation of 3(5)-nitropyrazoles by thermal rearrangement of *N*-nitropyrazoles. *Journal of Organic Chemistry*, **38**, 1777–1782.
(25) Hervé, G. (2008) FR Patent 2917409.
(26) Zaitsev, A.A., Dalinger, I.L. and Shevelev, S.A. (2009) Dinitropyrazoles (review). *Russian Chemical Reviews (Engl. Transl.)*, **78**, 589–627.

(27) Ek, E., Latypov, N. and Knutsson, M. (2010) Four Syntheses of 4-Amino-3,5-Dinitropyrazole. Proceedings of 13th Seminar on New Trends in Research of Energetic Materials, Czech. Republic, Pardubice, April 21-23, 2010.
(28) Wang, B.Z., Wang, Y.L., Zhang, Z.Z. *et al.* (2009) Synthesis and characterization of 4-amino-3,5-dinitropyrazole (LLM-116) condensation products. *Chinese Journal of Energetic Materials*, **17**, 293–295.
(29) Dalinger, I., Shevelev, S., Korolev, V. *et al.* (2011) Chemistry and thermal decomposition of trinitropyrazoles. *Journal of Thermal Analysis and Calorimetry*, **105**, 509–516.
(30) Zhang, Y., Guo, Y., Joo, Y-H. *et al.* (2010) 3,4,5-trinitropyrazole-based energetic salts. *Chemistry: A European Journal*, **16**, 10778–10784.
(31) Zhang, Y., Huang, Y. and Shreeve, J.M. (2011) 4-Amino-3,5-dinitropyrazolate salts – highly insensitive energetic materials. *Journal of Materials Chemistry*, **21**, 6891–6897 [10].
(32) Comte, S. and Jacob, G. (2011) Synthesis of di- and Trinitropyrazoles. High Energetic Materials Conference (HEMs2011), France, La Rochelle, October 3-4, 2011.
(33) Ravi, P., Gore, G.M., Tewari, S.P. and Sikder, A.K. (2012) Quantum chemical studies on the structure and performance properties of 1,3,4,5,-tetranitropyrazole: a stable new high energy density molecule. *Propellants, Explosives, Pyrotechnics*, **37**, 52–58.
(34) Pagoria, P.F., Lee, G.S., Mitchell, A.R. and Schmidt, R.D. (2002) A review of energetic materials synthesis. *Thermochimica Acta*, **384**, 187–204.
(35) Dalinger, I.L., Cherkasova, T.I. and Shevelev, S.A. (1997) Synthesis of 4-diazo-3,5-dinitropyrazole and characteristic features of its behavior towards nucleophiles. *Mendeleev Communications*, 58–59.
(36) Dalinger, I.L., Vatsadze, I.A., Shkineva, T.K. *et al.* (2010) Nitropyrazoles 18. *Russian Chemical Bulletin International Edition*, **59**, 1631–1638.
(37) Lebedev, V.P., Matyushin, Y.N., Inozemtcev, Y.O. *et al.* (1998) Thermochemical and Explosive Properties of Nitropyrazoles. Proceedings of 29th International Annual Conference of ICT, Karlsruhe, FRG, 1998.
(38) Dalinger, I.L., Shkinyova, T.K., Shevelev, S.A. *et al.* (1998) Synthesis and Physical-Chemical Properties of Polycyclic Nitropyrazoles. Proceedings of 29th International Annual Conference of ICT, Karlsruhe, FRG, 1998.
(39) Dippold, A.A., Klapotke, T.G., Martin, F.A. and Wiedebronk, S. (2012) Nitraminotriazoles based on ANTA A Comprehensive study of structural and Energetic Properties. *Eu. J. Inorg. Chem.*, 2429–2443.
(40) Hammerl, A., Klapotke, T.M., North, H. and Warchold, M. (2003) Synthesis, Structure, Molecular Orbital & Valence Bond Calculations for Tetrazolyl Azide. *Propel. Explos. & Pyrotech.*, **28**, 165–173.
(41) Hammerl, A. and Kaplotke, T.M. (2002) Tetrazolylpentazoles. Nitrogen Rich Compounds. *Inorg., Chem.*, **41**, 906–912.
(42) Substituted pentazoles.
(43) Chavez, D.E. and Hiskey, M.A. (1998) High-Nitrogen Pyrotechnics. *J. Pyrotech*, (7), 1–6.
(44) Kaihoh, T., Itoh, T., Yamaguchi, K., Ohsawa (1988) first Synthesis of a 1,2,3,4, Tetrazine. *J. Chem. Soc. Chem. Comm.*, 1608–1609.
(45) Churakov, A.M., Tartakovsky, V.A. (2004) Progress in 1,2,3,4-Tetrazine Chemistry. *Chem. Rev.*, **104**, 2601–2616.
(46) Zhu, X. and Tian, Y. (1987) Synthesis and [properties of Tetrazine explosives. *Proc. Int. Symp on Pyrotech. & Explosives, Bejing China, China Acad. Publ.*, 241–244.

6 Explosion Processes

6.1 Introduction

When explosives are initiated, their energy is liberated by one of two processes – burning or detonation. Most explosives are capable of either process, depending on the mode of initiation and the conditions under which it occurs. In practice, these factors are so regulated as to ensure that the explosive behaves in the desired manner. However, in the accident scenario, the material may be subject to an unusual stimulus, such as mechanical damage and fire. Under these circumstances, they may behave in the opposite mode from their design mode. The properties of each mode of decomposition are quite different, but they are driven by the same thermodynamics for the reaction, with the proviso that the products may be different under the different conditions prevailing in the two processes.

6.2 Burning

Virtually all explosives burn vigorously when ignited in a dry, unconfined state. Burning can also occur in a confined state, since explosives do not rely on an external supply of oxygen for their combustion. Typical gun propellant does not require outside oxygen, and will be in a totally enclosed system. Burning comprises a series of complex chemical reactions which take place at, or just above, the surface of the explosive. When the solid material is being converted into gases, its surface (provided it is non-porous) can be regarded as receding layer by layer. The temperature of each layer is brought in turn to the ignition point, and the heat necessary to do this is radiated and conducted from the reaction zone(s) into the solid material. Some heat is also evolved by the slow decomposition of the explosive before it reaches ignition point. The rate at which the surface recedes depends on the rate of heat transfer into the material. This, in turn, depends on the temperature at the burning surface, the thermal conductivity of the material, its transparency to radiation and, inversely, on its thermal stability. The thermal transfers that occur during the burning process are shown

Laser Ignition of Energetic Materials, First Edition. S Rafi Ahmad and Michael Cartwright.

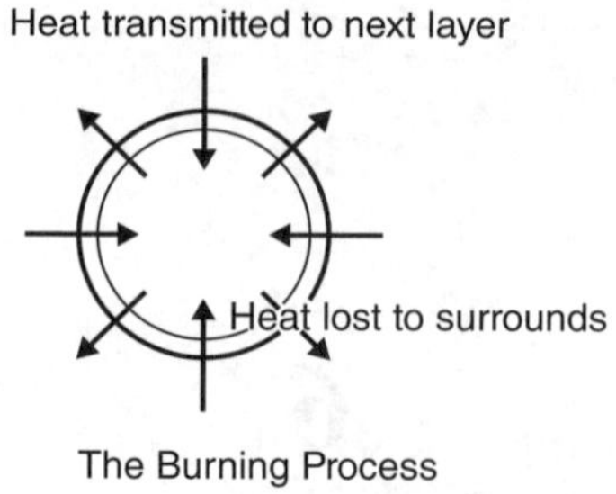

Figure 6.1 *Simple schematic of the burning process.*

schematically in the Figure 6.1. The heat lost is by a combination of radiative transfer from a hot body and convection by the hot gases moving away from the surface. The burning process moves steadily in towards the centre of the particle.

Heat is lost to the surroundings by radiation from the hot material, but also by the convection of hot gases from the surface, brought about by the pressure differential between these hot gases and the atmosphere. The rate of regression, or "linear" burning rate, is designated r. For a given explosive, the main factor determining r is the burn rate coefficient, identified as β, which is dependent on the rate of the chemical reaction occurring. The rate of reaction is a function of the explosive under consideration. Nitro ester explosives such as nitroglycerine burn much more rapidly than nitro compounds such as TNT or nitromethane. Thus, β is different for each energetic material. There is also an effect due to the ambient pressure P existing at the surface at a given instant. The effect of pressure can be interpreted by considering the burning process in more detail. There is more to the burning process than a simple single layer. The diagram below, Figure 6.2, depicts an idealized burning

Figure 6.2 *Detailed cross section through burning particle.*

process for a spherical particle which has been ignited on the surface. It shows the burning process in a little more detail.

The five layers identified have slightly different processes occurring. Starting from the centre, there is a region of unconsumed explosive, which is receiving heat released by the decomposition processes occurring above the solid surface. This will cause the material to melt, and this is the next region indicated in the diagram. As it melts, some of the explosive material is decomposing, releasing gas which bubbles through the molten material, causing a foam region in which some of the liquid material is being turned into vapour. In the gas phase, the decomposition of the energetic material becomes quite violent, causing the appearance of the fizz zone. At this stage, the reaction is largely the conversion into sub-molecular species and it is the recombination of these into the final products of the reaction which causes the light output seen to accompany the burning process. These stages can be clearly demonstrated by passing an inert gas stream over the burning surface and lifting the flame zone away from the surface, exposing the other zones.

The effect of system pressure on the burn rate can be clearly understood by examination of these gas-producing zones. When the sample is unconfined, the gases move away from the burning surface, taking away considerable quantities of heat; hence, the heat transfer to the unconsumed explosive will be comparatively low. However, as the gases are not free to move away from the surface, then less heat will be lost from the burning process and more heat will be transmitted into the system. There is also another factor, responsible for greater heat transfer into the unconsumed material. The Ideal Gas Law states that the volume occupied by a fixed volume of gas is inversely proportional to the pressure operating on it, so as the pressure rises, the liberated gases are reduced in volume, bringing the heat-generating processes closer to the unconsumed material and again increasing the heat transfer into the unconsumed material.

A typical temperature profile across a burning grain is shown in Figure 6.3, starting from the centre of the grain on the right hand side and moving towards the surface on the left hand side.

Therefore, for a non-porous explosive, the linear rate of burning r is given by Vielle's Law as:

$$r = \beta . P^{\alpha}$$

where β is the burning rate coefficient and α is the pressure index. The burning rate coefficient is specific to the explosive being burnt. The pressure index is different for each energetic material and is quite sensitive to the ingredients in mixtures and their proportions. It is also dependent on the physical parameters of the explosive, such as crystallinity and crystal defects.

Considering the burning rate law identified above, the shape of the curve for burning rate as a function of pressure depends on the value of the pressure index, α, for the particular explosive. For systems which have α values of less than 1.0 then as the pressure increases the rate asymptotes towards a maximum value, so their curve is convex Figure 6.4 below.

Black powder is in this category and most propellants are designed to come into this category. A typical gun propellant burns non-explosively in the open at 5 mm s^{-1} but, in a gun, where the pressure is about 4000 times higher, the rate can be as high as 400 m s^{-1}. Porous explosives under heavy confinement may burn at up to 500 m s^{-1}.

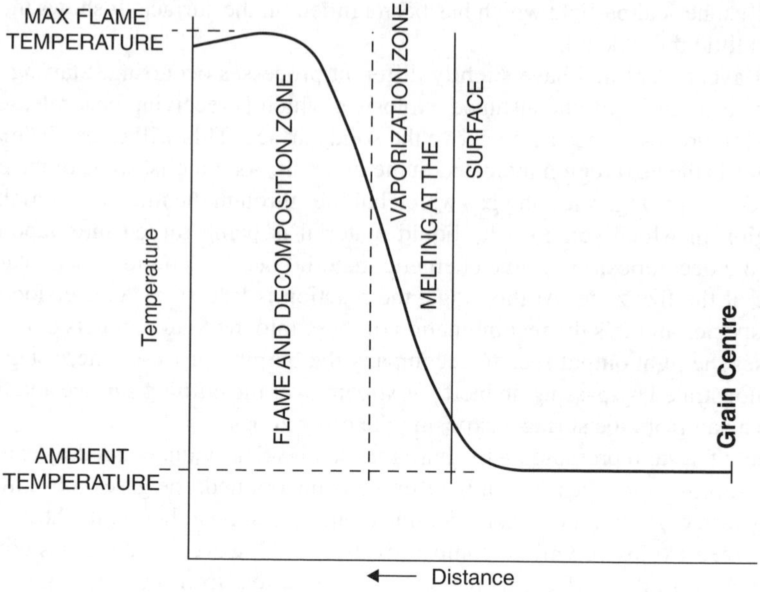

Figure 6.3 *Schematic temperature profile for a burning spherical particle of explosive.*

If it is considered that burning proceeds by parallel layers, it follows that the mass rate of burning, i.e. the mass of explosive consumed in unit time, designated dm/dt or m, is given by the equation

$$dm/dt = r \times A \times \rho$$

where r is the linear burn rate, A is the area of burning surface and ρ is the applied density.

If the burn process needs to be rapid, as in a gun, the design is to make A as large as possible and r and ρ rise interdependently to high values, then m becomes very large and the rapid generation of gas gives rise to a more or less explosive effect known as deflagration. Thus, all the propellant can be burnt in the gun before the projectile leaves the barrel, maximizing the propellant performance. To achieve this aim, propellant charges are made of a large number of small particles or using large, porous particles characterized by the multitubular arrangement shown in Figure 6.5. This has an added advantage in that, since the burn occurs within the hollow tube as well as on the outer surface, the burning

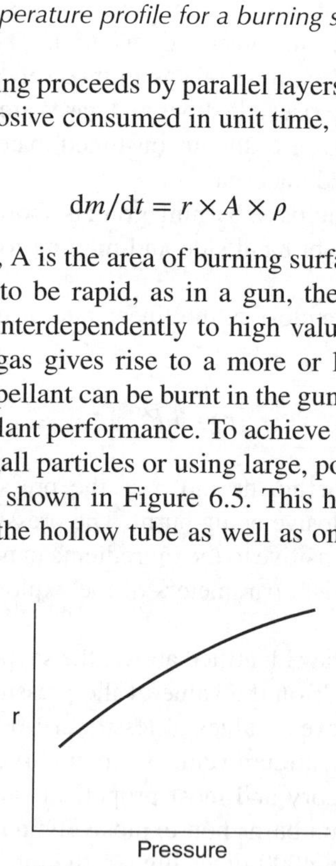

Figure 6.4 *Effect of increased pressure on the linear burn rate of a material with $\alpha < 1$.*

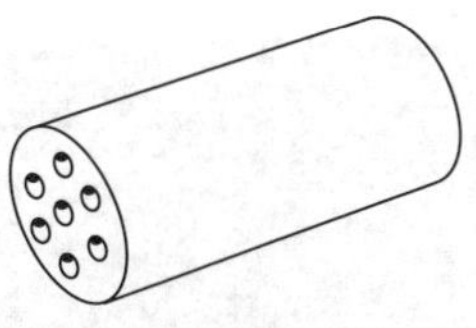

Figure 6.5 *Propellant charges are made of a large number of small particles or using large, porous particles characterized by the multitubular arrangement shown here.*

surface area increases as the burn proceeds. The rate of gas production increase just as the projectile starts to move in the barrel, and the volume available for the gases increases, evening out the pressure profile.

Certain other explosives, however, have α values greater than 1.0 and, as a result, the pressure and burn rate continue to increase, as indicated in Figure 6.6.

Under these circumstances, the pressure continues to increase until one of two things occurs. If the pressure exceeds the tensile strength of the container, then the container simply ruptures. The violence of the event depends on the strength of the container. If simple cardboard tubes or other low-strength materials are used, this results in a comparatively mild event with, in the case of cardboard, some non-injurious fragments being produced. However, strong steel cases with much higher tensile strengths only fail at high pressures, giving rise to a violent decomposition called deflagration, which produces a number of large fragments propelled with considerable velocity. Typical fragment produced by a deflagration are shown in the accompanying picture (Figure 6.7), which shows the fragments produced by inducing a deflagration in a $\approx$ 450 Kg (1000 lb) bomb filled with a high explosive composition.

Notice in the figure that the large fragment in the foreground has travelled about 10 m from the site of the event, as indicated by the crater near the base plate of the bomb. Other fragments have travelled much further.

Some explosives have the complex curve shown in Figure 6.8. The rate of burning increases as if it is asymptoting to a maximum typical for explosives with $\alpha < 1$ and then, as the pressure increases, the value of α changes and there is in effect almost a plateau region, in which increases in pressure cause very minor increases in burn rate.

Further increase in pressure causes the burn rate to accelerate again as if α is >1. This type of behaviour exhibited in Figure 6.8 is often designed into rocket propellants systems, where the requirement is for constant thrust and the effects of variation in homogeneity of the motor arising during manufacture on thrust can be minimized. There is a small region

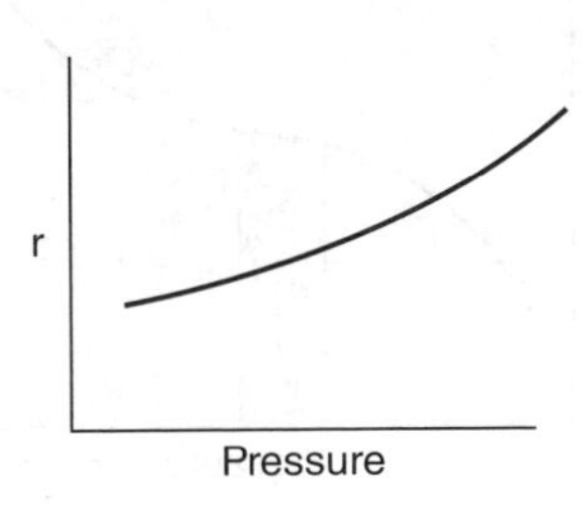

Figure 6.6 *Linear burn rate as a function of pressure for α >1.0.*

Figure 6.7 *Fragments produced from deflagrating of a 450 Kg bomb.*

on the curve, the shaded area in the figure, where the rate of burning is almost independent of pressure. This is an attempt to eliminate the pressure pulses which tend to be produced by the burning process for the materials with $\alpha > 1$ due to inconsistencies in the manufactured propellant. In the second and third cases, when $\alpha > 1$, there will be a tendency for increased pressure to be generated at the burning surface; these may then accelerate the flame front, which again increases the pressure.

When the velocity of the flame front matches that of sound in the explosive material, under the conditions of pressure and temperature prevailing, then transition to the alternative mode of explosive decomposition – detonation – occurs. The burning process which results in detonation is known as **D**eflagration to **D**etonation **T**ransition (DDT). Such systems show an appreciable delay, called run-up time, between the initiation of burning and the onset of detonation. The delay varies with the nature of the explosive, the particle size, charge density and porosity, together with conditions of confinement. The burning to detonation

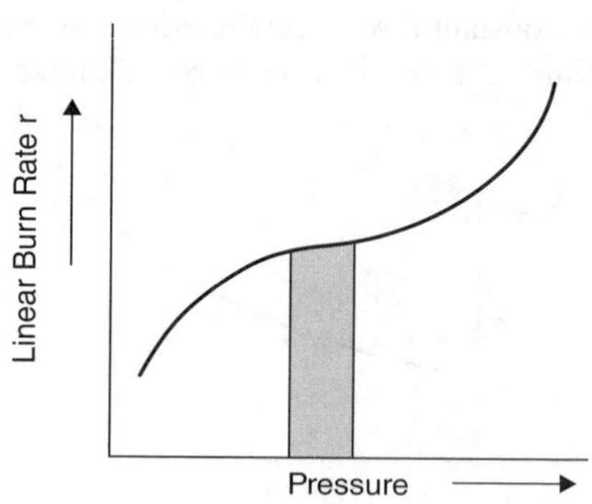

Figure 6.8 *Burn rate with pressure profile for some 'platonised' propellants.*

principle is utilized in detonators and delay fuzes. It must also be borne in mind, when disposing of bulk explosives by burning, that suitable precautions must be taken against an unintended detonation.

6.3 Detonation

The burning rate, travelling at the velocity of sound, generates a shockwave which passes through the charge. The velocity of the shockwave in solid or liquid explosives is between 1800 and 9000 m s^{-1}, an order of magnitude higher than that of a fast burning process. The rate at which the material decomposes is governed not by the rate of heat transfer, but by velocity at which the material will transmit the shockwave. The steady state velocity of the shockwave occurs when the energy delivered by the chemistry matches the energy required to compress the explosive ahead of the shockwave. A section through a detonating explosive stick is shown in Figure 6.9, together with the pressure profile along the stick.

In order to initiate a detonation wave and propagate it throughout the charge, it is necessary to provide conditions which favour the detonation process. As we have seen, initiation to detonation can occur in one of two ways:

1. Burning to detonation.
2. Shock to detonation.

These will now be considered in turn.

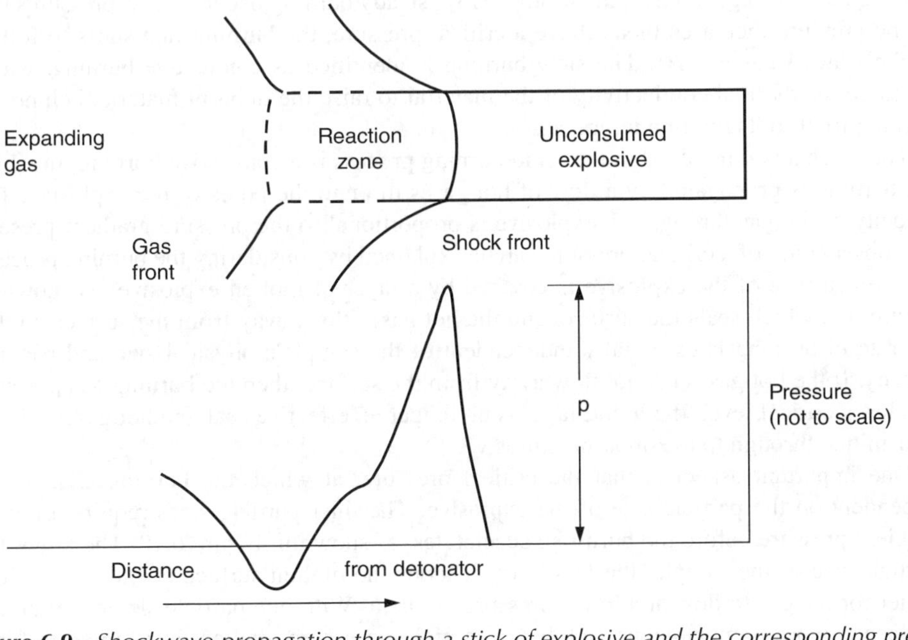

Figure 6.9 *Shockwave propagation through a stick of explosive and the corresponding pressure profile along the stick.*

6.4 Mechanism of Deflagration to Detonation Transition

So far, the two decomposition processes have been considered separately, because their properties and effects are so different. The differences are highlighted at the end of the chapter. The object of a high explosive train is to produce a detonation which will perform certain functions. Two processes are available for inducing detonation. The first of these is shock to detonation, in which a violent mechanical blow is delivered to the high explosive, while the second is burn to detonation, which starts with a burning system, which is then transformed into a detonation. The latter process is a very significant problem area for gun propellants, where the transition to detonation can be catastrophic for the gun and the crew. This is particularly important when high-performance propellants include materials such as RDX in their composition. The propellant system can be effectively sealed by the fibre optics used to deliver the pulse laser initiation sequence.

The burn to detonation process involves, at some stage, a material which consistently undergoes the burn to detonation transformation. This may be either in the primary explosive, which responds to the initiator stimulus by burning to detonation, or in a booster material which converts the primer output burn into a reliable detonation. In this aspect, lead azide primer is ideal. It is almost impossible to prevent lead azide detonating once it has started burning. Unfortunately, the detonation is insufficient to set off high-performance explosives and a detonation booster is required. The difference between primary and secondary explosives is that primers undergo a much faster deflagration to detonation transition. Also, as in the case of lead azide, they do not require special experimental conditions. The difference between primers and secondaries is one of degree rather than kind.

Two burning regimes have been identified; a slow, regular, steady burning, often identified as a regular burning, and a rapid burning. Slow, steady burning occurs at low pressures but, as the pressure increases then, above a critical pressure, the burning rate starts to follow Vielle's law. Rate $= \beta.P^{\alpha}$. The slow burning is identified as conductive burning, which relies on the thermal conductivity of the material to raise the unburnt material behind the burning front to its ignition temperature.

The mechanism involved in the rapid burning process is a convective burning, in which the burning is propagated by a flow of hot gases through the pores of the explosive. The velocity of the gas through the explosive is proportional to the pressure gradient present. The observation of a critical pressure can be explained by considering the burning process. The top surface of the explosive is covered by a layer of molten explosive, as shown in Figure 6.2, which seals the surface, and the hot gases flow away from the surface so that the rate of burning is essentially independent of the sample's physical size and packing density. If the hot gases cannot flow away from the surface, then the burning gas pressure reaches a critical level, the liquid layer is no longer an effective seal and hot gases will be transmitted through to the unburnt explosive.

One important aspect is that the critical pressure, at which the burning changes, is dependent on the particle size of the explosive. The finer particle sizes require a higher ambient pressure before the burn rate accelerates, as shown in Figure 6.10. The larger the particle size of the sample, the less homogeneous the molten surface is and, hence, it is easier for the gas to flow at a lower pressure gradient. With fine particle size material, the surface is more homogeneous and, hence, a higher pressure gradient is required to force the hot gases back into the unburnt sample.

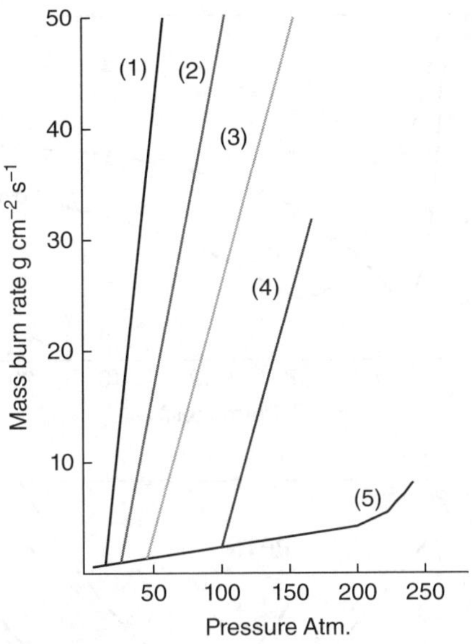

Figure 6.10 *Burning rate of HMX pressed powders as a function of pressure for various particle size materials [1].*

Key *1 ps 200–600 μm, density 1.20 g cm^{-3}*
2 ps 104–124 density 1.05 g cm^{-3}
3 ps 64–76 μm, density 1.07 g cm^{-3}
4 ps 53–64 μm, density 1.07 g cm^{-3}
5 ps 5 μm, density 1.02 g cm^{-3}

In order to explain the observation that some explosives, such as PETN, undergo a transition to detonation while some, such as TNT, do not, there must be another component to the DDT process. The introduction of the idea of crystal fractures playing a role can explain this difference. PETN is more easily fractured when compared to TNT – hence the difference in behaviour. There is an alternative explanation, in that the thickness of the molten layer in PETN is comparatively thin because the ignition temperature is close to the melting point, whereas the molten layer in TNT would be much thicker because of its low melting point and high ignition temperature producing a greater barrier to gas penetration and, hence, a lower burn rate, which would retard the DDT process.

The dominant parameters in this mechanism for pressed charges are permeability and confinement. Permeability allows the flow of hot gases through the charge, and related to this must be the reactive surface of the charge exposed to these hot gases. Permeability is determined by particle size, pore size, crystallite shapes and distribution. These are determined by the initial particle size and shape and the degree of compaction, which can be expressed as a percentage of theoretical maximum density, TMD (TMD is the single crystal density). The compaction process produces the final effective particle size of the charge. This parameter is impossible to measure, other than by permeability, which provides an average apparent size value. One of the parameters measured in experiments is the pre-detonation length. This is the distance travelled by the burning front before detonation occurs. Values of this, a function of gas permeability, expressed as percentage TMD, vary

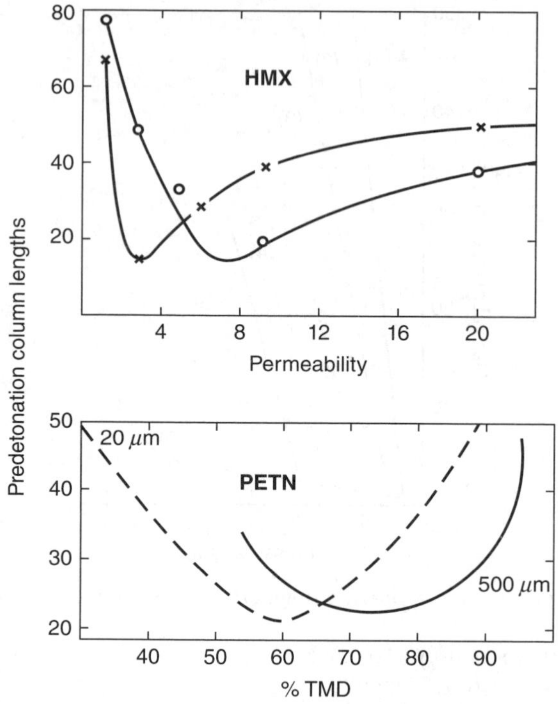

Figure 6.11 *Variation of predetonation column length as a function of permeability/particle size and TMD for HMX and PETN powders samples.*

with particle size of the original explosive, as shown in Figure 6.11 for HMX and PETN samples [2].

For both of these materials, there is a minimum in the pre-detonation column length, the run-up distance, which is dependent on the compaction and permeability.

In a series of tests for various packing densities of different particle size in RDX and HMX samples, a region can be identified in which detonation transition will be observed. However, when the density and the particle size are outside this region, then transition to detonation does not occur. Figure 6.12 below shows these regions. Note that this region is more restricted for HMX than for RDX. Small particle size materials must be compacted less than coarse particle size materials. The detonation that initially occurs is a low-velocity detonation which, if the conditions are right, will then further develop into the full velocity of detonation characteristic of the explosive material.

The acceleration of the detonation wave up to the full VoD is dependent on a number of parameters, the most significant of which is the confinement. There is also a critical minimum diameter, above which the explosive is self-confining and the wall strength of the container is unimportant. If the charge diameter is below the critical value, then the strength of the container is important. The greater the wall strength, the greater the shockwave reflection from the wall, which will add to the acceleration of the shockwave to the full VoD.

An area where DDT is vitally important is gun propellants, particularly high-energy systems. These are normally used as porous beds, because the rapid burn is required to

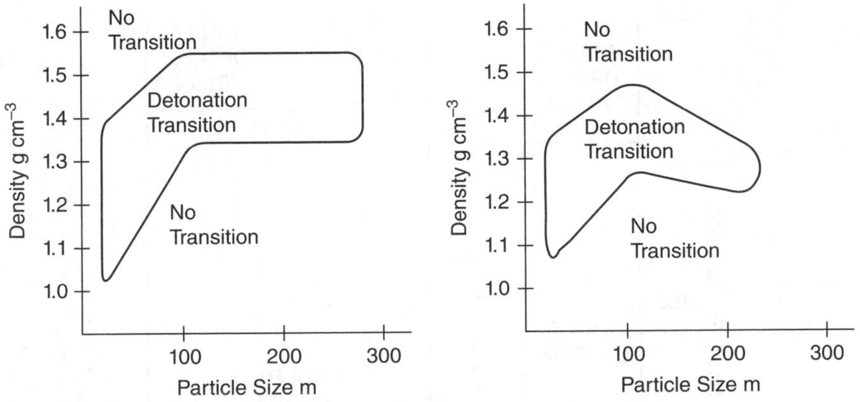

Figure 6.12 *Detonation transition regions for RDX (l) and HMX (r) showing variations with density and particle size.*

deliver most of its energy to the projectile before it exits the barrel. When the chemical energy released is greater than 4.18 kJ g^{-1}, then there is a strong possibility that DDT can occur under suitable confinement [3]. Decreasing the propellant constituent particle size increases the burn rate but decreases the run-up length and, thus, encourages DDT.

The alternative mechanism for DDT occurs when the bed is non-porous. In this situation, the pressure increase develops a plug of high-density material, which is accelerated by gas pressure and then delivers a shock to the unconsumed explosive, creating an STD wave which initiates a full VoD in the material. If the explosive is close to its theoretical maximum density, then the DDT process cannot occur and detonation is induced by shockwave impact on the explosive from either a donor charge or a high-velocity mechanical plate impact.

6.5 Shock-to-Detonation

When a high-velocity shockwave strikes an explosive charge and travels through it, two situations can arise. If the initiating shock velocity is too low, or other conditions are unfavourable, then the initiating shockwave will merely die out, leaving the bulk of the explosive chemically unchanged. If the shock velocity is above a threshold, the explosive undergoes compression and adiabatic heating in the shock front. This process liberates energy by exothermic decomposition of the explosive, thereby accelerating the shockwave and restoring the detonation velocity to that of the host explosive, accompanied by the emission of light. This is shown schematically in Figure 6.13.

If the velocity of the incident shockwave exceeds the velocity of detonation in the explosive medium, the shockwave velocity will decay to that of the host, with emission of additional light. Given the proper conditions, the shock-to-detonation method is the surest and most convenient way in which to initiate a main charge to detonation. The effect is not quite instantaneous; the impinging shockwave may have to run a distance of millimetres or centimetres into the receptor charge before it becomes self-supporting in the explosive material. The delay, however, is only of microseconds, and is practically negligible.

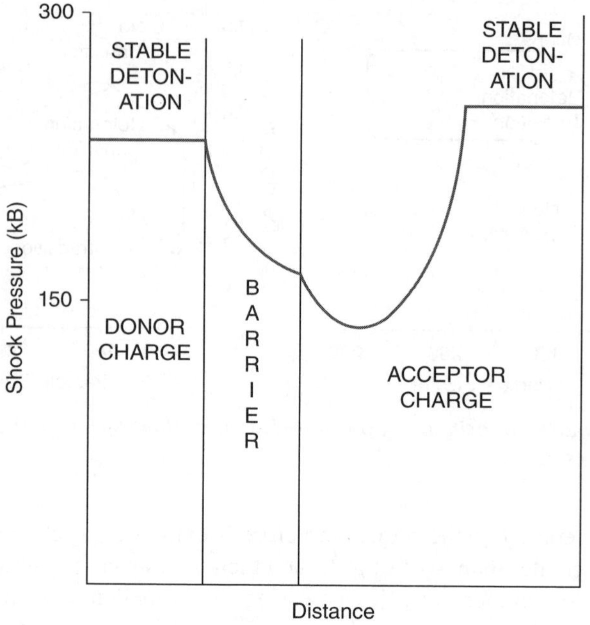

Figure 6.13 *Schematic pressure profile of a shock to detonation from a donor charge passing through a barrier, which is typically a metal casing.*

6.6 The Propagation of Detonation

When a column or length of high explosive is initiated to detonation at one end, the detonation wave will tend to propagate along the charge at its characteristic detonation velocity. To achieve propagation for considerable distances, two conditions are necessary:

- The charge, if it is only millimetres wide, must be confined in the form of a tube or channel, otherwise the detonation wave tends to dissipate laterally.
- The diameter of the column of explosive in the tube must exceed a certain critical value characteristic of the explosive, in order to avoid distortion of the wave front and reduction of velocity, with consequent fading of the detonation.

The commonest application of the linear charge is detonating fuze (or cord). The explosive filling is required to have fairly high sensitivity, a high detonation velocity and a low critical diameter. It must be packed in granular form at optimum particle size and low filling density. Cordtex, the well-known British-made detonating cord, consists of a PETN core in a flexible plastic tube, and detonates at about 6000 m s^{-1}. More recently Nobel's Explosives Co. (NEC) has marketed a non-disruptive type of detonating cord called Nonel. This consists of a plastic tube lined, but not filled, with a finely powdered explosive. Initiated from the breech of a starting pistol, it detonates at 1900 m s^{-1} and gives out a flash at the far end of the tube, sufficiently intense to ignite a subsequent item such as a detonator. Nonel tube, unlike Cordtex, does not disintegrate when it functions, nor is it even damaged.

Other applications of the detonating charge of limited width but unlimited length are linear cutting charge and fracture tape.

In a properly designed and assembled system, the detonation wave, once established either from initial burning or from a detonating donor charge, proceeds through the full volume of the charge at a velocity characteristic of the particular explosive. This process, however, is dependent on ideal conditions of:

- adequate intensity of entering wave (for shock initiation);
- adequate charge diameter;
- adequate confinement (of small charges);
- absence of sharp bends in linear charges.

Should any of these conditions not be met, the explosion of the charge will be less violent than that produced by a complete, 'high order' detonation. There are several distinct mechanisms by which these intermediate phenomena may proceed. A detonation may convert back to burning, but this is not usual, because the initial detonation may be too fast to set up the temperature gradient upon which the burning process relies. Another phenomenon is 'partial detonation', in which part of the charge detonates but the remainder is chemically unchanged and still visible in the vicinity. A 'low order detonation' is said to have occurred when the detonation wave passes right through the explosive but at a velocity well below its maximum one, giving rise to a correspondingly less destructive explosion, but more so than a deflagration. Low order detonation is commonly associated with nitroglycerine-based explosives. Authorities responsible for preparing reports on accidents or trials are expected to differentiate between these various occurrences.

To sum up this section on burning and detonation, some distinctions between the two phenomena are set out in Table 6.1. In two important respects, however, there is no distinction between the two. Regardless of whether a given charge burns or detonates, the molar quantities of the respective gases evolved will be practically the same, even though the consecutive reactions by which the final products are formed may differ. Explosives, like all other reacting substances, obey the general laws of chemistry, including Hess's Law, which states that the heat change accompanying a chemical reaction depends only on the initial and final states of the system and is independent of all intermediate states. We can, therefore assume that the heat generated by the above charge will also be the same whether the charge burns or detonates.

6.7 Velocity of Detonation

The velocity at which the detonation shockwave proceeds through a given charge (VoD) is an important parameter of the explosive material, because it largely determines the dynamic shock effect which the explosive exerts on its immediate surroundings. Detonation velocity is a variable property. It can be predicted by calculation and measured experimentally and, although a maximum value can be assigned to a particular explosive, practical results may be considerably lower, for reasons outlined below.

Table 6.1 *Some distinctions between explosive burning and detonation.*

Burning	Detonation
All explosives burn at the outset when initiated by igniferous means (exceptions: water-based compositions).	Most explosives are capable of detonation if sufficient stimulus is applied.
Burning slow compared with detonation. Burning rates lie between 0.001 m s^{-1} and 500 m s^{-1}.	Detonation is much faster than burning. Velocities of detonation lie between 1800 m s^{-1} and 9000 m s^{-1} for solid explosives.
Because of the comparatively low linear rate and because of conductive and radiative factors, there is often a tendency for the flame to spread along unburnt surfaces faster than it passes into the bulk of the explosive. Burning is a surface reaction.	The mechanism of detonation causes propagation to occur throughout the bulk of the explosive, proceeding radially from the point of initiation within the charge. The surface of the charge is normally reached from within by the wave, and is merely the boundary at which the wave ceases to be self-supporting. Detonation is not a surface reaction – it is a shockwave mechanism.
The rate of burning increases with ambient pressure (*P*) according to the law $r = \vartheta P^{l}$	The velocity of detonation has a limiting value for a given explosive. It is virtually independent of ambient pressure.
The rate of burning in a container depends on the cumulative pressure within the container, which in turn depends on the degree and rate of venting. The burning rate is not normally affected by the strength of the container.	When detonation occurs in a container, the velocity (for small charges) is affected by the strength of the container.
Damage to inert surfaces is limited to distortion of bursting.	Detonation damages an inert surface by distortion, perforation, shattering, denting, deep erosion and any other effects predicable by sudden, extreme pressure.
The linear rate of burning is not dependent on the size of the burning charge. There is virtually no critical diameter effect.	The velocity of detonation is dependent on charge diameter (for small charges). Detonation in a linear charge fails below a critical diameter of significant dimension.
Burning is initiated by direct heat or flame, not usually by explosive shock. Burning may convert to detonation if conditions are favourable.	Detonation is initiated by shock or by transition from burning. It does not usually revert to burning; if propagation fails, the charge remains chemically unchanged.
Granular explosives which have become wet cannot be ignited (with few exceptions).	Granular explosives can, with adequate shock stimulus, be detonated in a wet state.
Burning does not, intrinsically, create reports or shock phenomena, but noise and air blast may occur when cumulative overpressure is vented into the atmosphere.	Detonation creates a report which is caused by the shockwave emerging from the charge into the surrounding air.

6.7.1 *Effect of Density of Loading*

The detonation velocity of a given explosive is determined primarily by its thermochemical properties. However, in practice, the density of the particular charge is important. In order to achieve the maximum velocity of detonation for a given explosive, as well as compactness, it is necessary to consolidate the charge to its maximum density. For a crystalline explosive, the density achieved will depend on the crystal properties and the consolidation technique employed (e.g. compacted powder, casting or sintering). The limiting density will be the crystal density (Ψ) of the explosive, corresponding to the employment of a single crystal, and this is identified as the maximum theoretical density (TMD). In fact, the crystal densities of organic explosives do not exceed 2.0 g cm^{-3}, with very few exceptions. Charge densities (ρ) are invariably lower than the TMD, and this is usually expressed as a percentage of TMD. Providing the charge is of reasonable diameter and well confined, the velocity of detonation appears to be almost exactly proportional to the loading density. To calculate the approximate Velocity of Detonation, Marshall's formula is used:

$$\text{D (metres s}^{-1}\text{)} = 4430\sqrt{nT_d} + 3500\,(\rho - 1)$$

where:

n is number of moles of gaseous products per gram of explosive detonated
T_d is Temperature of Detonation (K)
ρ is charge density (g cm^{-3}).

This formula gives good agreement with observed results, though the calculation of T_d has to be of a rather approximate nature. If two charges of the same explosive are prepared with different densities, then their velocities of detonations are covered by the relationship:

$$D_1 = D_2 + 3500(\rho_1 - \rho_2)$$

where:

D_1 = velocity of detonation at density ρ_1
D_2 = velocity of detonation at density ρ_2.

6.7.2 *Effect of Diameter of Charge*

The velocity of detonation of a given substance falls when the diameter of the charge is reduced below a certain value if the degree of confinement is small or self-confining (i.e. 'bare' charges) is low as shown in Figure 6.14. It is not possible to give a firm figure for this diameter, as it depends on the nature of explosive. Values range from a few mm for nitroglycerine and PETN, hence the use of the latter in detonating cord to very approximately it is 10–20 cm for explosives like TNT, rising to higher values for certain mixed explosives. It is over 1 m for pure ammonium nitrate.

6.7.3 *Degree of Confinement*

Generally speaking, the greater the confinement, the easier it is to obtain high velocities of detonation. This is well illustrated by experimental results recorded for Minol 2 (ammonium nitrate/TNT/aluminium 40/40/20), using charges of differing diameter and differing confinement, i.e. in cardboard cartons, in lead and in steel as shown in Figure 6.15. The

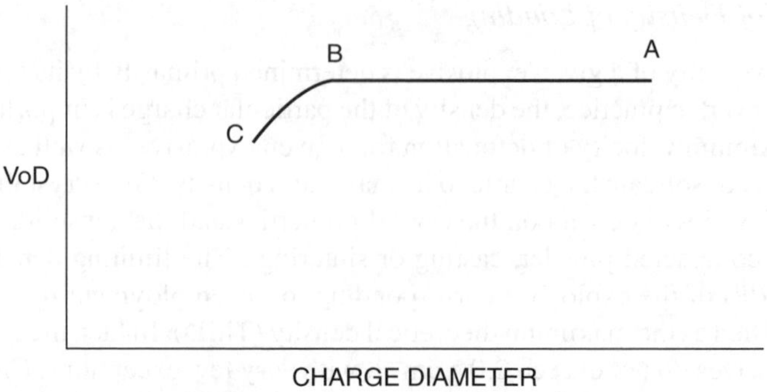

Figure 6.14 Indicating the critical diameter and effect of diameter on the velocity of detonation for an unconfined charge.

observations are plotted as $\frac{1}{D} \times 10^{12}$ cm^{-1} s against $\frac{1}{y}$ cm^{-1}, where D is the observed velocity of detonation and y is the radius of the charge (note that the use of reciprocals means that high values of D show as low points on the graphs). The limiting velocities for each type of confinement are clearly the same, and correspond to 5720 m s^{-1}. As the charge diameter gets smaller, the effect of the confining tube becomes increasingly noticeable, and D falls at a more rapid rate for the lowest degree of confinement.

It is generally accepted, from experimental results that, with adequate confinement of a charge, full velocity of detonation is reached if the charge diameter is greater than about 10 cm.

6.7.4 Effect of Strength of Detonator

Explosive charges and initiating systems used in military munitions are designed so as to ensure that the full velocity of detonation is achieved every time for maximum effect.

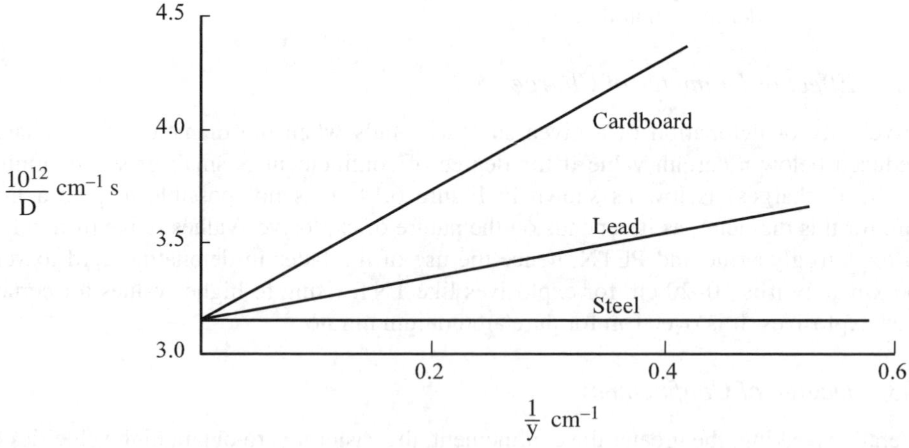

Figure 6.15 Effect of confinement strength on velocity of detonation, D, as a function of charge diameter y (plotted as 1/D against 1/y).

However, the more *ad hoc* combinations of detonator and main charge used in commercial blasting operations and by terrorists sometimes demonstrate the fact that the full velocity is not attained if the detonator does not give a sufficiently strong initiating shockwave. Nitroglycerine-based explosives, commonly used for rock-blasting but not in military detonating munitions, are capable of two very different velocities, and which one results is largely dependent on the strength of the detonator.

6.8 The Measurement of Detonation Velocity

The detonation velocity of a high explosive is a fundamental parameter for assessing its performance in a detonating device or demolition charge. The application of detonation theory provides a basis for calculating velocities, but such calculations are of limited value because they are approximate and they are not easily applicable to all types and conditions of high explosive charges. Therefore, experimental methods for determining velocities of detonation are important and have been successively improved in the last hundred years. Details of the current procedures are given in appendix 6.1.

6.9 Classifications of Explosives and Pyrotechnics by Functions and Sensitivity

In practice, explosives are employed to fill certain roles. The nature of the role determines which mode of explosion (i.e. burning or detonation) is required. The explosive will then be caused to function under conditions regulated to ensure that it behaves in the desired manner. Explosives can therefore be classified under the role in which they are *normally* employed. Explosives which are normally caused to detonate are identified as high explosives (HEs). Explosives which normally function in a burning mode are sometimes called 'low explosives'. However, this term is now discouraged because of the low-order explosions with reduced VoD. Because such compositions are mostly propellants of various kinds, they are nowadays designated as 'propellant explosives' or simply 'propellants'.

We can include in our scheme of functions the class of materials known as pyrotechnics, although they do not share all the characteristics of explosives which are set out in this book. Pyrotechnics are mixtures of oxidizing and reducing solids, capable of self-sustained combustion at rates that differ greatly from one composition to another. They are designed to produce special effects that supplement or simulate those produced by conventional explosives, and are outlined in Table 6.2.

While some explosive compounds can be assigned exclusively to the category of high explosives, others are more versatile. Nitrocellulose is present in many commercial high explosives, but it is also the universal ingredient of all conventional gun propellants. Nitroglycerine is also present in many commercial high explosives and in many propellants while RDX, hitherto regarded simply as a high explosive, is also being incorporated in more and more propellant compositions.

Another way of classifying explosives is by the readiness with which they are ignited and exploded, i.e. their sensitiveness to initiation (the difference in meaning between the

Table 6.2 Classification of explosives by function.

High explosives detonate to:
- create shockwaves
- burst
- shatter
- penetrate
- lift and heave
- create air blast
- create underwater pulses.

Propellants burn to:
- propel projectiles and rockets
- start internal combustion engines and pressurize other piston devices
- rotate turbines and gyroscopes
- inflate airbags in motor vehicles and deploy safety escape devices in aircraft.

Pyrotechnics burn to:
- create special effects
- ignite propellants
- produce delays
- produce heat, smoke, light and/or noise.

words 'sensitivity' and 'sensitiveness' is a fine one and is used to define the response to the planned initiating stimulus and the response to unplanned accidental stimuli such as fire). Substances which are readily ignited or detonated by a small mechanical or electrical stimulus are called primary explosives; those which are not readily initiated thus, and therefore require the influence of an impinging shockwave to initiate them, are called secondary explosives (propellants are not normally initiated by either of these methods in practice, but by the application of flame). Combining the two methods of classification, we can describe both the input normally applied to an explosive and the response expected of it:

- A primary high explosive can be detonated easily.
- A secondary high explosive can be detonated, but less easily.
- A propellant explosive is not required to detonate at all.

While it is probably true that most primary explosives are capable of detonation, some are not required to do so in use, but are only required to burn or deflagrate.

It must be stressed that the above statements are requirements, not firm predictions as to the behaviour of explosive materials under all conditions. Given the wrong kind of initiatory stimulus, some propellants may detonate, and some secondary explosives may burn instead of detonating. The requirements presuppose that the explosive system will function as it is designed to and will be subjected to the appropriate initiatory stimulus.

If we consider the military use of explosives, as exemplified by a complete round of gun ammunition (Figure 6.16 below), the roles of different types of explosive are seen.

The cap will contain a primary explosive (1) which burns rapidly (deflagrates) under the influence of a light blow or small electrical stimulus. The flame from the cap then ignites a pyrotechnic (2), which boosts the output of the primer and generates enough flame to ignite

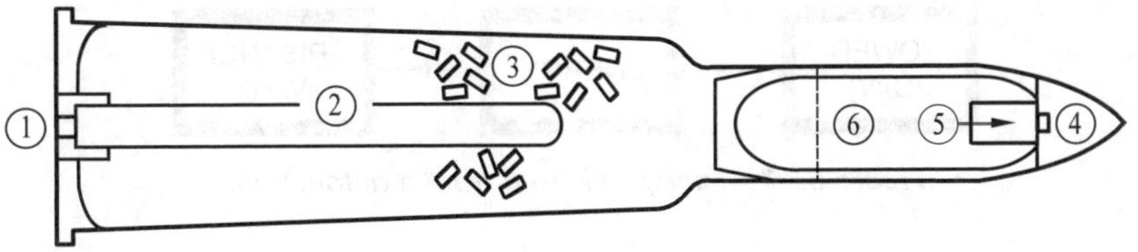

Figure 6.16 *Typical component propellant and warhead systems in an artillery shell.*

the propellant (3) in the cartridge case, and the projectile is ejected from the gun. When the projectile strikes its target, the impact causes a primary high explosive (4) to detonate, and the shockwave thus generated is enhanced through a series of secondary high explosive components (5) until it reaches the main charge of secondary high explosive (6) and the shell bursts.

Each of these explosive components has its own requirements to fulfil and, in some cases (particularly (1), (2) and (3)), no single explosive compound exists which can satisfactorily fulfil the role, so a carefully formulated mixture is used. These will be discussed in Chapter 7 in the section on initiation. In the field of commercial explosives, the requirements of explosives in, for example, rock blasting, are also very diverse; a careful assessment of the effect required is made and translated into terms of explosive properties, so that a choice of product can be made from the large number of blended explosives available on the market, many of them still based on Nobel's pioneering work.

In the chemistry of both military and commercial explosives, much of the expertise lies in blending explosive compounds with other such compounds, or with other compatible materials, to obtain an explosive mixture which has the right properties of safety, reliability and performance. This is particularly true of the ammonium nitrate systems widely used in commercial blasting operations. Some of the most useful explosive compounds are of little or no practical use in an unmixed state but, when mixed with other substances, they can fulfil one or more requirements to near perfection. Such compounds include RDX and nitroglycerine and, as research continues, more materials are discovered which give improved safety while maintaining performance – or, in the case of CL20, improving performance.

6.10 The Effects of High Explosives

6.10.1 Energy Distribution in Explosions

The total chemical energy released by an explosive charge can be calculated, and it can also be verified experimentally. This is the total energy available to do useful work, and it is independent of the mode of decomposition either burning or detonation. Thermodynamics dictates that the energy released is the difference between the enthalpy of the products and the enthalpy of formation of the starting compound:

$$\Delta H_c = \Sigma\Delta H_f(\text{products}) - \Delta H_f\,(\text{explosive})$$

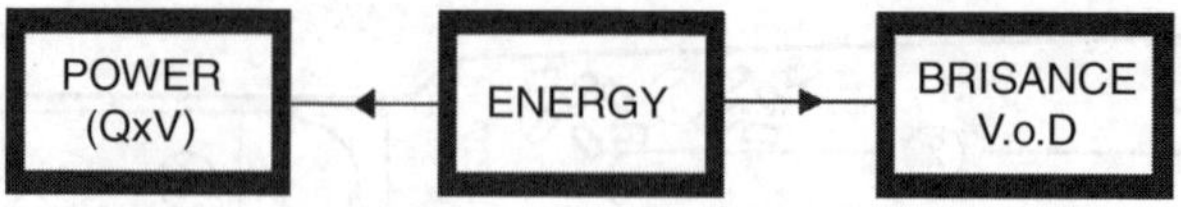

Figure 6.17 *Energy released from a detonation.*

The total chemical energy per unit mass of a given explosive material is called its heat of explosion, sometimes referred to as Q. The symbol Q denotes a modulus value of the large negative heat change ($-\Delta H$) which occurs when the explosive material is converted into gases:

$$Q = | - \Delta H_c |$$

When the charge detonates, this energy is released in the form of kinetic energy and heat over a very short but measurable time, of the order of microseconds. Two different processes utilize this energy – the shockwave and the expanding gases, as indicated in Figure 6.17.

The division of the energy between these two phenomena is impossible to calculate. An indication can be gained by measuring the detonation of a charge of the explosive in question under water. A supersonic wave is formed near the initiation point and travels through the medium of the explosive, sustained by the exothermic decomposition of the charge behind it. On reaching the periphery of the charge, the detonation wave passes, unsupported, into the surrounding medium, and exerts on it a sudden, intense pressure, equivalent to a violent mechanical blow. The shockwave moves away from the event very rapidly, as indicated by the expanding ring of dark, denser, water. The resultant expanding gas bubble moves away at a lower velocity and, thus, measurements can be made of the separate energies associated with the detonation shockwave and the expanding gases. These experiments indicate that the maximum proportion of energy in the shockwave is less than 50% of the total, and those closest to this limit are the explosives with the highest VoD. The VoD in water is higher than in air because of its higher density and, consequently, the velocity of the gas bubble is lower in water – again because of water's higher density.

If the medium is a solid, the stresses that are set up may exceed its mechanical strength, causing effects such as the fragmentation of a high explosive shell or the creation of multiple cracks in a mass of homogeneous rock. Thus, in the first phase of the explosion, a portion of the available energy has been dissipated in the form of the detonation shockwave, producing a crushing, shattering, disintegrating effect on the medium exposed to it. This property of explosives is called 'brisance'. The brisance depends on the velocity of the detonation shockwave. Some of the shockwave energy may be dissipated by propelling fragments with high kinetic energy.

After the shockwave has moved away from the charge. the remaining energy is expended in the heat of the gases, which are then beginning to expand and, in so doing, push out the surrounding medium (e.g. earth, water or air). In earth, a crater is formed (i.e. lift and heave). In water, the expansion causes the formation of a gas bubble; in air, a blast wave is formed. These effects are derived from the work capacity of the explosive – that is, its 'power'. Power is the useful chemical energy contained in an explosive; in the terminology of commercial explosives, it is more often referred to as 'strength'. It is proportional to two fundamental parameters, namely the number of moles of gas produced per unit mass of explosive, and the amount of heat (Q) available to expand the gas (neglecting the unknown proportion of the original energy that is dissipated in the shockwave). This concept of explosive power should not be confused with the more general scientific definition of power, i.e. work done per unit time. Explosive power involves the total work output of the charge over the full period of the explosion. The time scale will be of the order of milliseconds, and is therefore much longer than the detonation time (microseconds). Furthermore, the full time scale of the explosion will vary widely according to the weight of the charge and the nature of its environment. It is therefore not amenable to calculation.

6.11 Explosive Power

This function of the total work capacity of the explosive is based on the Ideal Gas Law, and is valid even though the gases released in the explosion are far from the ideal conditions. Using the Ideal Gas Law and thermodynamics, the work done by an expanding gas at constant pressure P is:

$$P \times \Delta V$$

where ΔV is the change in volume of the gas.

Since the volume occupied by the original explosive is negligible (<1% of the gas volume produced by the explosion), ΔV can be approximated to V, the volume of gases produced. From the Ideal Gas Law,

$$\text{system force capability} = \mathbf{power} = \mathbf{P \times V = nRT}$$

where:

n = number of moles of gas produced
R is the gas constant
T is the absolute temperature in degrees Kelvin.

The number of moles of gas is given by the expression

$$n = V_g/22.4$$

where:

V_g is the volume of gases produced
22.4 dm^3 is the volume occupied by one mole of any gas.

The temperature of the gases produced is given by the relationship

$$T = \Delta H_c/\Sigma n_i C_I$$

where the summation is over the number of moles of gas i with its thermal capacity at constant volume of C_i. If the energy liberated by a gram of explosive and the moles of gases per gram are used, then the temperature is given by:

$$T = Q/\Sigma\, \dot{C}$$

where $\dot{C}$ is the mean thermal capacity of the products.

Substituting for n and T in the Ideal Gas Equation gives:

$$\text{Force} = V_g/22.4 \times R \times Q/\Sigma\dot{C}$$

$$\therefore \text{Force} = K \times Q \times V_g \qquad \text{where } K = R/(\Sigma\dot{C} \times 22.4)$$

$$\therefore \mathbf{Power} \propto \mathbf{Q \times V_g}$$

If we can calculate Q and V_g, then we should be able to assess the power of an explosive. Note that this will not be absolute, because of the energy partition between shockwave and expanding gases shown above.

6.12 Calculation of Q and V from Thermochemistry of Explosives

6.12.1 General Considerations

Having seen some examples of explosive molecules, there is a need to look at the type of chemical reactions that occur when a material undergoes an explosion. The speed of the chemical reaction during the detonation process causes difficulties. In burning, for example, the process is relatively slow, giving time to measure the reaction taking place; typically, propellants will burn in milliseconds or longer. When looking at the detonation reaction, the time scale is very short – a matter of microseconds – and the actual events occurring at the molecular level are not visible. Thus, indirect observation of the 'before' and 'after' type are often used.

Explosive reactions may be regarded as a breaking of the explosive molecule into its component atoms followed by a rearrangement of the atoms into a series of small, stable molecules. The main molecules found in the reaction products are those of water (H_2O), carbon dioxide (CO_2), carbon monoxide (CO) and nitrogen (N_2) from the nitro group. Their enthalpies of formation are determined from the following reaction equations:

$$H_2\,(g) + {}^1\!/_2O_2\,(g) = H_2O(g) \qquad \Delta H_f(H_2O) = -241\ \text{kJ mol}^{-1}.$$

$$C\,(s) + O_2\,(g) = CO_2\,(g) \qquad \Delta H_f\,(CO_2) = -394\ \text{kJ mol}^{-1}.$$

$$C(s) + {}^1\!/_2O_2(g) = CO(g) \qquad \Delta H_f(CO) = -110\ \text{kJ mol}^{-1}.$$

6.12.2 Energy of Decomposition

When a chemical reaction takes place, be it the simple rusting of iron or an explosive detonation, the final products are more stable than the starting materials, unless we have forced energy into the system. Thus, a spontaneous reaction is one that generates materials of higher stability and, in the process, energy is released. Most explosives give out large amounts of energy as heat, which we can calculate as the enthalpy of detonation, ΔH_d. Energy liberated in the decomposition is given by the initial and final states and is independent of the route.

The energy released on explosion can be calculated by consideration of the enthalpy (heat energy) of formation data for the explosive and its products. The enthalpy of formation is the heat energy change when a material is formed from its constituent elements in their natural states. This data is readily available for most materials but can be calculated, if necessary, from (for example) bond energies. Chemical equations for the decomposition of the explosive provide product composition details, enabling the calculations needed to be undertaken.

6.12.3 Products of the Explosion Process

The decomposition of an explosive to produce products will depend on the ratio of fuel to the oxidizer. If we examine EGDN, then, because the compound is exactly oxygen balanced, the products we would expect are carbon dioxide, water and nitrogen. The balanced equation can be written as:

$$\begin{array}{c} CH_2ONO_2 \\ | \\ CH_2ONO_2 \end{array} = 2H_2O + 2CO_2 + N_2$$

The balanced chemical equation for nitroglycerine decomposition can be similarly written – and note the production of free oxygen.

$$\begin{array}{c} CH_2ONO_2 \\ | \\ CHONO_2 \\ | \\ CH_2ONO_2 \end{array} = 2.5H_2O + 3CO_2 + 1.5N_2 + 0.25O_2$$

When the explosive contains insufficient oxygen, to convert all carbon to carbon dioxide and all the hydrogen to water (i.e. is oxygen deficient), then we require some rules to enable the product composition to be predicted. A good example of this difficulty is the decomposition of TNT. According to Figure 6.18 below, there are three possible decomposition schemes, all of which are chemically balanced and possible. Thermodynamics can help to predict the output of the decomposition, but it can only predict the most energetically favoured, without defining which will occur. Simple rules based on experimental observations can be used.

The data under each equation is the enthalpy of formation of the compounds produced. The heat evolved by the decomposition ($\Delta H_c = \Sigma\ \Delta H_{f(products)} - \Delta H_{f\ (starting)}$) will depend on which route the decomposition follows. There is almost a factor of two between the two extreme cases. Application of the rules devised by Kistiakowsky and Wilson during the Second World War enables an approximate answer for the products of the detonation for

$$C_7H_5N_3O_6 \rightarrow 3\,CO_2 + 4\,C + 2.5\,H_2 + 1.5\,N_2$$
$$3 \times -394 \qquad -1182\ \text{kJ Mole}^{-1}$$

$$C_7H_5N_3O_6 \rightarrow 6\,CO + C + 2.5\,H_2 + 1.5\,N_2$$
$$6 \times -110 \qquad -660\ \text{kJ Mole}^{-1}$$

$$C_7H_5N_3O_6 \rightarrow 3.5\,CO + 3.5\,C + 2.5\,H_2O + 1.5\,N_2$$
$$3.5 \times -110 \qquad 2.5 \times -242 \qquad -990\ \text{kJ Mole}^{-1}$$

Figure 6.18 *Possible decomposition schemes for TNT.*

materials, which are oxygen-rich or moderately oxygen-deficient, $\Omega < -40$. Even complex mixtures of materials can be evaluated by their rules to give approximate decomposition equations and, hence, an estimate for explosive power.

6.13 Kistiakowsky - Wilson Rules

i. All carbon atoms converted to carbon monoxide.
ii. Any remaining oxygen converts hydrogen to water.
iii. Any remaining oxygen converts carbon monoxide to carbon dioxide.
iv. Nitrogen atoms are converted to molecular nitrogen.

Applying KW rules to some examples:

RDX formula $C_3H_6N_6O_6$ Oxygen balance -22

Applying the KW rules:

i. $3C \rightarrow 3CO$
ii. $6H \rightarrow 3H_2O$

Since there is no unused oxygen, rule iii does not apply. Rule iv states nitrogen atoms to nitrogen molecules:

iv. $6N \rightarrow 3N_2$.

therefore the equation is:

$$C_3H_6N_6O_6 \leftrightarrow 3CO + 3H_2O + 3N_2$$

PETN $C_5H_8N_4O_{12}$ Oxygen Balance -10

Rule i $5C \leftrightarrow 5CO$ Oxygen left $= 12 - 5 = 7$

Rule ii $8H \leftrightarrow 4H_2O$ Oxygen left $= 12 - 5 - 4 = 3$

Rule iii $3CO \leftrightarrow 3CO_2$ Oxygen left $= 12 - 5 - 4 - 3 = 0$

Rule iv $4N \leftrightarrow 2N_2$

Therefore the equation for the decomposition of the PETN is:

$$C_5H_8N_4O_{12} = 4\,H_2O + 3\,CO_2 + 2\,CO + 2N_2$$

For TNT, the rules would predict the equation to be:

$$C_7H_5N_3O_6 = 6\,CO + C + 2.5\,H_2 + 1.5N_2$$

This equation is the middle, lowest energy, option in the scheme given above, and it gives a calculated Q of 2728 J g^{-1}. The agreement with the experimental value of 4306 J g^{-1} is very poor. Generally, for oxygen-deficient explosives, KW rules i and ii should be reversed, and thermodynamics based on total energy changes supports the formation of water ($\Delta G_f =$ –228.57 kJ mol^{-1}), compared to carbon monoxide ($\Delta G_f = -137.17$ kJ mol^{-1}).

Revised (or reversed) Kistiakowsky-Wilson rules

i. All hydrogen converted to water.
ii. Remaining oxygen converts C atoms to carbon monoxide.

Rules iii and iv are the same as the original

Therefore, applying these modified rules, the equation for the decomposition of TNT is:

$$C_7H_5N_3O_6 \leftrightarrow 3.5CO + 3.5C + 2.5H_2O + 1.5N_2 \ \mathbf{\Delta H_c} \ -990 \text{ kJ Mole}^{-1}$$

The experimental value is 4306 J g^{-1}. The reversed KW rules give 4352 J g^{-1}. Before perfect agreement between theory and experimental Q values for many explosives can be obtained, additional reactions need to be taken into account.

6.14 Additional Equilibria

The above equations assume that the compounds CO and H_2O and C are produced independently, but unfortunately they interact at the high temperatures of the explosion. Two important reactions are the disproportionation of CO into CO_2 and C, according to the equation:

$$2CO(g) = CO_2\,(g) + C(s) \quad \Delta H_R = -170.6 \text{ kJ mole}^{-1}$$

and the reaction of water with carbon, the water gas equilibrium, according to the equation:

$$CO + H_2 = H_2O + C \quad \Delta H_R = -129.4 \text{ kJ mole}^{-1}$$

Springall and Roberts produced two empirical rules governing both of these reactions. However, both of the reactions are, in fact, equilibria, and the relative compositions are given by the following expressions for their equilibrium constants:

For the carbon with carbon dioxide:

$$K_c = \frac{[CO_2][C]}{[CO]^2}$$

And for the carbon with water reaction:

$$K_c = \frac{[H_2O] \cdot [C]}{[CO] \cdot [H_2]}$$

Tables of the values of these equilibrium constants as a function of temperature and pressure are available, and incorporation of thermodynamic equilibrium data into the calculations enables a better agreement between theoretical and experimental values for TNT to be obtained by an iterative process. The simple rules enable a starting set of values to be obtained for the temperature and composition of the products, which then act as a basis for introducing the equilibria and a re-determination of the composition and, hence, the heat of reaction and then the temperature of detonation. The process is repeated until constant values are obtained.

There is a third additional reaction occurring, involving the products CO and H_2O, which participate in the gaseous equilibrium given by the equation:

$$\begin{array}{llll} CO_2 + H_2 = & CO + & H_2O & \\ -394 & -110 & -240 & \mathbf{\Delta H_r} + 44 \text{ kJ Mole}^{-1} \end{array}$$

An equilibrium constant can be written for this reaction as:

$$K = \frac{[CO] \cdot [H_2O]}{[CO_2] \cdot [H_2]}$$

The reaction is almost perfectly balanced at 1000 K, but favours the left hand side at low temperatures and the right hand side at high temperatures. For most explosions, this reaction is a minor correction.

6.15 Energy Released on Detonation

Using equations for the detonation of the explosives, the energy released during an explosion can be calculated from the enthalpy (heat energy) of formation data for the explosive and its products. The enthalpy of formation is the heat energy change when a material is formed from its constituent elements, in their natural states. This data is readily available for most materials, and can be calculated if necessary from other data (for example, average bond energies), which assumes that the effect of the NO_2 groups on the bond energies of the rest of the molecule is negligible.

Consider the EGDN case:

$$\begin{array}{c} CH_2ONO_2 \\ | \\ CH_2ONO_2 \end{array} = 2H_2O + 2CO_2 + N_2$$

Substituting the enthalpy values gives:

$$\begin{array}{llll} \Delta H_{f(EGDN)} & 2 \times \Delta H_f(H_2O) & 2 \times \Delta H_f(CO_2) & \\ -259 & 2 \times -241 & 2 \times -394 \therefore \Delta H_c = 1001\ \text{kJ mol}^{-1} & Q = 6.63\ \text{kJ g}^{-1} \end{array}$$

For the oxygen-rich NG, the product of the decomposition contains some excess oxygen and the reaction equation and the thermodynamics reflect this.

$$\begin{array}{l} CH_2ONO_2 \\ | \\ CHONO_2 \\ | \\ CH_2ONO_2 \end{array} = 2.5H_2O + 3CO_2 + 1.5N_2 + 0.25O_2$$

$$\begin{array}{lllll} \Delta H_{f(NG)} & 2.5\Delta Hf_{(H_2O)} & +3\Delta Hf_{(CO_2)} & & \\ -380 & 2.5 \times -241 & 3 \times 394 & \Delta H_C = 1404\ \text{kJ mol}^{-1} & Q = 6.18\ \text{kJ g}^{-1} \end{array}$$

When RDX detonates, an approximation for the overall equation is:

$$\begin{array}{llll} C_3H_6N_6O_6 & \rightarrow 3CO & + 3H_2O & + 3N_2 \\ +61.5 & 3 \times (-110) & 3 \times (-242) & \end{array}$$

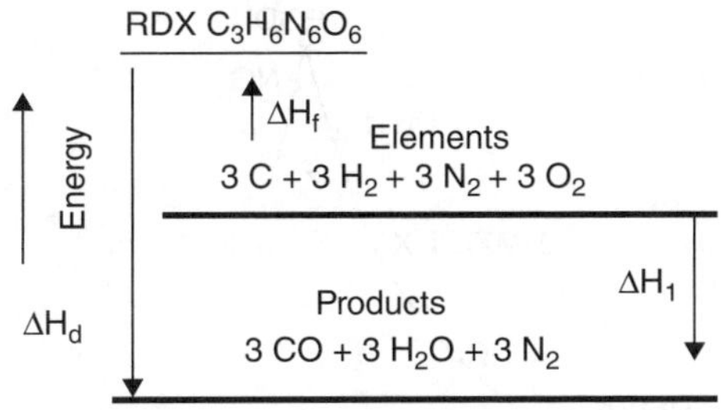

Figure 6.19 *Thermodynamic Cycle for formation and decomposition of RDX.*

Note that RDX has an endothermic enthalpy of formation, as shown in Figure 6.19 below, and this will be added to the enthalpy of formation of the products.

$$\Delta H_d = -1056 + (-61.5) = -1117.5 \text{ kJ mol}^{-1} \quad Q = 5.034 \text{ kJ g}^{-1}$$

The accepted experimental heat of explosion, Q, for RDX is 5056 J g^{-1}.

The higher the value of Q for an explosive, the more heat energy generated when the explosive is decomposed. Table 6.3 lists the accepted Q values for a number of explosives. All low oxygen balanced explosives produce either carbon monoxide or, at worst, carbon. Any carbon will not contribute to the overall energy change, resulting in a relatively low Q value. The production of carbon monoxide liberates less heat than the production of carbon dioxide, 111 kJ $mole^{-1}$ for CO and 394 kJ $mole^{-1}$ for CO_2. This represents an inefficient combustion of the carbon fuel.

The effect of oxygen balance on Q is shown in Figure 6.20. Calculations of Q assume that any water formed on detonation is present as gas.

Table 6.3 *Calculated heats of explosion (Q), and gas volumes (water gaseous).*

Secondary explosives	J g^{-1}	Secondary explosives	cm^3 g^{-1}
Nitroglycol (EGDN)	6730	Nitroguanidine (picrite)	1077
Nitroglycerine	6275	DATB	1015
PETN	5940	HMX	I910
RDX	5130	RDX	908
HMX	5130	Tetryl (CE)	845
RDX/TNT 60/40	4500	RDX/TNT 60/40	796
Pentolite (PETN/TNT 50/50)	4475	PETN	780
Tetryl (CE)	4396	TNT	740
TNT	4308	Picric acid	790
DATB	3805	Nitroclycerine	714
Picric acid	3745	Nitroglycol (EGDN)	789
Picrite	2876		
Primary explosives		**Primary explosives**	
Lead styphnate	325	Lead styphnate	1885
Mercury fulminate	235	Mercury fulminate	1755
Lead azide	230	Lead azide	1610

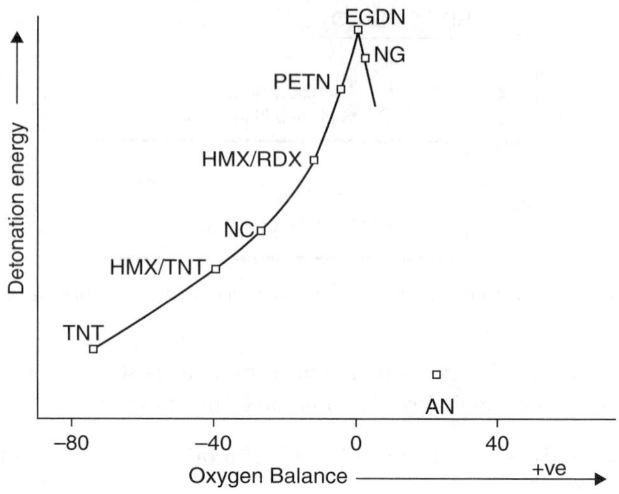

Figure 6.20 Heat of explosion as a function of oxygen balance.

6.16 Volume of Gases Produced during Explosion

The second parameter required before explosive power is calculated is the gas volume (V). This is the volume of gas generated by unit mass of explosive. Such volumes are as measured at standard temperature and pressure (STP), and because one mole of any gas occupies (ideally) a fixed volume (22.414 litres) under those conditions, the values of V are directly related to the number of moles of gas generated in the explosion.

Although the volume of gas generated depends, according to the Ideal Gas Law, on the temperature and pressure, measurements are corrected to a standard temperature and pressure (STP) 273 °K, 0 °C and normal atmospheric pressure, 101.3 kPa (kiloPascals) or 1 atmosphere. This allows an easy comparison between explosives. Under these conditions, a mole of gas will occupy 22.4 dm^3 (litres), the molar gas volume. Once the equation for the detonation is known, a calculation of the volume of gas liberated can be made. Examination of the equation for detonation shows which materials are gaseous; often the only non-gaseous product is carbon. The number of moles of gaseous product is calculated, divided by the relative molecular mass of the explosive and finally multiplied by the molar volume to give the volume of gas liberated at STP by one gram of explosive. This value is sometimes known as 'V', the gas volume, but V_g prevents any confusion with the velocity of detonation.

For the simple case of EGDN with the equation below:

$$\begin{array}{c} CH_2NO_2 \\ | \\ CH_2NO_2 \end{array} \quad (l) = 2H_2O + 2CO_2 + N_2$$

All of the products are gaseous and the number of gas moles liberated is 5, giving a volume of $5 \times 22.4\ dm^3\ mol^{-1}$ and dividing by the molar mass of EGDN, 152.1. Similarly, for nitroglycerine (NG), then there are 7.25 moles of gas produced. Note that the free oxygen produced, 0.25 moles, adds to the effective gas volume even though it is wasted thermal output. The gas volume is $7.25 \times 22.4\ dm^3\ mol^{-1}$.

RDX is an example of how the Vg can be evaluated:

$$C_3H_6N_6O_6(s) \rightarrow 3\,CO(g) + 3\,H_2O(g) + 3\,N_2(g) \text{ (K-W rules)}$$

Nine moles of gas from a single mole of RDX; total volume of gas is 9 × 22.4, i.e. 201.6 dm^3 which, when divided by the mass of one mole of RDX (222), gives an answer of 0.908 $dm^3 g^{-1}$ (908 $cm^3 g^{-1}$).

TNT can also be treated in a similar fashion. In this case, the carbon will be a solid product and contributes nothing to the gas volume. The equation for detonation:

$$C_7H_5N_3O_6 \rightarrow 3\tfrac{1}{2}CO(g) + 3\tfrac{1}{2}C(s) + 2\tfrac{1}{2}H_2O(g) + 1\tfrac{1}{2}N_2\,(g) \text{ (modified K-W)}$$

gives a total of $7^1/_2$ moles of gaseous products. This results in a total of 168 dm^3 of gas from the $7^1/_2$ moles, or 740 cm^3 of gas for a gram of material. Table 6.3 contains accepted values for the heats of explosion and the gas volume for a number of common explosives and some mixed formulations.

There is a wide gap between the Q and V_g values of the secondary high explosives used in the main charges of munitions and as the basis of commercial blasting explosives, on the one hand, and those of the primary explosives used in igniters and detonators, on the other. Note also the inversion of the order between Q and V. High Q materials tend to have low V values.

6.17 Explosive Power

Explosive power is calculated as the product of Q and V_g, and these values for the common explosives, normalized to picric acid having a power of 100, are given in Table 6.4. It is clear that the explosive power of primary explosives is well below that of secondary explosives.

Table 6.4 *Calculated power indexes, based on picric acid = 100.*

Secondary explosives	Power index
Nitroglycol	170
PETN	161
HMX	160
Nitroglycerine	159
RDX	159
RDX/TNT/ 60/40	138
DATB	132
Pentolite 50/50	129
Tetryl	123
TNT	117
Picric acid (obsolete)	100
Primary explosives	
Lead styphnate	21
Mercury fulminate	14
Lead azide	13

Table 6.5 *Enthalpies of combustion for some light elements.*

Element	Atomic mass	Enthalpy of combustion (kJ mol^{-1})	Enthalpy of combustion kJ g^{-1}	Reaction		
Hydrogen	1.01	–121	120	$\frac{1}{2}H_2 + \frac{1}{4}O_2$	→	$\frac{1}{2}H_2O$
Lithium	6.94	–296	42.7	$Li + \frac{1}{2}O_2$	→	LiO
Beryllium	9.02	–611	67.7	$Be + \frac{1}{2}O_2$	→	BeO
Boron	10.82	–636	58.7	$B + 1\frac{1}{2}O_2$	→	$\frac{1}{2}B_2O_3$
Carbon	12.01	–394	32.8	$C + O_2$	→	CO_2
Magnesium	24.32	–603	24.8	$Mg + \frac{1}{2}O_2$	→	MgO
Aluminium	26.97	–839	31.1	$Al + 1\frac{1}{2}O_2$	→	$\frac{1}{2}Al_2O_3$
Silicon	28.09	–909	32.4	$Si + 2O_2$	→	SiO_2
Sulfur	32.06	–297	9.3	$S + 2O_2$	→	SO_2

Even if the high sensitivity of the primary explosives did not preclude their use in large charges, they would be ineffective in at least some of the roles of high explosives, namely those which rely on high power (i.e. lifting, heaving and creating air blast and underwater bubbles). The importance of primary explosives lies in the ease with which they are initiated and the reliability with which some of them burn to detonation, even in very small quantities of 0.1 g or less.

6.17.1 Improving Explosives Power

To produce a "better" explosive, the power of an explosive can be increased if either Q, the heat of the reaction, can be increased, while retaining the gas volume, or V, the gas volume, can be increased, or by increasing both. In practice, the required Q performance increase can be achieved by the use of a fuel that produces a higher Q value. Other light elements offer higher enthalpies of combustion than carbon and hydrogen. Table 6.5 tabulates this data for likely elements.

Combustion of aluminium can achieve this goal. The burning of aluminium to produce the oxide is very exothermic.

$$2Al(s) + 1.5O_2 = Al_2O_3 \quad \Delta H_c = -839\,\text{kJ mole}^{-1}\ \text{(of Al combusted)}$$

Note that, in the table, although hydrogen is the most efficient in terms of enthalpy of combustion, several materials – notably the lighter elements, lithium, beryllium and boron – produce greater enthalpy of combustion per gram of material combusted than carbon. Also note that, apart from hydrogen and carbon, only one other material – sulfur – produces an oxide which is gaseous at normal combustion temperatures. Gas production and, hence, power will be essentially zero for those materials producing solid products.

As an alternative, the products of the explosive decomposition, CO_2 H_2O and CO, can react with aluminium as shown in the equations:

$$3CO_2(g) + 2Al(s) = 3CO(g) + Al_2O_3(s) \qquad \Delta H = -821\ \text{kJ}$$
$$3H_2O(g) + 2Al(s) = 3H_2(g) + Al_2O_3(s) \qquad \Delta H = -950\ \text{kJ}$$
$$3CO(g) + 2Al(s) = 3C(s) + Al_2O_3(s) \qquad \Delta H = -1344\ \text{kJ}$$

Note that for the first two reactions, the gas volume is maintained, but for the third reaction, the gas is removed. Aluminium is, therefore, added to mixtures until the first two reactions are essentially complete. This increases the heat output and, hence, the power of the explosive, but note that it is impossible to prevent some of the third reaction occurring once some carbon monoxide is produced. When the reaction with carbon monoxide occurs, the power of the explosive will diminish.

The optimum amount of aluminium added depends on the oxygen balance of the composition. Calculations indicate that the ideal composition for pure TNT is with 18% added Al and, for improved oxygen-balanced amatol (50 : 50 TNT : AN), the optimum occurs at 25% added aluminium. This mixture gives a 50% improvement in explosive power over pure TNT, and a 25% improvement in explosive power over the optimum TNT/Al mixture. The addition of aluminium is used practically to enhance the blast effects of explosives. Other elements could be utilized in a similar fashion. When the energy available, oxygen requirement and densities are all considered, aluminium currently provides the greatest enhancement in performance. The addition of aluminium will reduce the velocity of detonation of the mixture.

6.18 Shockwave Effects

The shattering effect, or brisance, of a high explosive is largely due to the very high dynamic pressure in the shock front of the detonation wave which passes through the charge and out into the surrounding medium. This pressure can be as much as 390 kilobars, far higher than the static pressure produced by the gases occupying the volume of the charge at the temperature of detonation.

In solid explosives, two practical methods exist for creating the shock wave; we can either ignite the explosive under conditions that cause the burning surface to recede at sonic velocity, or we can expose the charge to a shockwave from an adjacent charge. In either case, the wave is accelerated by an increasing rate of energy release occurring within and behind it, caused in turn by increasing temperature and pressure within the explosive material. This acceleration process does not continue indefinitely, however; it is countered by the loss of energy in the compression process and by the finite time of the chemical reaction. A balance is soon reached between the energy released by the reaction and that used in compression, radiation and other heat losses. Consequently, the velocity of the wave becomes stabilized at a value characteristic of the particular system, dependent on the chemical identity of the explosive, its physical state and so on.

Calculations can be made which enable all the hydrodynamic properties of ideal detonation waves to be predicted. The velocity of such waves, D, is found to be equal to the velocity of sound (c) within the explosive, at the temperature and pressure existing in the shock front, plus the velocity of the material (w) as it moves forward in the wave. Therefore:

$$\mathrm{D} = c + w$$

For a typical high explosive, say, TNT, c is about 5400 ms^{-1} in the conditions of the wave and w is about 1500 ms^{-1}, so D is about 6900 ms^{-1}.

Behind the shock front is a short (1–10 mm) reaction zone in which the explosive material is decomposing into atoms, which then recombine into gases. The rear of this zone is marked by negative pressure, because atoms and molecules are moving backwards relative to the forward movement of the shock front. Behind the reaction zone, the gases immediately move outwards if confinement is absent (as in a bare charge in air) or too weak to withstand the pressure (as in a munition) (in the case of rock blasting, there may be a comparatively long delay of milliseconds before the cracks in the rock spread sufficiently for the rock mass to disintegrate and allow the gases to expand).

High-speed photography of a stick of explosive detonating in air shows the roughly conical detonation shockwave radiating outwards ahead of the expanding gases. This can be depicted diagrammatically and juxtaposed with the pressure profile along the stick as it is detonated from the left-hand end (Figure 6.9).

The peak dynamic pressure in the shock front is called the detonation pressure (p) of the explosive. An empirical method of calculating it is due to Cook, as follows:

$$p\,(\text{kbar}) = \rho D^2 \times 2.50 \times 10^{-6}$$

where:

ρ is the charge density in g cm^{-3}
D is the velocity of detonation in m s^{-1}.

Example Given the velocity of detonation of RDX at 1.50 g cm^{-3} as 7400 m s^{-1}, calculate its detonation pressure (p) at the same density.

$$p = 1.50 \times 7400^2 \times 10^{-6}\ \text{kbar}$$
$$= 205\ \text{kbar}$$

The calculated detonation pressures of some high explosives are given in Table 6.6 below.

It is seen from Table 6.6 that primary explosives are less inferior as regards detonation pressure than they are in terms of power; in fact lead azide has a higher calculated detonation pressure than TNT has, due to the high density of lead azide. The comparatively high detonation pressure of lead azide, combined with its readiness to burn to detonation, enables

Table 6.6 *Detonation pressure of some explosives.*

Explosive	D(m s^{-1})	at ρ (g cm^{-3})	p (kilobar) calc.
Secondary explosives			
HMX	9110	1.89	392
RDX	8440	1.70	300
PETN	8300	1.56	269
RDX/TNT 60/40	7900	1.72	268
DATB	7520	1.79	253
Nitroglycol	8100	1.50	246
Nitroglycerine	7700	1.60	237
Tetryl	7160	1.50	192
TNT	6950	1.57	190
Primary explosives			
Lead azide	4500	3.8	192
Mercury fulminate	4500	3.3	167
Lead styphnate	4500	2.6	157

very small quantities of it to bring about detonation in moderately sensitive secondary explosives such as RDX or PETN when they are juxtaposed in fuze or demolition detonators. In turn, RDX and PETN, by virtue of their higher detonation pressures, are able to initiate less sensitive explosives such as RDX/TNT fillings in munitions, and nitroglycerine- or TNT-based commercial explosives. In addition to being a good guide to the initiating ability of an explosive in a train, detonation pressure is the main factor in determining brisance, the lesser being the power index. In single explosive compounds of the secondary type, brisance and power are roughly parallel properties. The same is not true in mixed explosives; for example, it is common practice to mix RDX or HMX with aluminium to increase their power, and TNT with ammonium nitrate for the same purpose, but in each case the velocity of detonation is lowered and therefore so is the detonation pressure and the brisance.

The property of brisance enables a high explosive to break its container (e.g. a shell or bomb) into small fragments which fly at high velocities. There is no close analogy between this process and the ejection of a projectile from a gun. The latter depends on the relatively steady production of hot gases by the propellant, whereas fragmentation relies on the shockwave to break up the casing and to begin the acceleration, subsequently aided by the expanding gas.

In contrast, the production of air blast by an explosive relies little on the detonation shockwave. It is mainly caused by the expansion of the gaseous products, which sets up a secondary shockwave in the surrounding air by a process analogous to the original formation of the detonation shockwave. The overpressure of the blast wave at some distance from a charge is proportional to the power index.

The phenomenon of shockwave reflection varies with the relative densities of the two media concerned. A shockwave in air, when reflected from a solid surface, produces a much higher overpressure. Conversely, a shockwave generated through a solid medium (e.g. metal or rock) is partially reflected in the form of a negative pressure wave, or tension wave, when it reaches an interface with air. This latter effect causes, within the solid medium, a rapid alternation between extremes of compression and tension near the reflecting interface. Even the strongest material is unable to withstand the stress thus produced, and tensile fracture therefore occurs in the plane of the reflected wave near the interface. This phenomenon is of great importance in the effective use of explosives. The principle is employed in the 'scabbing' of armour by squash-head projectiles, and in the breaking of concrete, rock and minerals by commercial explosives. In the former case, the effect has to be maximized, so a high detonation pressure in the filling of a squash-head shell is a prime requirement. In the blasting of rocks and minerals, on the other hand, an optimal effect is aimed at; the fragments will be required in a particular range of sizes for ease of handling and processing, so the explosive is selected for its characteristic degree of brisance.

6.19 Appendices: Measurement of Velocity of Detonation

Appendix 6.A: Dautriche Method

The simplest method of measuring the velocities of detonation is by a comparative method with reference to a standard explosive whose velocity of detonation is known, usually a standard detonating cord 'cordtex'. The method due to Dautriche is shown in the figure below

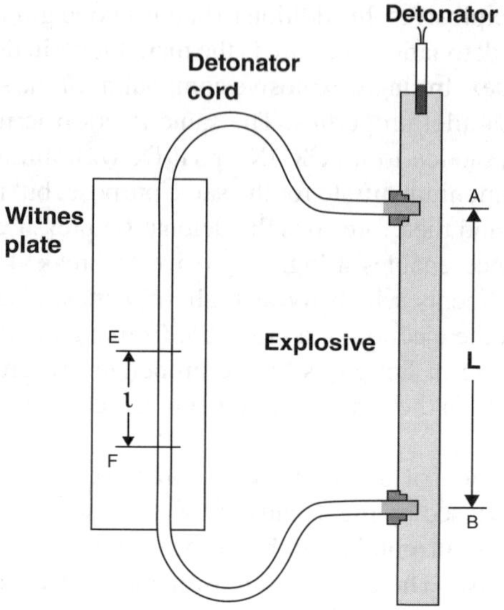

The Dautriche Method for VoD determination.

The explosive under test is pressed or cast at the required density of loading into a steel tube which contains two junctions into the side of the tube at a measured distance apart L. These junctions are fitted with a loop of standard det cord and produce a gas-tight fit to the cord (If measuring low VoD materials, the det cord may not pick up from the main explosive filling, so inserted into the junctions are two open-ended plain detonators which then guarantee to start the det cord). The standard detonator at the end of the charge is fired, and the shockwave travels down the tube and fires the ends of the det cord in sequence. This means that there are two detonation waves travelling in opposite directions in the cord and, where they meet, they produce an indentation on the witness plate at the point F in the figure. If E is the centre point of the cordtex loop, then the VoD of the unknown Du and Dc is that of the 'cordtex' they are related by the expression:

$$\mathrm{Du} = \frac{\mathrm{Dc\,L}}{2\mathrm{EF}}$$

Thus, the displacement is a direct measure of the ratio of the VoD of the explosive and the standard. The Dautriche Method is convenient for routine measurements and requires no costly timing equipment, but its accuracy is limited to about 3%. Furthermore, it relies on an accurately calibrated detonating fuze, which must be checked periodically by an absolute method. Other disadvantages are that the method gives only a mean velocity over a relatively long length of explosive, and that the experimental error increases with increasing velocities in the test explosives. In contrast, the following methods are both absolute and precise.

Appendix 6.B: The Rotating Mirror Streak Camera Method

The velocity of detonation of an uncased explosive charge can be measured with high precision using a rotating mirror streak camera, as shown below. The method is essentially simple and it provides an absolute determination, but the experimental technique is laborious and time-consuming. An image of the luminous region associated with the detonation shock front is projected via a rapidly rotating mirror onto a cylindrical photographic film.

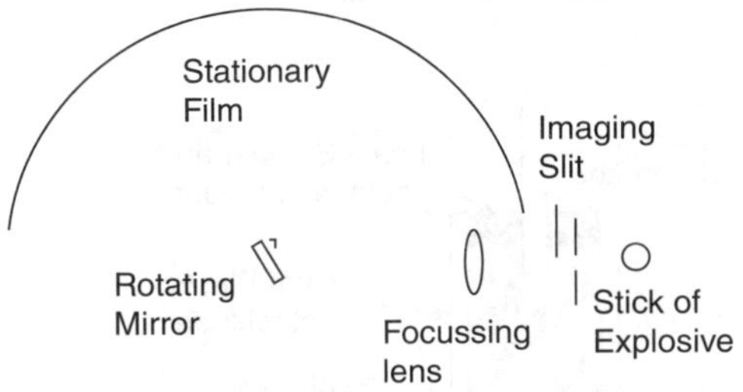

Schematic of the basic rotating mirror streak camera.

The idealized film trace produced from the streak camera is shown in the figure below.

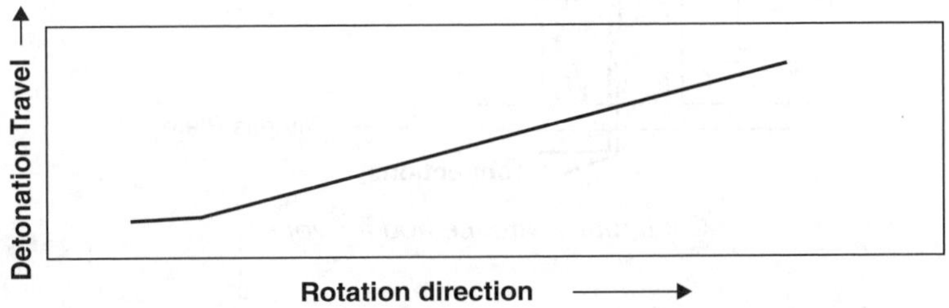

Idealized photographic trace from rotating mirror streak camera.

A straight line film trace indicates a uniform velocity of detonation, while the cases of accelerating or decelerating waves give, respectively, a concave or convex trace. The slope of the line measures the VoD with respect to the angular velocity of the mirror and the magnification of the image produced by the lens system. The firing sequence has to synchronize the initiation of the charge with the rotation of the mirror to ensure a complete trace. It is then possible to calculate not only the velocity of detonation but also, for example, the length of the reaction zone immediately behind the shock front, which corresponds to the thickness of the luminous trace.

Appendix 6.C: The Continuous Wire Method

This involves inserting a resistance wire axially into a cylindrical charge. A current is then passed through the wire and, when the charge is detonated, the wire is short-circuited progressively along its length by the detonation wave, which is of high conductivity. Thus, suitable instrumentation gives a record of voltage drop against time, from which the velocity of detonation can be calculated. Again, the firing sequence has to be synchronized so that the arrival of the shockwave at the end of the resistive loop triggers the oscilloscope trace. This is easier to do than with the streak camera.

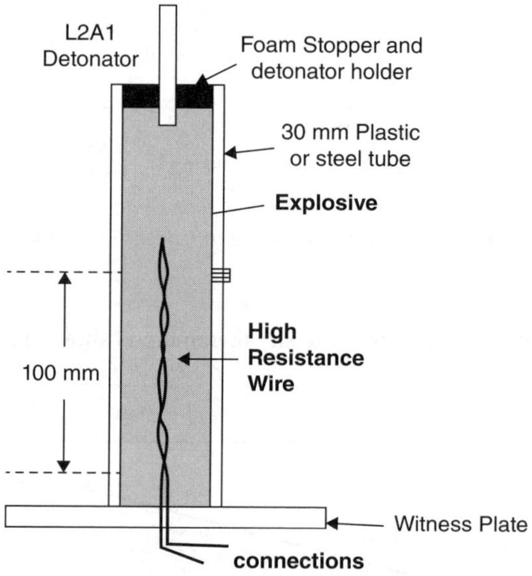

Continuous wire method for VoD.

Appendix 6.D: The Event Circuit

This method also relies on the high conductivity of the detonation shock front, compared with that of the undetonated explosive. A flexible printed circuit, which includes a number of ionization probes, is attached to the charge, as shown in the figure below. The probes are connected to charged capacitors and, when the circuits are successively closed by the passage of the detonation wave, the capacitors discharge into a load resistor, causing voltage pulses to appear across the load resistor. These are fed into a counter-timer. This system, although it uses advanced equipment, is easy to operate, uses only small charges of explosive and gives accurate results. Modern digital timing devices, capable of measuring time accurately to better than a nanosecond, are readily available

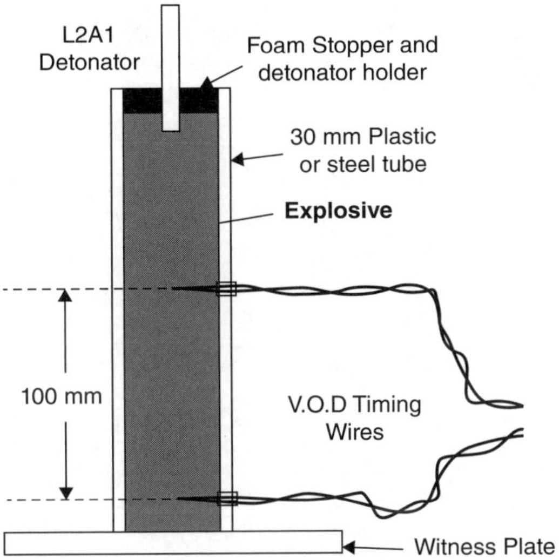

Schematic arrangement for event system of VoD measurement.

Bibliography

Baker, W.E. (1973) *Explosions in Air*, Univ. of Texas Press.

Baker, W.E., Cox, P.A., Westine, P.S. *et al.* (1983) *Explosion Hazards and Evaluation*, Elsevier Sci. publishing, Amsterdam.

Cole, R.H. (1948) *Underwater Explosions*, Princeton Univ. Press (Repub. Denver pubs. 1965).

Cooper, P.W. (1997) *Explosives Engineering*, Wiley-VCH.

Zukas, J.A. *et al.* (1997) *Explosives Effects and Applications*, Springer, New York.

Johansson, C.H. and Persson, P.A. (1970) *Detonics of High Explosives*, Academic Press, London.

Kinney, G.F. and Graham, K.J. (1985) *Explosive Shocks in Air*, 2nd edn, Springer Verlag, Berlin, Walters and Zukas.

Mott, N.F. (1947) Fragmentation of H.E. Shells. Proceedings of the Royal Society, **A187**, 300.

Cook, M.A. (1971) *The Science of High Explosives*, Reinhold Pub. Corp., New York.

Politzer, P. and Murray, J.S. (eds) (2003) *Energetic Materials Part 1 Decomposition Crystal Structure and Molecular Properties, Part 2 Detonation and Combustion*, Elsevier, Amsterdam, ISBN 978-3-540-87952-7.

References

(1) Taylor, J.W. (1962) Burning of secondary Explosive Powders by a Convective Mechanism. *Transactions of the Faraday Society*, **58**, 561–568.

(2) Korotkov, A.I., Sulimov, A.A. Obmenin, A.V. *et al.* (1969) Transition from combustion to detonation in porous explosives. *Combustion Explosions and Shockwaves* (Eng Trans), **5**, 216.

(3) Butler, P.B., Lembeck, M.F. and Krier, H. (1982) Modelling of shock development and transition to detonation initiated by burning in porous propellant beds. *Combustion and Flame*, **46**, 75.

7

Decomposition Processes and Initiation of Energetic Materials

7.1 Effect of Heat on Explosives

The basis of all energetic materials is the rapid chemical decomposition of unstable chemical formulations. If the materials were stable, then they would not be useful as energetic materials. That is not to say that some chemical compounds are not stable in their own right, but that there are more stable states for some mixtures of chemical compounds. That is, sugar is stable indefinitely, as is sodium chlorate, but the mixture is very reactive to yield more stable compounds. All energetic materials are decomposing, even at room temperatures, but the rate of decomposition is miniscule, allowing materials to be stored for several years in controlled environments and still function when required. Examining the energy profile for a typical chemical reaction, as shown in Figure 7.1 below, the starting material has to be provided with energy E to arrive at the energy maximum or transition state. At this point, the molecules can either lose energy by collision with their neighbours and return to a lower-energy unreacted state, or some of the bonds in the molecule can fracture, absorbing this energy and moving in the reacting direction, undergoing further decomposition and liberating energy. Note that the energy required to get to the transition state E_a is liberated again in the decomposition, so that the net energy change in the reaction Q is independent of the magnitude of E.

Arrhenius showed that the rate of chemical reactions, particularly decompositions, is controlled by the magnitude of E_a. The energy of the molecules is given by a Maxwell Boltzman distribution law. In the diagram in Figure 7.2, the molecular energy distribution is shown at two temperatures. If the activation energy is E_a, denoted by the dashed line on the figure, then the proportion of molecules with energy greater than E_a – the threshold value for decomposition, depicted by the curves to the right of E_a – increases with temperature.

Two conditions are applicable to the decomposition, namely the reaction is first order (i.e. only involves the explosive and no outside chemical agency), and the rate of decomposition

Laser Ignition of Energetic Materials, First Edition. S Rafi Ahmad and Michael Cartwright.

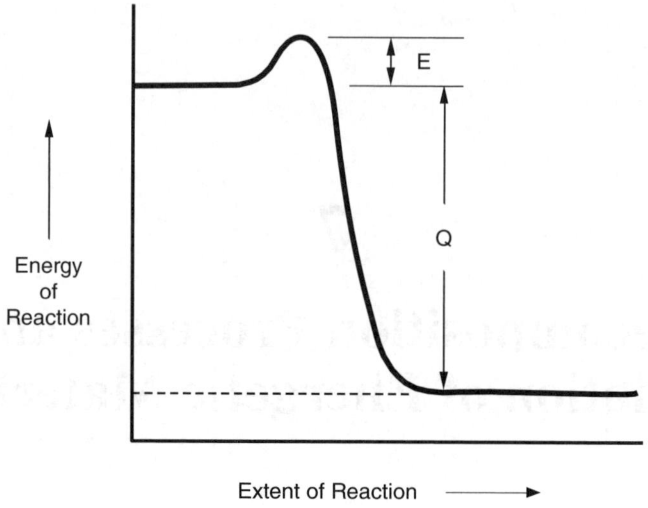

Figure 7.1 Typical energy profile of an energetic material decomposition reaction.

is proportional to the active mass of the explosive, which can be approximated to the mass of explosive present according to the equation below. Starting with a mole of compound, then at any time t when x moles of product have been produced, the rate of product production is proportional to the quantity of compound remaining, which is the starting concentration minus the product concentration x moles.

$$\text{Explosive} \rightarrow \text{gaseous products}$$
$$(a - x)\,\text{mol} \quad x\,\text{mol}$$
$$\text{The rate of decomposition} = \frac{dx}{dt} = k(a - x)$$

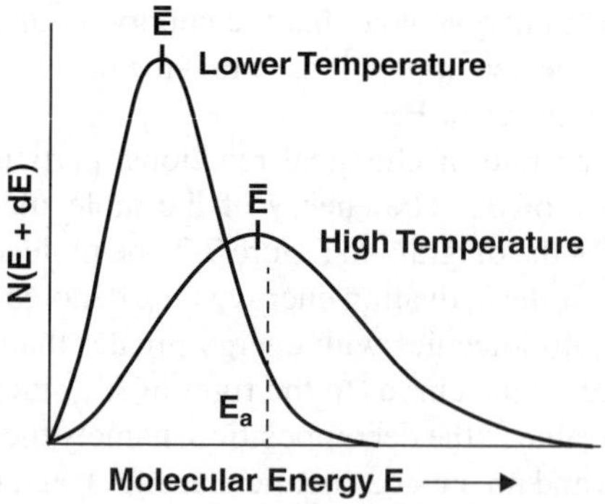

Figure 7.2 The Maxwell-Boltzman distribution of molecular energy levels at two arbitrary temperatures. E is the mean energy and E_a the activation energy for the decomposition reaction.

where k is the first order rate constant. The value of k is obtained from the Arrhenius equation:

$$k = \mathrm{A}e^{-\mathrm{E}/\mathrm{RT}}$$

where:

E is the activation energy for the decomposition
T is the temperature of the explosion in degrees Kelvin
R is the universal gas constant
A is a constant for the particular explosive and varies with the explosive.

This equation shows that if there is an increase in temperature, then the rate of reaction will increase. A very approximate rule is a 10-degree increase in temperature doubles the rate of reaction.

Figure 7.3 shows the effect of activation energy on the relative rate of reaction. If the activation energy is 1 J mole^{-1}, then the reaction is complete at 100 K. The thermal vibration of the molecules, even at 100 K, is sufficient to cause the molecule to decompose very rapidly whereas, at E_a 100 kJ mole^{-1}, the reaction is negligible by comparison, even at 1100 °K. It follows that all explosives are susceptible to heat to a greater or lesser degree. This will become self-evident when it is considered that all the various systems which are used to initiate explosives consist of applying heat energy in some form to the explosive; a material which is completely thermally stable could not function as a chemical explosive. The reliability of explosives depends on thermal instability, whereas their safety, which is

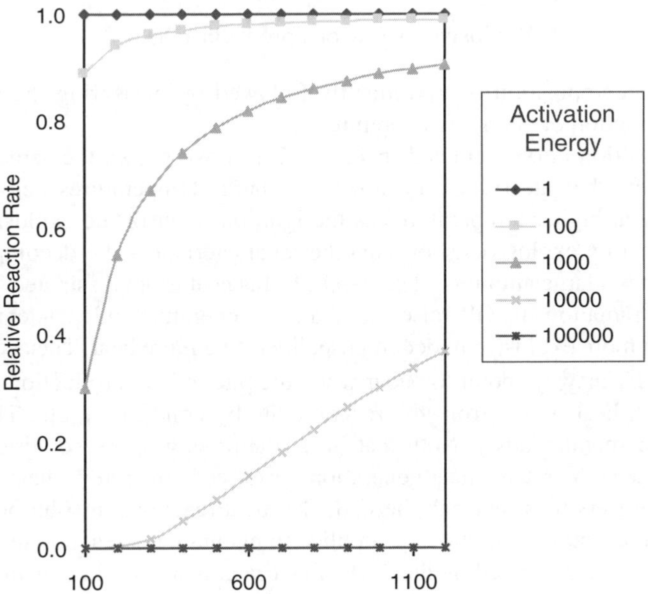

Figure 7.3 Effect of activation energy value on the rate of reaction.

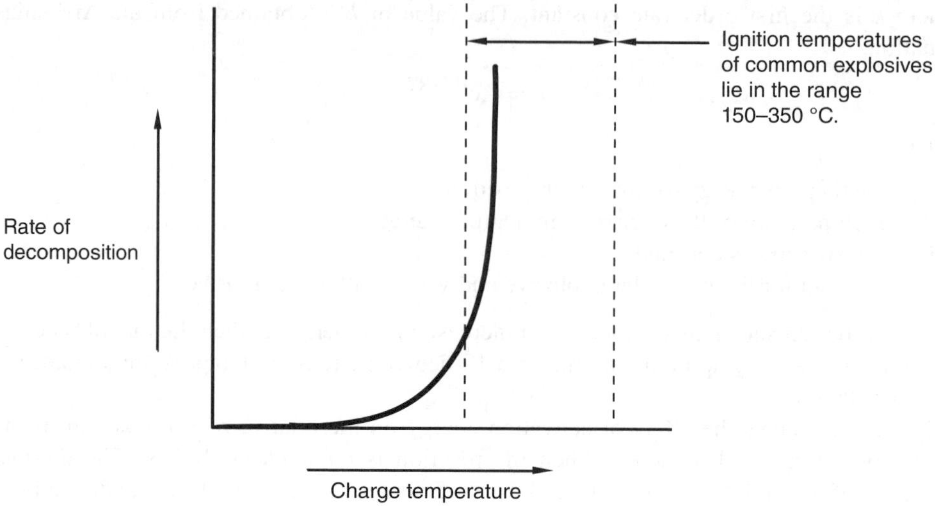

Figure 7.4 *Rate of charge decomposition as a function of charge temperature.*

the obverse quality, depends inversely on it. An optimum value of thermal stability must clearly be aimed at, and an understanding of the effect of heat on explosives is vital to the user as well as to the chemist and the manufacturer.

As the temperature of a charge of explosive is raised above normal (25 °C or 198 °K), slow decomposition sets in, according to the equation:

$$\text{Explosive} \rightarrow \text{gaseous products} + \text{heat}$$

The *rate* of decomposition is conveniently followed by measuring the volume of gas produced as a function of charge temperature.

The decomposition curve depicted in Figure 7.4 above shows the same general form for all explosives; the rate rises very slowly at charge temperatures below 100 °C, but steepens rapidly at higher temperatures as the ignition point of the explosive is reached. The effect of heat on explosives shows another characteristic – the decomposition of the explosive liberates a large amount of heat (−Q). If this heat is not dissipated by conduction, convection and radiation, it will raise the charge temperature independently of external heating, causing more explosive to decompose, liberating more heat. The accelerating loop can then cause a 'runaway' point to occur when the rate of heat production $\frac{(d\,\Delta Q)}{dt}$ exceeds the rate at which heat is lost from the reaction site by conduction, etc. The charge will then decompose spontaneously. Note that in the latter stages, the conditions are almost 'adiabatic', because the rate of heat generation far exceeds the rate of heat transfer out of the system and differs from normal, thermal, chemical reactions in solution ('isothermal' conditions), where heat is continuously supplied to maintain the reaction temperature. The heat generated can be evaluated by the Arrhenius/Frank Kamenetski equation:

$$\frac{E}{Tm} = R \ln \left[\frac{\alpha \rho Q Z E}{Tm 2 \lambda \delta R} \right]$$

where:

R is the gas constant (see above),
α is the radius of a sphere or cylinder or the half-thickness of a slab,
ρ is the density,
Q is the heat of reaction liberated during the self-heating process,
Z is the pre-exponential,
E (or E_a) is the activation energy from the Arrhenius expression,
λ is the thermal conductivity,
δ is the shape factor.

The rate of heat loss, dH/dt, is determined by Newton's Law of cooling:

$$\frac{dH}{dt} = k\Delta T$$

where:

k is a constant
ΔT is the temperature difference between the sample and its surroundings.

The value of k is different for different explosive materials, and also depends on the sample shape and container material, thickness and conductivity. If these two functions are plotted, then where the curves cross indicates the temperature at which the reaction will be self-sustaining or spontaneous, i.e. the 'ignition temperature' (Figure 7.5).

In bulk explosive subject to prolonged high temperatures, thermal decomposition tends to start at some point within the mass of explosive. Explosives have low thermal conductivity. As a result, the dissipation of heat ($-Q^4$) around the point of decomposition is small, and the reaction may speed up and rapidly ignite the whole charge. Heat dissipation in bulk charges depends on the mass, volume and shape of the charge and on its degree of compaction. The

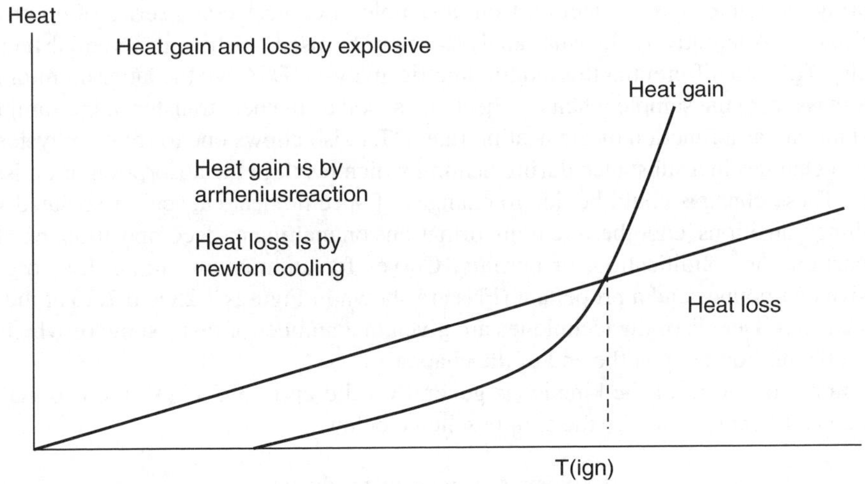

Figure 7.5 *Plots of heat generated and heat lost by a system as a function of temperature.*

Table 7.1 *Ignition temperatures (°C) of some explosives.*

Tetrazene	433 (160 °C)	RDX	486 (213 °C)
Mercury fulminate	443 (170 °C)	TNT	513 (240 °C)
CE (tetryl)	453 (180 °C)	Lead styphnate	523 (250 °C)
Nitroguanidine	458 (185 °C)	β-HMX	573 (300 °C)
Nitrocellulose	460 (187 °C)	DATB	578 (305 °C)
Nitroglycerine	461 (188 °C)	Lead azide	623 (350 °C)
PETN	478 (205 °C)	TATB	632 (359 °C)

temperature at which the runaway reaction occurs is identified as the ignition temperature, $T_{(ign)}$. Table 7.1 shows that those explosives of low stability to heat, such as the nitrate esters (NG, NC and PETN), have lower ignition temperatures than most of the nitramines (e.g. RDX and HMX) and nitro-compounds (e.g. TNT).

Note that the temperature of ignition is only one useful parameter. Lead azide is more easily set off than TNT, yet its T_{ign} is over 100 °K higher than TNT. Clearly, other factors, such as sensitivity to impact, have to be taken into account. In addition to mechanisms of deliberate initiation, the effect of heat on explosives is important in the following situations:

- Explosives kept in hot environments; for example, in unsuitable storehouses in hot climates or in deep boreholes for seismic, geophysical and mineral extraction.
- Explosives in missiles subjected to aerodynamic heating in flight. 'Cook-off' of propellants in hot gun barrels.
- The incorporation of explosives in hot melts for filling munitions.

The kinetics of the thermal decomposition of explosives at temperatures below 150 °C (i.e. in the pre-ignition phase) provides useful information on the susceptibility of explosives to heat. The kinetic data are normally determined under isothermal conditions, by measuring the rate of gas production from a sample of explosive at a series of controlled temperatures. Alternatively, thermal analysers may be used, such as differential thermal analysis, DTA and differential thermogravimetric analysis (DTGA); the latter monitors the loss of mass from the sample when it is heated, as well as the heat transfer in the sample at a constant rate as a function of the heating rate. DTA also allows one to follow physical or chemical changes in a substance during heating which result in the absorption or emission of heat. These changes could be due to changes of specific heat capacity associated with crystalline transitions, endothermic transformations on melting or decomposition, exothermic reactions on solidification, or burning. Curves for ammonium nitrate, four organic explosive compounds and a propellant (FN) are shown in Figures 7.25 and 7.26 at the end of this chapter. Details of the techniques are given in a number of texts, some of which are given in the bibliography at the end of the chapter.

The first order form of the kinetics is generally valid up to 393 K (120 °C) so that the reaction can be represented by the simple scheme below.

$$\text{Explosive} \rightarrow \text{gaseous products}$$

And may be quantified by applying simple first order kinetic laws. If the initial quantity of explosive is a moles and if x moles of gas produced at time t we have:

	Explosive	products
quantities	$(a-x)$ mol	x mol

The reaction kinetics will follow the equation for the rate of decomposition $da/dt = dx/dt = k(a-x)$ where k is the first order rate constant.

The Arrhenius equation determines the form of the (isothermal) rate-temperature curves. These are similar in form to the adiabatic curves mentioned earlier (bulk charge conditions) but they rise less steeply. In the lower range of temperatures (T = 298–373 K, 25–100 °C), the rate-temperature curves for Groups I and II show different forms (Figure 7.6).

It is evident that those explosives in Group I (e.g. mercury fulminate, tetrazene and certain propellants) show a higher rate of thermal decomposition at lower temperatures than those in Group II, but at higher temperatures the rates of reaction are reversed. Group I are considered primary explosives, more easily set off, and Group II are considered secondary explosives, less easily initiated. Table 7.2 shows Arrhenius factors of activation energy, E_a and the pre-exponential factor A for various explosives. Values of A and E found fall broadly into two groups:

1. For certain primary explosives and propellants, $A \approx 10^{11}$ and $E \approx 100$ kJ mol^{-1}
2. For most secondary high explosives, $A \approx 10^{20}$ and $E \approx 200$ kJ mol^{-1}

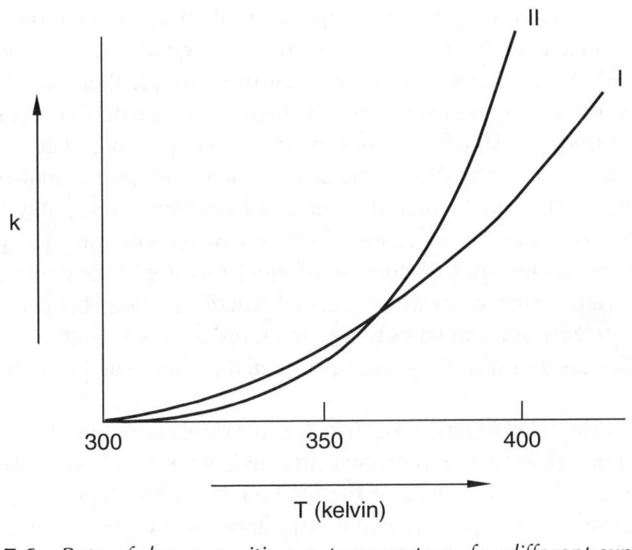

Figure 7.6 *Rate of decomposition vs temperature, for different explosives.*

Table 7.2 *Arrhenius Factors for the thermal decomposition of some explosives.*

Explosive	E (kJ mol^{-1})	A
Tetryl (liquid)	251	$20^{27.5}$
Tetryl (liquid)	230	$10^{24.5}$
Tetryl (crystalline)	217	$10^{22.5}$
Picric acid	242	$10^{22.5}$
TNT	222	10^{19}
Nitrocellulose (13%N)		
90–135 °C	205	10^{21}}
140–155 °C	201	10^{20}}
155–175 °C	234	10^{24}}
RDX	199	$10^{18.5}$
PETN	196	$10^{19.8}$
Nitroglycerine		
90–125 °C	176	$10^{19.2}$}
125–150 °C	188	$10^{23.5}$}
Silver axide	167	—
Lead azide	160	—
Trinitrotriazidobenzene	134	—
Copper (I) azide	111	—
Mercury fulminate	105	10^{11}

7.2 Decomposition Mechanisms

The mechanism operating during the decomposition of energetic materials is very dependent on two important parameters: the nature of the chemical species involved and the conditions under which the decomposition is occurring. Simple thermally induced decomposition generally proceeds by a different mechanism from the reaction occurring during the decomposition induced by a shockwave present in some explosions. The elucidation of the latter is nowhere near as clear as the mechanism occurring under thermal-induced decomposition. Reaction mechanisms, which dominate at low temperature, may be swamped by different mechanisms under the extreme conditions of temperature and pressure occurring during detonation. Our brief discussion of mechanism will concentrate on thermally induced reactions rather than shockwave-induced reactions, since this is closer to the conditions under which laser ignition takes place, and more data is available on these chemical reactions. For fuller discussions, the reader is directed towards the bibliography at the end of this chapter.

Only thermally stimulated reaction mechanisms for single compounds will be considered here, and no attempt will be made to discuss mechanisms of reaction between mixtures of different compounds. These introduce different chemical reaction possibilities, depending on the exact composition. The mechanism is very dependent on the chemical nature of the explosive compound, but almost all CHNO compounds involve the activation of the NO_2 group, which may involve formation of an NO_2 radical as an important part of the first stage in the decomposition. Nitroesters follow a different mechanism from nitramines, and both are different from nitrocompounds.

Most of the proposed mechanisms have been based on the kinetics determined for decomposition reactions, and the determination and identification of the various molecules produced during these reactions. The basic assumption is that there is a trigger linkage in the molecule, which is the first bond to split, and this step is the activation energy of the reaction, since the product of the bond fission will be much less stable and will spontaneously react faster. The major problem is to identify which is the critical trigger linkage. It is possible that there is more than one trigger linkage in the molecule and, hence, the complex kinetics. Although most energetic materials follow first-order kinetics, which means the rate of reaction is only dependent on the quantity of material present, but there are a number of anomalies in that some of the reactions are self-catalytic, and the products of the first stages of decomposition catalyse further decomposition. This often leads to an induction period before the onset of appreciable decomposition reaction.

DSC, when coupled to a GC/MS machine, has provided considerable data on the decomposition mechanism, particularly as a function of treatment temperature. The effect of temperature is often reflected in the change of state associated with both melting and vaporisation. This represents the enhanced freedom for molecules as the state changes form solid to liquid to vapour. Note that the rate of decomposition of RDX increases by an order of magnitude as it transforms between the solid and liquid state at its M.Pt. of 477 K (204 °C). Comparison of the N—N bond length between solid and liquid phases shows a dramatic increase in this bond length accompanying the melting process, which suggests that this bond is the trigger linkage.

Discussion will consider TNT as the first example, since this material has been studied for the longest time and is the least hazardous type of explosive compound to be investigated. Unfortunately, the sheer complexity of the mechanism, due to the presence of a high-stability benzene ring, has meant that the mechanism is quite involved, with multiple decomposition mechanisms.

7.2.1 Thermal Decomposition Mechanism of TNT

The thermal stability of pure TNT has been studied by the variation of temperature with time during adiabatic self-heating in a Dewar vessel, in the unstable range from 150 to 175 °C [1]. The first part of the decomposition was a heat-generating step of simple order with high activation energy (> 210 kJ mole^{-1}). This is unusual, since the rate-determining step has always been assumed as endothermic bond schism. The second step was an autocatalytic reaction with low activation energy (82 kJ/mole). Thermal decomposition product mixtures contain at least 25 discrete species, as well as large amounts of telomeric or polymeric materials (tars) [2]. The following major products appear in the mixtures: 2,4,6-trinitrobenzyl alcohol, 4,6-dinitroanthranil, 1,3,5-trinitrobenzene, 2,4,6-trinitrobenzoic acid and traces of unidentified compounds [3]. The thermal decomposition activation energy varies from 58.6 to ≈192.5 kJ mol, demonstrating that the process is far from well understood.

The suggestion that thermal decomposition of TNT is controlled by a homolytic, covalent C—H bond cleavage to its methyl moiety could explain the second order kinetics discussed later. This C—H bond cleavage, with an intra-molecular transfer of a hydrogen atom to the ortho-nitro group, followed by the elimination of H_2O, would provide a route to the isolated

4,6-dinitroanthranil product [4]. An intermolecular hydrogen atom transfer between two TNT molecules could provides a 2,4,6-trinitrobenzyl radical and a protonated TNT radical where the para-nitro group picks up the hydrogen atom.

Samples of TNT were partially decomposed (≈10–25%) by heating for 16-hour periods at 200 °C, and the products were identified as 4,6-dinitroanthranil, 2,4,6-trinitrobenzaldehyde, and 2,4,6-trinitrobenzyl alcohol. The absence in decomposition residues of 1,3,5-trinitrobenzene, a decomposition product found in others, indicates difficulties over the reaction mechanism. This observation has possible implications regarding activation parameters for TNT thermal decomposition.

The activation energy was given as 144.5 kJ and 222 kJ $mole^{-1}$. The differences were in terms of precision of the measurements, and also the role of TNT sublimation. In sealed gas phase systems, this is not a problem, but in open systems it may well obscure the fundamental kinetics by removing some of the TNT from the observation system without any decomposition. Modern theory suggests that the higher value is more consistent with a secondary explosive, and is more in agreement with the bond strengths of the likely trigger linkage.

There is considerable debate about which linkages break during the reaction. The assumption is also that the decomposition reaction of TNT is a monomolecular process. However, it was also proposed that bimolecular reactions also participate in the initial stage of thermal decomposition of poly-nitroaromatics [5]. These bimolecular reactions are absent in the primary fragmentation of polynitroaromates, due to heat, if oxidizable substituents (methyl-, hydroxy-, or amino groups) are not contained within the molecules of the compounds [6]. The problems of participation of bimolecular reactions in the measured data are also linked with the detection method of secondary phenomena occurring in the initial stage of thermolysis.

There is no doubt that there is an induction period before the reaction assumes its final kinetics, and the suggestion is that the production of some autocatalyst, or a fragment which performs the oxidation (i.e. NO_2), is the rate determining step (RDS). Some of these options are indicated in the scheme in Figure 7.7(a) and (b). Since addition of 10% of 4,6-dinitroanthranil the lower molecule in the scheme to TNT sharply increased the reaction velocity and the content of CO_2 and N_2 in the gaseous decomposition products, it was assumed that II was the autocatalyst in the thermal decomposition of TNT [7]. The difficulty is compounded by the effect of pressure on the system. In completely confined apparatus, the reaction rate is increased; this may be due to the reduction in the heat loss via escaping components, not least of which could be sublimating TNT vapour, since TNT is easily sublimed below its ignition temperature.

7.2.2 Non-Aromatic Nitro Compounds

The thermal decomposition of TNT is unusually complex because of the presence of the aromatic ring and its propensity to undergo substitution reactions rather than addition reactions, so that some of the initial products then substitute in the aromatic ring. With straight-chain alkyl and fully saturated ring nitro compounds, the absence of an aromatic ring simplifies the chemistry involved.

The simplest molecule in this category is nitromethane, CH_3NO_2. This compound is not a particularly sensitive explosive and requires a minimum of 30 g of a booster such as tetryl

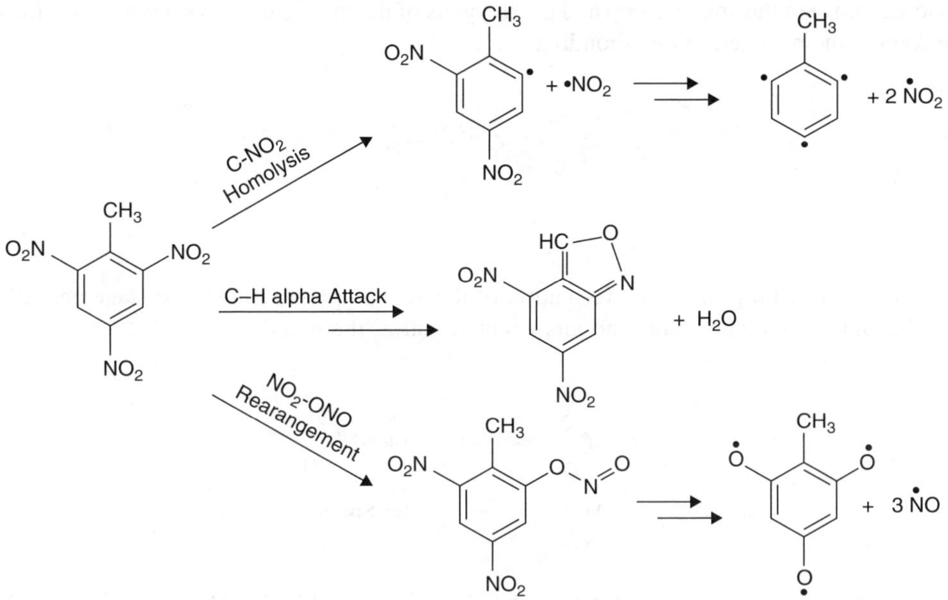

(a)

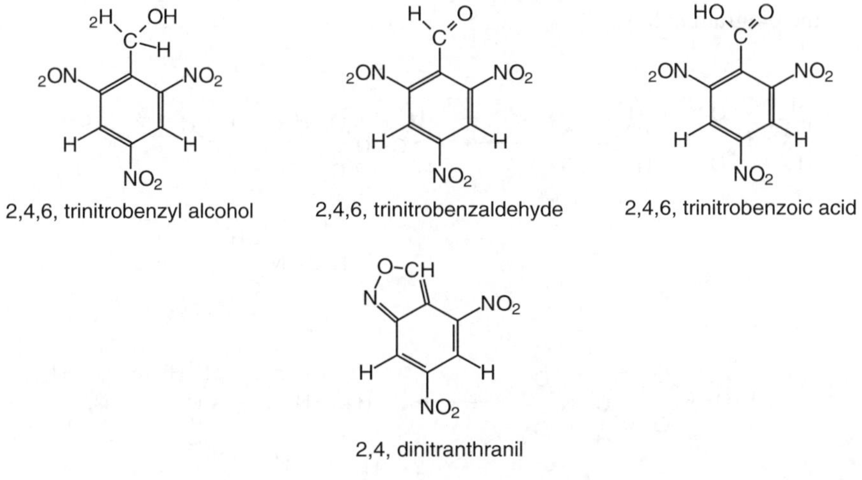

(b)

Figure 7.7 *(a) TNT decomposition scheme. (b) Some of the detected products.*

before it will detonate. Some indication of its decomposition mechanism can be given by the observation that the addition of nitrogen donor molecules, such as ethylene diamine (EDA) and diethylene triamine (DETA) render the liquid detonator sensitive but, over a period of time, this sensitivity increase disappears and is replaced with a detonator-insensitive, dark brown oil. The original suggestion was that the amine sensitized the NM by hydrogen bonding between the amine group and the oxygens of the nitro group, as shown below, thus weakening the nitrogen oxygen bonding.

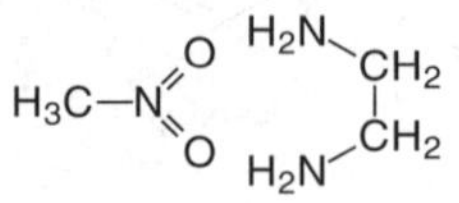

An alternative proposal is the formation of the aci compound as the first stage and this could then react with an amine and this would sensitize the liquid.

NM   Aci Species

NM decomposes on heating to give as products NO, N_2O, CO, CO_2, CH_4, H_2O, HCN and CH_2O (formaldehyde), which is indicative of the aci molecule but also of some other processes which may involve a bimolecular process. Some of these bimolecular options are shown in the diagram below.

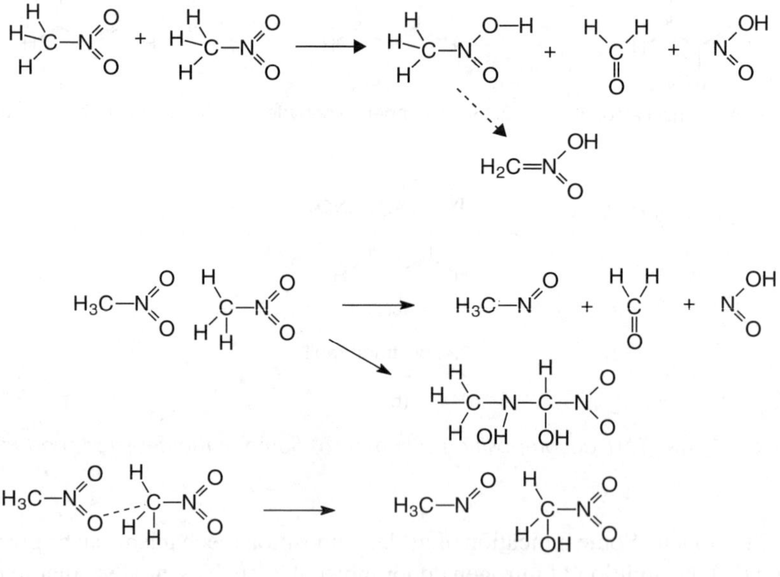

Note that one of the molecules forms the aci species and the other is split into formaldehyde and nitrous acid. Also note that one of the products of the bimolecular process is almost a simple addition of two molecules, which then undergoes further decomposition processes.

FOX 7, 1,1′ diamino, 2,2′ dinitro-ethylene has a structure dominated by the strong inter- and intramolecular hydrogen bonding, which stabilizes a number of polymorphs during thermal treatment.

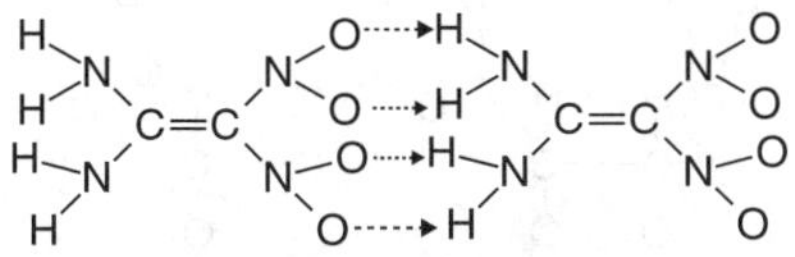

Nitrotriazalone is another important nitro compound which is being investigated. It contains a heterocyclic ring containing three nitrogens. As such, it is an interesting molecule to consider the decomposition mechanism, since it provides some indicators on how these polynitrogen ring compounds, which are being extensively researched as 'green' energetic materials, behave.

7.2.3 Nitro Ester Thermal Decomposition

Examining the nitroesters, which were incorrectly identified as nitrocompounds, the simplest is that formed from methanol with nitric acid – methyl nitrate. Few kinetic studies have been performed on this material, due to its extreme volatility at ambient and its low boiling point, 338 °K (65 °C). Also, it tends to be temperamental and often explodes without warning. The next member in the series, ethyl nitrate, is beset by similar properties, and the easiest nitrate ester to study is ethylene glycol dinitrate (EGDN). This molecule has only two carbon atoms and is safer to handle than nitroglycerine (NG), the next member in the series. The EGDN decomposition scheme is depicted in Figure 7.8.

The weakest linkage is the O—NO_2, hence splitting of this bond must be the first stage [8,9]. The two radicals produced, the first stage shown in the reaction scheme Figure 7.8, will be very reactive, and both can then result in chain reactions, accelerating the decomposition. This means that either of these radicals could react with unreacted molecules as well as simply decomposing themselves. Under these circumstances, the kinetics may deviate from simple first-order kinetics and there may well be an induction period. The ignition temperature for EGDN is very low, and thus the kinetics have to be studied at comparatively low temperatures, otherwise a rapid explosive reaction will occur. The activation energy for the decomposition has been quoted as 229 kJ mol^{-1}, but this is only an approximate value, due to the kinetics following non-linear curves as the extent of the decomposition increases.

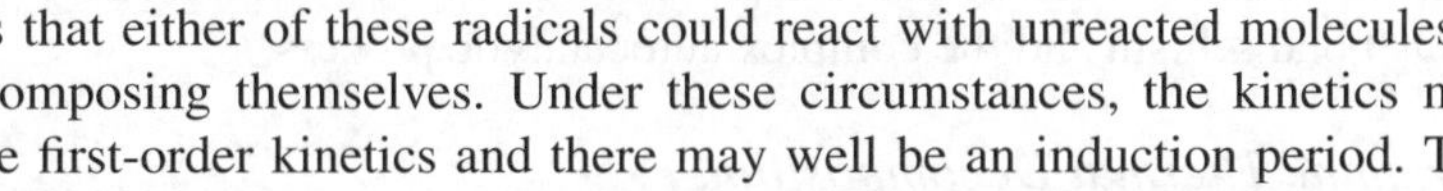

Figure 7.8 *Scheme for the decomposition of EGDN (ethyleneglycol dinitrate).*

Another very important nitro ester compound widely studied is nitrocellulose, an important ingredient in many propellants. The structure shows a long polymeric chain of glucoside rings, linked by oxygen bonding between the 1 and 4 positions in the glucopyranose rings.

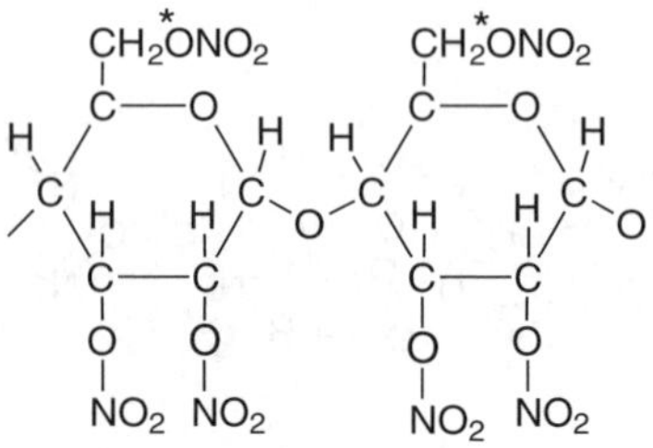

Nitro groups form ester linkages with the three alcoholic groups attached to the glucopyranose rings, as shown in the structure given in an earlier section. Different degrees of nitration are possible and this determines the chemistry involved in the decomposition. Because the nitro groups take two forms, one linked to an alkyl side group, while the other two are linked directly to the ring carbons, there is a difference in energy required to split the O—NO_2 bond between the two types, so the initial stage of loss of NO_2 radical displays some variation in activation energy. The weakest O—NO_2 bond is that attached to the 2 position carbon in the ring [10, 11]. There is again evidence of an autocatalytic reaction by this liberated NO_2 radical, occurring after an induction period giving a non-linear curve for the rate of reaction with time.

The attack on the glucodise ring in the polymer chain by the NO_2 radical is quite an involved process, because of the difference in availability of the hydrogen and the carbons. There is some conflicting evidence about the effect of additional NO_2; rather than increasing the rate of decomposition, it decreases in the presence of excess NO_2 [12]. If the released NO_2 is removed from the reaction vessel, the acceleration of decomposition is prevented [13, 14]. The kinetics are also very dependent on the temperatures used, and they fall into three groups <373 °K (100 °C), between 373 and 473 K (100 and 200 °C) and >473 °K (200 °C). This is deemed to be due to the presence of at least two reaction stages – the initial O—NO_2 bond schism and the complex autocatalytic process.

7.2.4 Nitramine Thermal Decomposition

Examining the nitramines as the next example, the N—NO_2 linkage is considered the trigger linkage. However, in both the RDX and HMX decomposition, there has been detected a 1 niroso 3,5 dinitrocyclohexane for RDX, and the corresponding nitroso compound from HMX. This indicates that a simple loss of oxygen by the NO_2 group in the 1 position has occurred. This molecule may have a higher stability than the parent compounds' decomposition via loss of the NO_2 radical, and hence its detection. Some studies [15] have indicated that the first-stage products are the gases N_2O and NO_2. These two species will have a very different behaviour, since the former has an even number of electrons and, as such, is a single molecular system, while the latter has an odd number of electrons and is more reactive. This is indicated by its brown colour and also the tendency to dimerize,

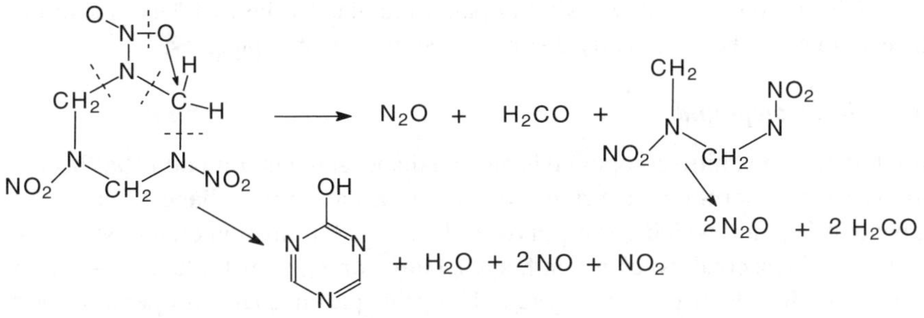

Figure 7.9 *Suggested scheme for the decomposition of RDX.*

forming the molecule dinitrogen tetroxide, N_2O_4, a stable liquid material widely used as an oxidizer in rocket propellants.

These two initial products show that there are two bond schisms occurring as shown in Figure 7.9. The predicted trigger linkage is the N-NO_2 thus creating the NO_2 species but the N_2O comes either from a splitting of the ring C—N bonds followed by loss of oxygen from the N—NO_2 radical or by a concerted reaction between two nitro groups either on adjacent positions in the molecule or from two adjacent molecules. The identity of the bond N—NO_2 trigger linkage for the decomposition is confirmed by the rapid increase in rate of decomposition when the temperature rises above the melting point of the solid. The length of this N—NO_2 increases dramatically from 1.37 Å in the solid state [16] to 1.45 Å [17] in the liquid, once the constraints of neighbouring molecules are eased. The appearance of CH_2O in the early products is an indication of simple splitting of the ring N—C—N bonds, with formation of the C=O with a neighbouring oxygen atom, perhaps formed by loss from N—NO_2 species during the formation of the N_2O species.

7.2.5 Photon-Induced Decomposition Mechanisms

So far, the discussion of decomposition mechanisms has focused on thermal decompositions but, since the main aim of the book is to highlight laser initiation, some discussion on photon-induced decomposition must be undertaken since, at some wavelengths, the mechanism of energy transfer can be different. For radiation in the IR region, it is thermal vibrations which are excited and, thus, the decomposition mechanism may be different from excitations using radiations in the blue end of the spectrum and the UV regions, where it is the electron excitation processes that are involved. Absorption in the UV region is due mainly to electron excitation from either the π orbital to π^* orbital in the NO_2, or of a lone pair electron on the oxygen being excited to the π^* orbital in the NO_2. In molecules with an aromatic ring, there is another absorption due to ring electron π orbital to π^* orbital excitation.

Although the first studies of photodecomposition were undertaken in 1936 by studying the end products, the lack of sufficiently rapid analytical instruments meant that considerable intuition was involved in any proposed mechanism [18]. The reaction occurring during photodecomposition is extremely rapid. Research has indicated that decomposition occurred within 5 picoseconds of irradiation, with a 5 μs pulse of laser irradiation [19]. The critical wavelengths for any compound depend on the possible electron transition processes for that compound. The mechanism operating will depend on which of these processes is occurring. Discussion will follow a similar pattern to that for thermal decompositions, and specific examples of the chemically distinguished classes of explosives.

7.2.5.1 Nitro Compounds

The simplest energetic nitro compound is nitromethane, and this compound has been widely studied. The UV absorption spectrum consists of two absorption bands. One is the very strong band at 198 nm, which corresponds to the $\pi \rightarrow \pi^*$ transition on the NO_2 moeity of the molecule. The second, weaker band, at 270 nm, corresponds to the $\pi^* \leftarrow$ n transition from a non-bonding loan pair on oxygen. The principal mechanism operating is that of simple radical formation:

$$CH_3NO_2 \rightarrow CH_3^\bullet + NO_2^\bullet$$

This homolytic splitting is supported by evidence of the presence of methyl nitrite, formed by a recombination of these radicals [20]. Studies using the longer wavelength radiation showed that nitrite formation may be involved in the primary process leading to elimination of the HON species, according to the equation:

$$CH_3NO_2 \rightarrow CH_2O + HON$$

The HON species will rapidly rearrange to the HNO species. All of these processes have been identified by fast spectroscopic techniques. Recent studies have shown that several components arising from reactions such as O elimination, OH formation and C—H cleavage have low quantum yields and, thus, are minor contributors to the process, and that the principal process occurring is C—N bond schism, following excitation of the $\pi \rightarrow \pi^*$ transition mentioned above [21]. Femtosecond pulses showed the nitromethane parent ion as the major product, and CH_3 and NO_2 were at a much lower level. The studies have shown that the photon-induced decomposition mechanism is dependent on the materials phase, the experimental temperature and the wavelength of the photon radiation.

7.2.5.2 Aromatic Nitrocompounds

TATB is a widely used insensitive explosive, and the strong hydrogen bonding between the nitro and the amine groups would be expected to modify the photon-induced decomposition, compared to the normal nitraromatic TNT. The first observation was that the colour of the TATB changed from yellow to green following exposure to UV light. The photoelectron spectrum shows that the N and O 1s band intensity is reduced by the irradiation, indicating that the simple C—NO_2 bond schism is the first step. Similar results were obtained with white light. Electron Paramagnetic Spin Resonance (EPSR) has shown that product of photodecomposition is a free radical which can, under the right circumstances, have a prolonged lifetime (two years) in the solid state. A mechanism based on simple splitting of the C—NO_2 bond would support this observation. Time of Flight (TOF) mass spectrometry suggests that simple loss of an oxygen atom from one of the NO_2 groups via a mono-nitroso derivative, and further oxygen abstraction, producing a di-nitroso compound, would also be a possible mechanism [22].

Recent photodecomposition studies using 532 nm light pulses have indicated that the products formed have a pressure dependence [23]. At higher pressures (8 GPa), the quantity of NO_2 produced is reduced and water is observed. At normal ambient pressure, there is apparently no water produced. The simple proton transfer arising from the strong inter-and intra-molecular hydrogen bonding may be encouraged by high pressure conditions. Further work is obviously required before there is a full understanding of the mechanism.

7.2.5.2.1 Nitramines

Dimethylnitramine DMNA is the simplest stable nitramine, which should demonstrate the fundamental mechanism for nitramine decomposition without the complexity of the heterocyclic polynitramines such as RDX. In the solid phase, exposure to 253.7 nm radiation appears to produce simple elimination of an O atom as the major process. Studies using 266 and 248 nm radiation on gas phase molecules indicate that the decomposition produces mainly NO_2 species in both the ground and excited states. There are also some products of nitro-nitrite isomerism, with the subsequent elimination of NO species. A simple hydrogen transfer between the methyl group and an oxygen of the NO_2 group leads to elimination of the HONO species elimination [24, 25]. The products appear to be very dependent on the wavelength of the irradiation. The products, other than simple O atom elimination, are summarized in Figure 7.10.

Three other important nitramines are RDX, HMX and CL20. These differ from DMNA in that there are more nitro groups in the molecule and there are complications due to the presence of cyclic rings. The major product would be expected to be simple nitrogen-to-nitrogen bond fission, eliminating an NO_2 radical. There is very strong EPSR evidence [26] that this is the major process for all three, and it is consistent with the N—N bond being the trigger linkage for the decomposition. A striking difference between RDX and HMX is in the number of NO_2 radicals produced in the molecule. β-HMX produces only one NO_2 radical by one N—N bond schism per molecule, while in single-crystal RDX under the same irradiation conditions, NO_2 radicals are produced at each N—N bond position in the molecule [27].

Both RDX and HMX show photoinduced decomposition with the elimination of HONO species, probably via proton transfer between the ring CH_2 group and the oxygen on the neighbouring NO_2 group. There may be an intermediate five-membered ring, formed initially by hydrogen bonding between the oxygen atom of the nitro group and the hydrogen of a ring CH_2 group. This process would be followed by N—N and C—H bond fission, as indicated in Figure 7.11 by the dashed lines with the formation of a C=N in the residual triaza-cyclohexane ring. The HONO can also undergo further splitting to produce NO and the OH radical observed in some experiments [28].

The NO molecule is a major product in the photodecomposition of all three nitramines. The origin of this molecule was either by nitro-nitroso isomeristion, with subsequent release of NO, or simple O atom loss from the nitro group, followed by N—N bond fission,

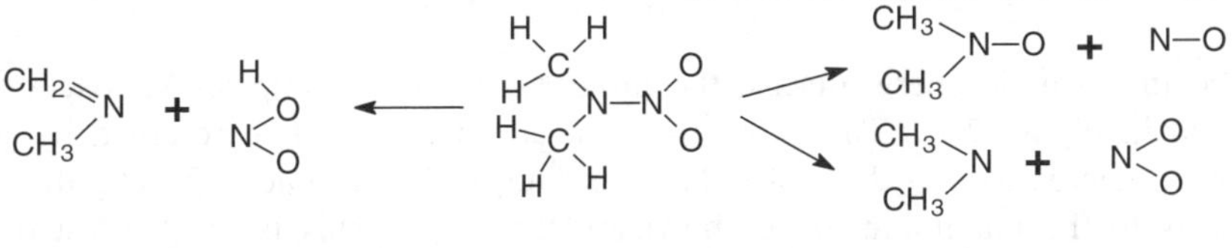

Figure 7.10 *Products of the photodecomposition of DMNA.*

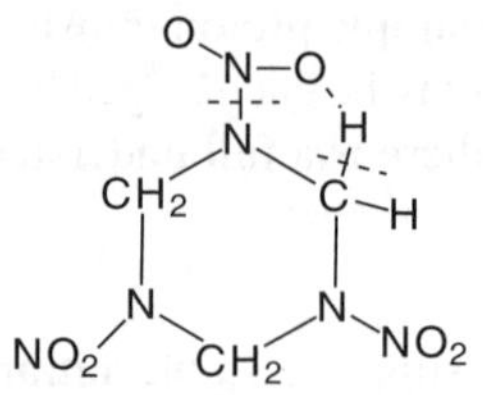

Figure 7.11 RDX, showing possible formation of the transient five-membered ring and elimination of the HONO species by bond schism across the dotted lines in the figure.

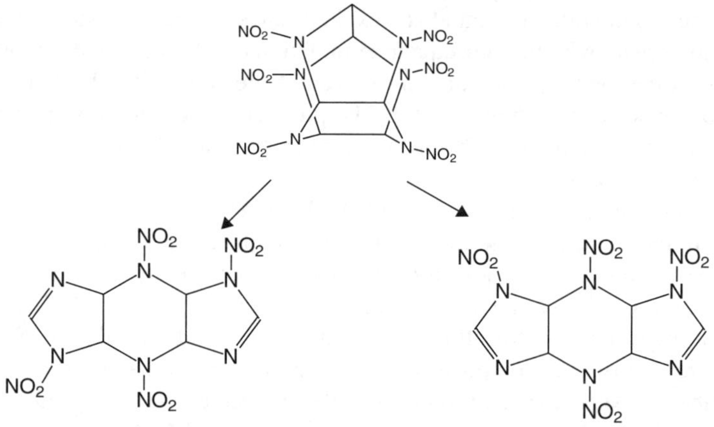

Figure 7.12 Two tetra-nitro isomers produced by photo induced loss of two NO_2 radicals from CL20.

releasing NO [29]. CL20 has a complicating factor, in that there are three component rings, bridged by carbon-carbon linkages. As well as the simple N—N bond rupture, there is also some electronic rearrangement, which leads to rupture of the carbon to carbon inter-ring linkage, particularly after the splitting of the second N—N linkage. The final products are the isomeric tri-cyclic structures, with four residual nitro groups, as shown in the diagram. One observation was that the quantity of NO produced appeared to be directly proportional to the number of NO_2 groups in the molecule. Hence, CL20, with six nitro groups, produced the most NO, and RDX, with only three nitro groups, produced the least number of NO molecules [30] (Figure 7.12).

7.3 Practical Initiation Techniques

Expressed in a simple energy terms, the progress of the decomposition reaction is as represented in Figure 7.13. The explosive is at an energy level represented by the flat line near the Energy axis. If the molecule gains energy, then its energy level will increase until it gets to the maximum value shown on the curve. This is the 'activated state'. The molecule can either lose energy by collision with its neighbours (like a snooker

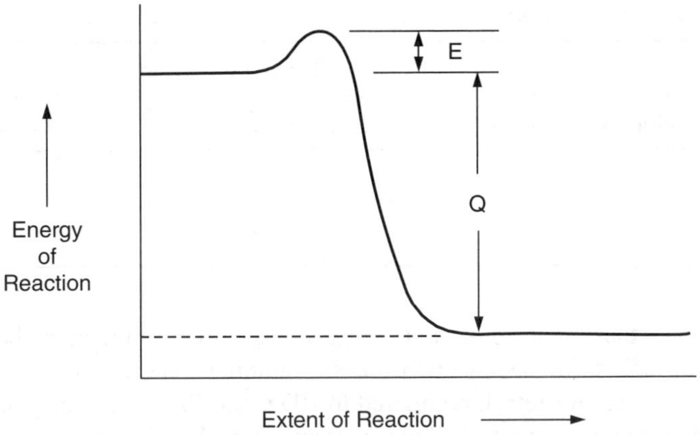

Figure 7.13 *Energy curve for the decomposition of an explosive.*

ball impact) and return to the original energy level, or it can decompose, following the curve to the right of the maximum, evolving heat. This may then be transferred to other explosives molecules, which may also decompose. If sufficient molecules decompose and evolve sufficient energy to overcome the heat losses, ignition will occur. This process is the ignition process mentioned earlier, and may continue rapidly until all the explosive material in the charge has decomposed. The energy required to raise the explosive to the activated state is the activation energy, E – or, better, E_a.

The net heat liberated in the reaction, Q, is independent of E, and hence E can not be measured from measurements of Q. There are a number of ways in which the activation energy can be supplied. Individual explosives can be particularly susceptible to one way and less so to another. It is usual for a particular explosive to be suited to one method of initiation in a predictable way, but inconveniently prone to initiation in a less predictable way by some other stimulus, which it may encounter by accident. The task of the explosives chemist and technologist is to match the selected explosive with the appropriate means of initiation, and to design the initiating system such that it is not prone to accidental functioning by any foreseeable hazard. The guiding principles are, as always, *safety and reliability*. The following account of the various means of initiation therefore looks at them from two standpoints, their practical usefulness and their accident potential.

7.3.1 Methods of Initiation

All methods of initiation are methods of raising the energy of a sufficient number of energetic material molecules to the activated state required for them to decompose, liberating sufficient energy to sustain the runaway reaction. This idea is encapsulated in the 'hot spot' concept introduced by Bowden and Yoffe. This means that not all the material has to be raised to the ignition temperature, but only that a sufficient number of molecules in one locality must be raised to the ignition temperature. Then they will decompose, liberating heat capable of heating the surroundings molecules, and the reaction will become self-sustaining. The temperature which the hot spot is required to achieve depends on the size of the hot spot. Some data for typical explosives are given in Table 7.3.

Table 7.3 *Calculated critical hot spots parameters for several explosives.*

	Critical Temperature K (°C)		
Explosive	Hot Spot Radii 10^{-2}	Hot Spot Radii 10^{-3}	Hot Spot Radii 10^{-4}
PETN	623(350)	713(440)	833(560)
RDX	(385)	(485)	893(620)
Tetryl	698(425)	(570)	1086(813)

The table shows that, as the hot spot size decreases, the temperature has to increase to counteract the effects of heat lost to the surrounding material. The hot spot size is marginally more critical for tetryl, compared to RDX and PETN, because of the presence of an aromatic benzene ring in tetryl. The development of these hot spots is a function both of the method used to generate them and of the material itself. During the discussion of the various initiation methods, where possible the mechanism of hot spot generation will be highlighted.

Each of these processes involves transferring energy into the solid to generate hot spots. Methods of initiation can be divided into the following categories, depending on the source of energy:

- Direct heating.
- Mechanical methods.
- Electrical sources.
- Chemical reaction.
- Shockwave.

7.3.2 Direct Heating

7.3.2.1 External Heat

The application of heat to an explosive through a container wall or barrier, or in an oven, may raise its temperature to the ignition point as shown in the graph, Figure 7.5 on page 159, When heat is supplied to a solid, it is transferred to the vibration of molecules. There is a Maxwell-Boltzman distribution of energy throughout the molecular population, as discussed earlier. If there is a sufficient number of molecules with vibration energies above the bond-breaking energy, the energy released by this decomposition will transfer to other molecules, thus sustaining the decomposition reaction. This method is not a practical initiation technique, being slow, of uncertain timing and unpredictable result. It is an accident hazard in the case of hot guns, outbreaks of fire, aerodynamic heating in missiles, overheated explosive storehouses and unsuitable filling processes.

Note that for some energetic materials, placing them in the EM field of a microwave beam can rapidly heat them to their ignition temperature, but this is again a potential hazard rather than a practical method. Often, lightning protection is modified to provide microwave shielding as well. The major heating accident is fire for explosive materials contained in building fires. Impingement of hot fragments from a fire can also provide a heat-generated accidental initiation.

7.3.2.2 Flame

This method is one of the essential methods for the initiation of propellants and pyrotechnics which are required to function in the burning mode. The flame transfers heat energy to a fairly localized region of the surface. It is also suitable for initiating certain explosives, which then burn to detonation in a reliable and predictable manner. It is of no use in the case of secondary high explosives, because it will cause them to burn at the outset and they may or may not convert to detonation, giving rise to a situation of no safety and no reliability. TNT will happily burn on the surface if ignited by direct flame, producing considerable quantities of dark, unburnt carbon soot. The luminosity of the flame resembles that of a candle. Standard safety fuse is a typical example of a flame initiation system. The gunpowder filling produces a small flame at the opposite end of the fuse from the end ignited with a pyrotechnic match 'fuzee' after a predictable period of time. The output is sufficient to initiate lead azide, which can burn to detonation and is used in plain detonators (see later).

7.3.2.3 Laser Heating

Lasers are capable of providing sufficient energy to raise steel to melting temperature (1800 °K) in seconds, and can be used for cutting metals. Direct impingement of the laser beam on secondary explosives can raise them to ignition temperature in milliseconds. See Figure 7.24. Sufficient applied power can lead to burn to detonation. This method is becoming more popular in costly missile systems because of the avoidance of sensitive primary explosives. It may be almost impossible to accidentally set of the system normally initiated by laser impingement. Use of cheap laser diodes with sensitizer materials has some civilian blasting applications. More will be said on this matter later in this book.

7.3.3 Mechanical Methods

There is a well-established principle in physics called the mechanical equivalent of heat. Performing mechanical work can be a source of heat output. Doing mechanical work on the energetic material can generate sufficient heat to raise the solid to its ignition temperature. Some of the mechanical methods used in munitions and commercial devices will now be discussed.

7.3.3.1 Friction

Rubbing two rough surfaces together generates heat by the frictional forces occurring between them. High points or asperities on the surface of one solid interact with similar high spots on the other surface. Overcoming the resistance to the motion requires force, and this applied force is the source of heat. Boy scouts have used this phenomenon to generate fire and flame by rapidly rubbing together two wooden sticks. The rubbing of an explosive or pyrotechnic on a rough surface, or between two such surfaces, can cause ignition by friction. The ordinary domestic match and the Christmas cracker are familiar examples of this. The principle is little used in the initiation of explosive munitions or commercial charges and, where it is used, it is little different to applying a hand-held match to the fuze (fuse).

The original thunderflash used friction to initiate the device. Pulling the firing strip caused a sandpaper strip to be dragged across the initiation composition. Like percussion initiation, however, it is an important hazard in the manufacture and handling of explosives. One major drawback of friction initiation is the effect of liquids and, particularly, water on the process. Water will lubricate the motion of the two surfaces and prevent build-up of temperature at the contact hot spots between the two surfaces and, hence, the initiation process will not work. Neither Christmas crackers nor matches will function when wet, and it is impossible to set off TNT by friction, since the solid melts at 80 °C, a process which absorbs heat, and the liquid then lubricates the roughened surfaces, dispersing any locally generated heat.

7.3.3.2 Percussion

This method delivers a sharp blow to the explosive, which is usually contained in a miniature device. By contrast, percussion initiation is an ever-present accident hazard wherever explosives are manufactured or handled. The threshold of danger is not easy to establish with certainty, and this is a good reason to treat all explosives with respect. It is believed that there are two mechanisms operating in which percussion imparts heat energy to the explosive. One is the friction between grains of the explosive, which can create 'hot spots' where the temperature exceeds the ignition temperature. The second is the adiabatic compression of air cavities within the explosive and between the crystallites of explosive. The effect of sudden compression of a small air cavity on the temperature within it is given by the equation:

$$\frac{T_2}{T_1} \left(\frac{P_2}{P_1}\right)^{\frac{\gamma-1}{\gamma}}$$

where:

T_1 and P_1 are the initial temperature (° Kelvin) and pressure
T_2 and P_2 are the final temperature (° Kelvin) and pressure
γ is the ratio of the specific heats at constant pressure, Cp and constant volume Cv, for air, having a value of 1.40 for the compression process.

If the pressure is increased by a factor of six from atmospheric and T_1 has a value of 293 °K, then the temperature of the air under the new conditions is given by the expression:

$$\begin{aligned} T_2 &= 293 \times 6\frac{0.4}{1.4}\ \text{K} \\ &= 489\ \text{K} \\ &= 216\ ^\circ\text{C} \end{aligned}$$

This temperature is sufficient to ignite some explosives. Sometimes, materials that rely on this gas compression for a major source of initiation can be dead-pressed by removing the majority of the trapped air during the pressing process. Both the inter-granular friction/shear and the trapped air content are dependent on the crystalline properties of the explosive material.

Percussion is a suitable method for most primary explosives but, in a miniature component such as a small arms cartridge, the degree of sensitivity required in the primary explosive is fairly critical to achieve both reliability and safety.

7.3.3.3 Stabbing

This is an important method for the initiation of both detonating munitions and propellants of several kinds. A small needle is caused to stab into a primary explosive to ignite it. The effect is a combination of percussion and friction between the particles and the needle, as it is forced into the explosive. There is little gas heating in this highly reliable method, often used in small arms initiation. Some materials are particularly sensitive to this method of initiation and, as will be discussed later, are often added to improve the stab sensitivity of materials which have the desired output but are stab-insensitive.

7.3.4 Electrical Systems

There are several ways of converting electrical energy into heat for the initiation of explosives, all of them based on the resistance heating effect. Practical electrical initiation systems are three in number. The most common is the bridgewire, in which a low-tension supply to a filament or gold leaf heats it sufficiently to ignite an attached bead of explosive. This method is commonly used in the electric matches which form the basis of many detonators. Often, the firing system is the discharge of a charged capacitor through the bridgewire. This is a safety feature, since the capacitor has to be charged by connection to the power source, and the switch is then moved to the discharge position for firing.

The second electrical method is the exploding bridgewire (see Figure 7.27 at the end of this chapter). In this, a high-tension pulse is used to explosively vaporize a filament, with the creation of a shockwave as well as high temperature. This is a safer system than the ordinary bridgewire, because it eliminates the use of a primary explosive, the shock effect being sufficient to bring a fairly sensitive secondary explosive directly to detonation. The necessary HT supply is not available in most munitions, but the warhead initiation system in large missiles is a possible application. The third method is the 'conducting cap', in which a low voltage supply is connected to a small quantity of primary explosive – sometimes mixed with graphite to render it conducting – trapped between two metal electrodes. The high resistance of the composition raises its temperature virtually instantly to the ignition point. The conducting cap method is favoured in the cartridges of guns down to 30 mm calibre. Below this size, the problem of short circuits due to grit and grime is quite high.

Passage of an electric discharge, 'sparking', is not used in practical systems, but it is a serious accident hazard in the manufacture and handling of primary explosives. Some materials are so sensitive to spark initiation that precautions against a build-up of static charge on a person, or in explosive material, are always necessary – usually, conducting floors and earthing wrist straps for the operator. Stores containing energetic materials must be protected by lightning conductors sunk into the water table.

7.3.5 Chemical Reaction

The initiation of explosives can be achieved by means of inflammatory chemical reactions. Exothermic reactions resulting in spontaneous ignition are associated with strong oxidizing agents in contact with flammable materials. One common example is the action of concentrated sulfuric acid on a mixture of sugar and potassium chlorate, from which chloric acid, $HClO_3$, is liberated and ignites the mixture. This mixture can be either explosive in its own right, or can be used to supply flame to a more highly explosive substance. The self-ignition (hypergolic ignition) of some liquid-fuelled rocket motors is another example of initiation by chemical reaction. Two examples of hypergolic ignition are anhydrous hydrazine, with either hydrogen peroxide (>90% H_2O_2) or inhibited red fuming nitric acid (IRFNA), which is concentrated nitric acid with excess dinitrogen tetroxide, (N_2O_4) dissolved in the liquid. Both the anhydrous hydrazine and the oxidants are very hazardous materials and, hence, this method is not widely used, since it often requires sophisticated machinery to give reliable performance and to ensure operator safety.

7.3.6 Initiation by Shockwave

As we have already seen, for secondary explosives, the optimum means of deliberate initiation is almost always to expose them to an intense shockwave generated by the detonation of another explosive, known as the donor charge. It is usually necessary for the receptor charge to be in close proximity to the donor charge. We have seen earlier that the exploding bridgewire is a shockwave-generating system, and the slapper detonator relies on the shock impact of a flier plate, which can be produced either by rapid electrical heating methods or by laser-driven flyer plates. The mechanism of shock initiation is, like other techniques, basically thermal in nature; the receptor charge is brought to detonation by the impact of the shockwave. Note that where the burn to detonation is not possible with wet materials, shock initiation works even for wet materials in sea mines. While for a reliable system, the two charges should preferably be in contact, accidents can be caused by shockwaves initiating charges at surprisingly long distances from their source – up to tens of metres from large charges. This is known as sympathetic detonation, and is a recognized hazard to be guarded against and tested for in insensitive munitions.

Summing up these various methods of deliberate initiation, and eliminating initiation by external heating as impractical, we can make a distinction between them. The following can be used for the first component of an explosive system:

- Percussion.
- Stabbing.
- Electrical.
- Friction (uncommon).
- Chemical reaction (uncommon).
- Laser (uncommon).

The other two methods – direct flame and shockwave – require a preceding pyrotechnic or explosive component to supply the appropriate stimulus in order for the process to occur.

7.4 Classification of Explosives by Ease of Initiation

Different energetic compounds have different responses to the different initiation process, and they can be broadly classified into four classes, based on the ease of initiation. This is not a rigid classification, because a material may be sensitive to one stimulus but insensitive to another stimulus. There is no definitive boundary for each class, since the various stimuli have different properties (e.g. impact is measured in Newtons, but electrostatic sensitity is measured in Joules).

- **Primary** explosives are easily initiated.
- **Intermediary** (boosters) are less easily initiated.
- **Secondary** explosives are more difficult to initiate.
- **Tertiary** explosives are very difficult to initiate.

Energetic materials can be placed in a particular category, depending on their response to a series of standard tests.

7.5 Initiatory Explosives

From the foregoing, it will be seen that the choice of an explosive for use in an igniter or detonator will depend primarily on:

- The type of initiating stimulus to be applied to the system.
- The characteristic type of output required.
- Considerations of safety and reliability.

Thus, the foremost requirements of a primary explosive are as follows:

1. High sensitivity to one or more of the applicable stimuli: percussion, stabbing, flash, shock, electrical energy or friction.
2. The production of flash (in igniters) or rapid burning to high-order detonation (in detonators).
3. Consistency of reaction time.
4. A good shelf life, necessitating stability and compatibility.

The number of single compounds approximating to these and other requirements is comparatively small, and the initiating components of current UK explosive stores rely mainly on the following compounds:

- Mercury fulminate.
- Lead azide and silver azide.
- Lead styphnate.
- Lead dinitroresorcinate.

Some details of these components are given below and their structures were given in Appendix 4A.

7.5.1 Primary Explosive Compounds

7.5.1.1 Mercury Fulminate. Mercury Isocyanate $Hg(ONC)_2$

This compound is easily manufactured by the reaction of ethanol with concentrated nitric acid in the presence of mercury. Product is grey crystals (unreacted mercury) with a density of 4.45 g cm^3.

Properties include:

- Poor shelf life (3 months at 40 °C in tropics).
- Incompatible with Al.
- VoD 4500 m s^{-1}at 3.3 g cm^3.
- Figure of insensitivity (F of I), (Rotter) 10.
- Good percussion sensitivity but 'dead' pressed at moderate to high pressures.
- Obsolete for munitions. IED usage because of ease of manufacture.

7.5.1.2 Lead and Silver Azides $Pb(N_3)_2$ and AgN_3

Both salts are easily prepared by precipitation on slow addition of lead or silver nitrate solutions, respectively, to an aqueous solution of sodium azide in the presence of added organic material (dextrin or similar material), which coats the solid as it is precipitated and reduces the sensitivity to accidental initiation. A reduced-sensitivity basic salt, $Pb(OH)N_3$, is also precipitated by controlling the solution pH, and this avoids the need for the added organic compounds.

Material exhibits a very good shelf life, when dry. Moisture lowers stability and can induce a reaction with copper salts to give very sensitive copper azide (incompatability). Hence, lead azide detonators use aluminium and not copper tubes to enclose the azide.

Properties include:

- White crystals, density 4.8 g cm^3.
- VOD 4500 m s^{-1} at 3.8 g cm^3.
- Figure of insensitivity (Rotter) 30.
- Cannot be dead pressed, which is a distinct advantage, since the VoD is a function of density and the higher the compaction of the charge as a percentage of theoretical maximum density, the greater the performance.
- Poor sensitivity to percussion and stabbing,
- Good friction sensitivity.
- Very reliable burn to detonation. It is almost impossible to prevent these azides burning to detonation. Only at the nano-particle size can these azides simply burn.
- Two polymorphs, α and β, with different impact sensitivities. The less sensitive β form is the normal polymorph, which crystallizes in the preparation, and the more sensitive α form is a metastable form which slowly reverts to the β form during ageing.

7.5.1.3 Silver Azide

As per lead salt, but more easily initiated. Lower density and VOD than lead salt. Incompatible with copper, tetrazene and some first fire compositions.

7.5.1.4 Lead Styphnate

Prepared by the nitration of resorcinol with HNO_3/H_2SO_4 to give styphnic acid.

Styphnic acid Treated with $MgCO_3$ to give magnesium styphnate solution. Add lead nitrate solution; lead styphnate precipitates.

Properties include:

- Basic salt (some $Pb(OH)_2$ impurities) is a poor initiator.
- Reddish to brown crystalline, solid density 3.09 g cm^3.

- Sensitive to flame initiation and spark (safety hazard).
- Figure of insensitivity of 20. Good percussion sensitivity for igniters.
- Constituent in conducting composition.
- High stability and good shelf life, particularly with metals.
- VOD of 5200 m sec^{-1} at p = 2.9 g cm^3.
- Poor transition to detonation on decomposition.

7.5.1.5 *Lead Dinitroresorcinate*

Similar preparation to the styphnate, but with the nitration stopped at the 2 nitro group substitution.

Properties include:

- Good burn characteristics and flash generator.
- F of I of 12. More sensitive then styphnate.
- Poor transition to detonation.
- Good shelf life.
- Widely used in first fire compositions

7.5.2 *Primer Usage*

Since a large variety of methods of initiation are in regular use, and all types depend on this small number of compounds, none of which is ideally suited to the purpose, some means has to be found in each case to adapt the chosen explosive to its intended use. Particularly important is the desired initiation method. These methods of adapting the primary explosives to the required use fall into two categories: mixtures and composites.

7.5.2.1 *Mixture*

The explosive primer can be intimately mixed with another explosive and/or some other material which will correct its deficiencies, so that the resulting mixture possesses both the required sensitivity and output. A typical example is the mixing of 2% tetrazene with lead azide, which reduces the stab energy required for reliable initiation of the azide from 50 mJ for ignition to 5 mJ without any deterioration in the output of the azide. Hard materials, such as grit and glass, are added to improve sensitivity to mechanical stimuli, and soft materials such as wax are added to materials to reduce their sensitivity to mechanical stimuli.

7.5.2.2 *The Composite Arrangement*

One layer of explosive material can be used in juxtaposition to another, one supplying the required sensitivity and the other supplying the required output. The explosives are pressed in two or three successive layers into a small metal container, which then contains a miniaturized explosive train in its own right. Each of the layers can be either pure compounds, or intimate mixture of two or more compounds, with or without additional inert materials.

These two methods are used in both the two main classes of initiators (igniters and detonators), and sometimes both methods are used in the same component.

Table 7.4 Initiation of igniters.

Percussion	Flash from another igniferous element
Stabbing	Friction (uncommon)
Electrical methods	Laser (uncommon)
(Hot wire and spark)	

7.6 Igniters and Detonators

Primary explosives are contained in two distinct types of component: igniters and detonators.

Igniters are used to ignite the next component in a sequence. The explosive used is designed to burn explosively, but not to detonate and not to disrupt its container. The output required of it is FLASH, together with some associated gas pressure and perhaps hot particles. Its role is said to be igniferous.

Igniters are used for the following:

(a) in cap form, to ignite propellant, either directly or via a cartridge primer;
(b) to ignite delay elements or gaines in fuzes.
(c) to ignite pyrotechnics.

Figures 7.14, 7.15 and 7.16 show some examples of igniters.

In each of the above cases, the igniter itself is only a few millimetres in diameter, but displays the results of great care in design and manufacture, which is necessary because the effectiveness of the munition as a whole depends on its efficient functioning. Percussion

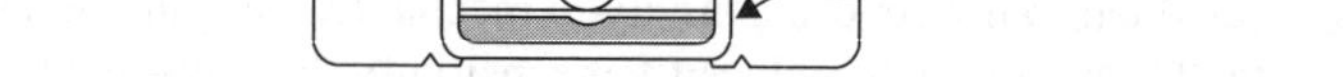

Figure 7.14 Percussion cap for small arms cartridge.

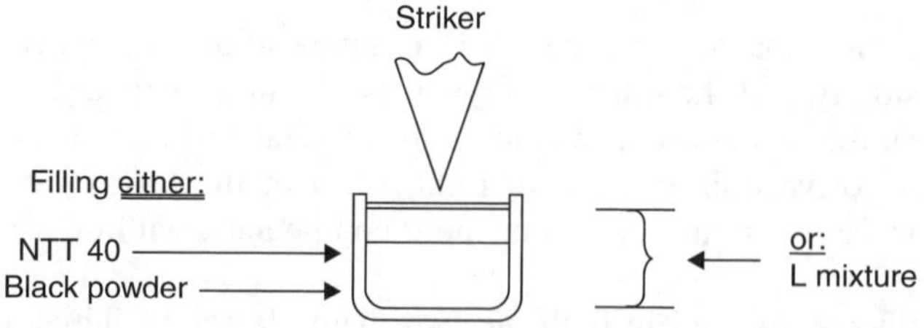

Figure 7.15 Stab igniter for shell fuze detailing alternative filling compositions.

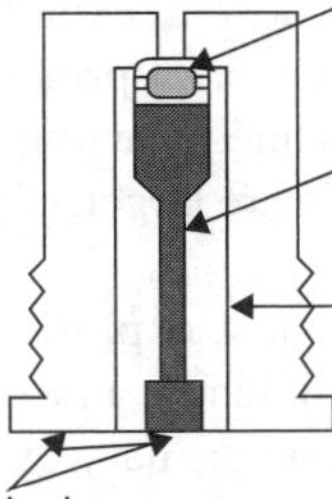

Figure 7.16 *Arrangement for conducting cap igniter.*

caps, in particular, contain complex mixtures of up to six ingredients, and continual research is aimed at obtaining 100% reliability.

Detonators are used to directly induce detonation in the next component of a train. To do this, the explosive used must itself detonate when initiated, so that it imparts a shockwave to the next component. In doing so, it normally disrupts its container. Its role is said to be disruptive. Detonators can be sub-divided into fuze detonators and demolition detonators.

Fuze detonators initiate the detonation train in a compact munition, such as a projectile, missile, bomb, mine or grenade. It may be initiated by:

- stabbing
- electrical means
- flash (in an igniter-and-detonator system)
- direct impact against a target (percussion)

A demolition detonator is the first explosive component in a hand-placed system for a demolition or any other blasting operation. It can be initiated either by safety fuze, in which case it is called a 'plain' detonator, or by an electric circuit. It is larger than a fuze detonator, being invariably of the 'composite' type (see section 7.5.2.2). It is inserted either directly into a demolition charge, or taped to a sample of cordtex detonating cord, or inserted into a pellet of boosting explosive. Note that in the USA, demolition detonators are called 'blasting caps', but this term is misleading when compared with UK usage, in which a cap is a percussion- or electrically-initiated igniferous device, i.e. an igniter.

7.7 Explosive Trains

Few energetic materials have the correct properties to be used on their own. From a safety point of view, we want the minimum quantity of materials which are easily set off – primaries – but from a user point of view, the maximum output is required. As a result, use is made of explosives trains. An arrangement of explosive components, based on the initiation methods already outlined, is called an explosive train. All explosive munitions, commercial blasting equipment and terrorist explosive devices comprise some form of explosive train. In all these differing applications, the design of the train will depend on the prime requirements of safety for the operator, reliability and effect, with economy in the bulk and cost of the device also in mind. Almost all explosive trains contain a primary explosive in their first component, and it is a universal principle that the amount of such explosive is kept to a minimum, consistent with adequate output, for reasons of safety and convenience. Succeeding components of the train are larger, of greater output but less prone to accidental initiation.

The sensitivity and output requirements of most explosives cannot be achieved by a single explosive material. To overcome this deficiency, an explosive train is used. In this arrangement, a number of materials are used. The primary explosive responds well to the initiatory stimulus, but does not have the output required for final performance, so its output is used to start the next component in the initiation train. The sensitivity to initiation steadily decreases, but the explosive output increases in the explosive train. The number of stages in the train depends on the requirements, but they always commence with a primary explosive contained in an igniter or a detonator. Trains are of two kinds:

- An igniferous train functions throughout in a burning mode, as distinct from a detonative one. Its first component must therefore be an igniter. Such trains are found in, for example, gun cartridges, carrier projectiles and pyrotechnic devices.
- A disruptive train is designed to terminate in a detonation. It may be designed to function in a detonative mode throughout, in which case its first component is a detonating material. Alternatively, it may commence in a burning mode, starting with an igniter and converting to detonation at some intermediate point, via a detonator. Note that if a delay is required between the operation of the firing sequence and the output, then it must occur in the burning part of the device. In either case, therefore, a disruptive train contains a *detonator*.

The various combinations of components that make up explosive trains are shown diagrammatically below in Figure 7.17.

Both commence with an igniter (though not all projectiles do); the primer in the cartridge and the exploder in the shell both boost the output from the igniter and detonator,

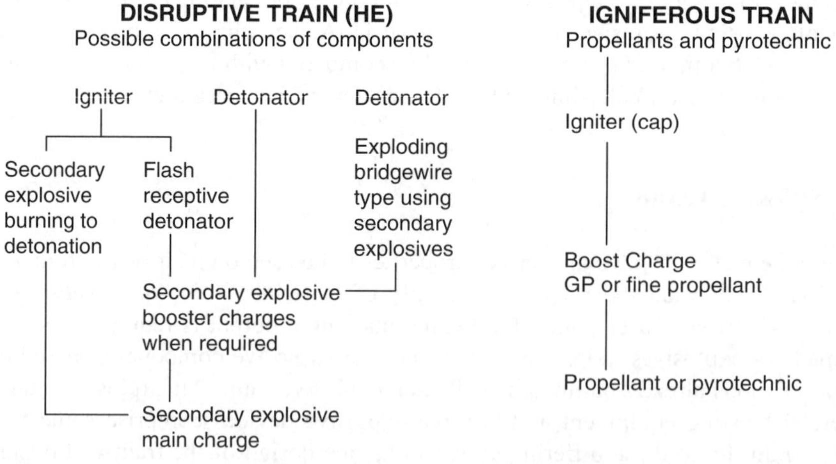

Figure 7.17 *Illustration of an artillery round containing two separate trains – one of each kind.*

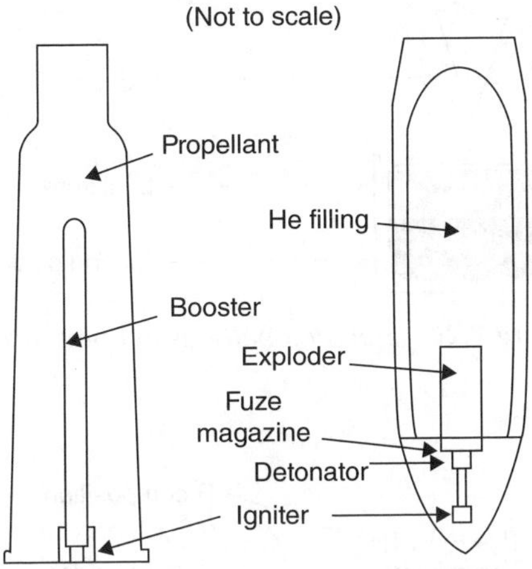

Figure 7.18 *Schematic explosive trains in HE shell ammunition.*

respectively, and initiate the respective main charges of propellant and secondary high explosive.

In the example of an explosive train illustrated in Figure 7.18, both the propellant train (igniferrous) and the HE train (disruptive) in the warhead start with an igniter in a burning mode, but that is not always necessary.

In the type of detonator shown in Figure 7.19, used in both military demolitions and commercial blasting, the electric bridgewire ignites the igniferous composition adhering to it and this, in turn, ignites the azide primary composition, which burns readily to detonation.

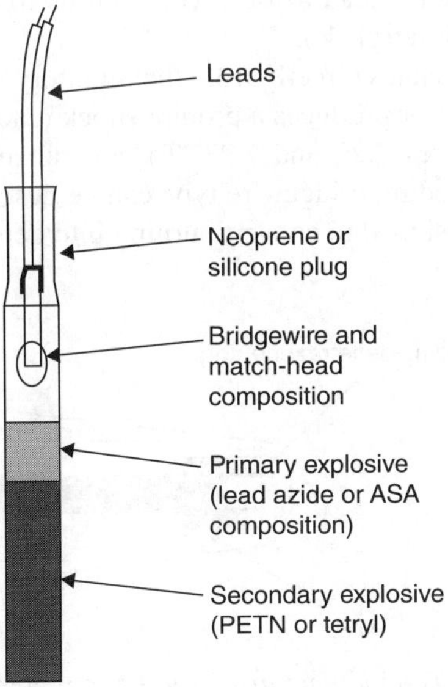

Figure 7.19 *Electric demolition detonator.*

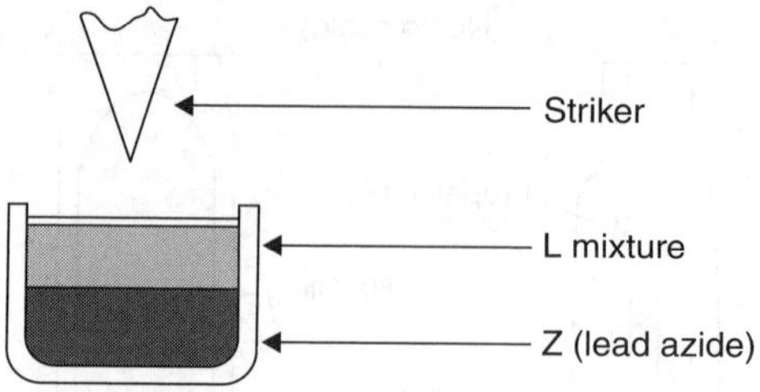

Figure 7.20 *Stab detonator as in bomb fuze.*

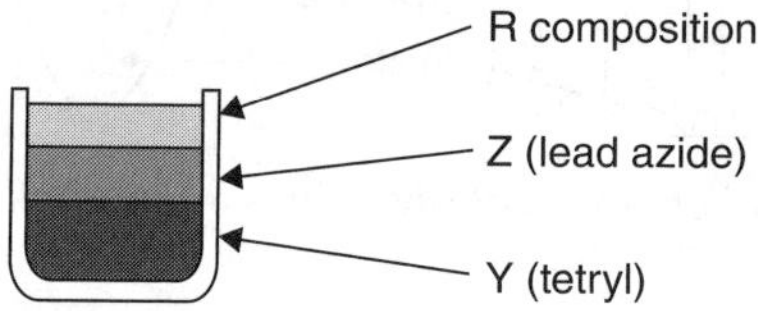

Figure 7.21 *Flash-receptive detonator as in 105 mm HE shell fuze.*

The base charge is a fairly sensitive secondary explosive which gives a disruptive output sufficient to cause detonation in a less sensitive main charge in contact with the detonator.

In the type of detonator shown in Figure 7.20, widely used by the military, 'L mixture' (mainly lead dinitroresorcinate) is reliably initiated by stabbing, but its output is igniferous rather than disruptive. It serves to ignite 'Z composition' (lead azide), which instantly burns to detonation and provides a disruptive output.

'R composition' (lead dinitroresorcinate – see Figure 7.21) is easily ignited by flash and, in turn, ignites 'Z composition' (lead azide), which burns to detonation and initiates by shockwave the base charge of tetryl (Y).

The direct impact fuse detonates directly from the initiatory stimulus. Also, the mechanical shock of a slapper detonator produces a prompt shock reaction.

This is illustrated in Figures 7.22 and 7.23. To generalize, all detonators except the percussion type and the exploding bridgewire type can be described as devices containing miniature explosive trains designed to convert burning into detonation.

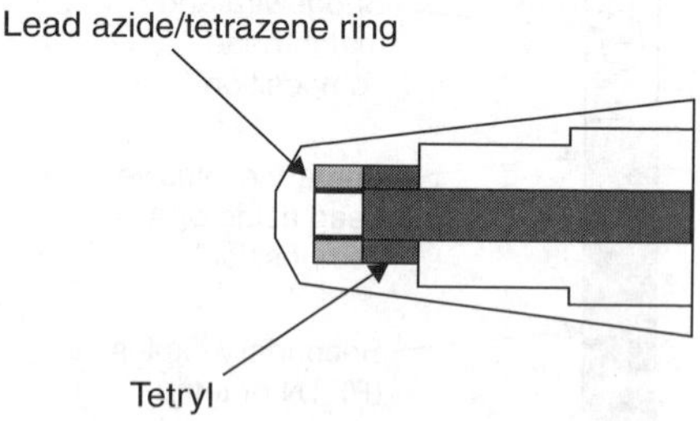

Figure 7.22 *Direct impact fuze as in 84 mm HEAT projectile.*

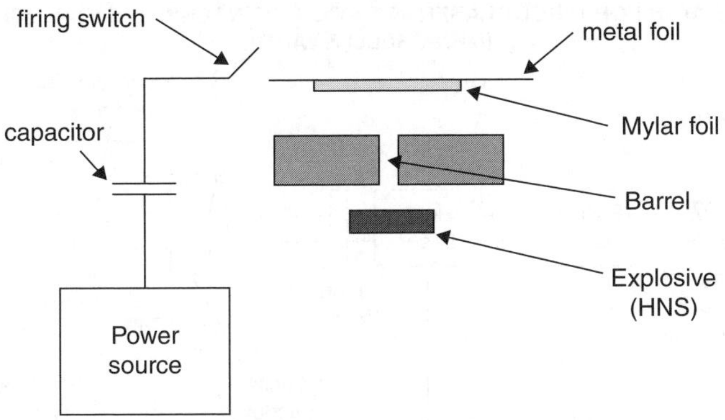

Figure 7.23 *Schematic of slapper detonator.*

In the system shown in Figure 7.22, the lead azide is here sensitized by a small percentage of tetrazene (≈4%) and, when the fuze strikes the target, the percussion impact causes the azide to detonate directly and impart a shockwave to the tetryl.

In a disruptive train designed for a munition or a demolition process, there are usually several explosive components between the detonator and the main charge. Each of these receives a shockwave from the preceding component and, on detonating, imparts a stronger shockwave to the next one, until the main charge is initiated with adequate stimulus. These intermediate components are traditionally filled with tetryl but, because of the high toxicity of tetryl, there is now a move towards RDX desensitized by wax – a more environmentally friendly material. The most successful of these mixtures in the UK is manufactured and marketed under the name of 'debrix' ('desensitized Bridgwater explosive').

7.7.1 Explosive Trains in Commercial Blasting

Explosive trains used in commercial blasting are simpler in detail than the compact, sophisticated and precision-made devices used in the fuzes of munitions, but they utilize the same principles. A typical system for commercial demolitions or quarrying is initiated by a portable exploder combining a dynamo and a capacitor. This device is connected to one end of a **reusable** shot-firing cable, the other end of which is connected, usually in series, with expendable wires, to individual detonators embedded in their respective explosive charges in holes drilled in masonry or rock. Electric detonators can have a pyrotechnic delay interposed between the match-head and the primary charge, designed to stagger a series of explosions by fractions of a second to limit the magnitude of the ground shock. Effectively, the detonation shockwave ripples across the rock surface.

Alternatively, charges in different shot-holes can be linked by detonating cord instead of being separately wired up. Some highly insensitive blasting explosives, unless containing a sensitizer such as glass microballoons, may need to be primed with a cartridge of nitroglycerine- or TNT-based explosive. Secondary blasting, i.e. the breaking of scattered boulders by separate charges, employs 'plain' detonators initiated by lengths of safety fuse (fuze) which, in turn, are ignited by hand or linked to a fast-burning pyrotechnic cord, known as igniter cord or quarry cord. This system is not permitted in underground mining,

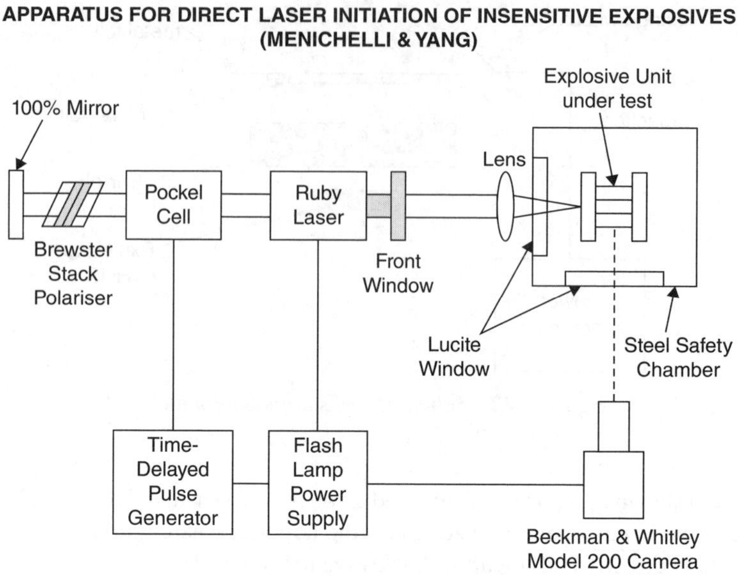

Figure 7.24 Schematic of early laser ignition experiments adapted from [30].

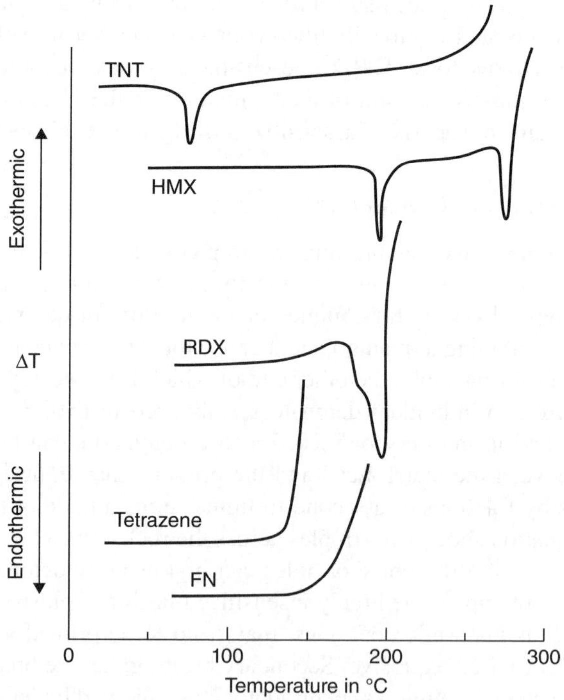

Figure 7.25 DSC scans for various energetic materials showing phase transitions and decomposition.

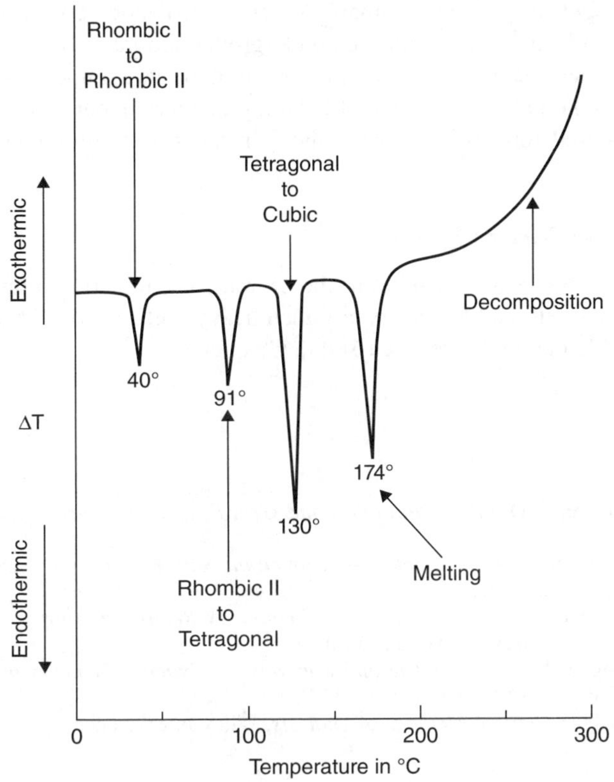

Figure 7.26 DSC of ammonium nitrate, showing phase transitions and decomposition.

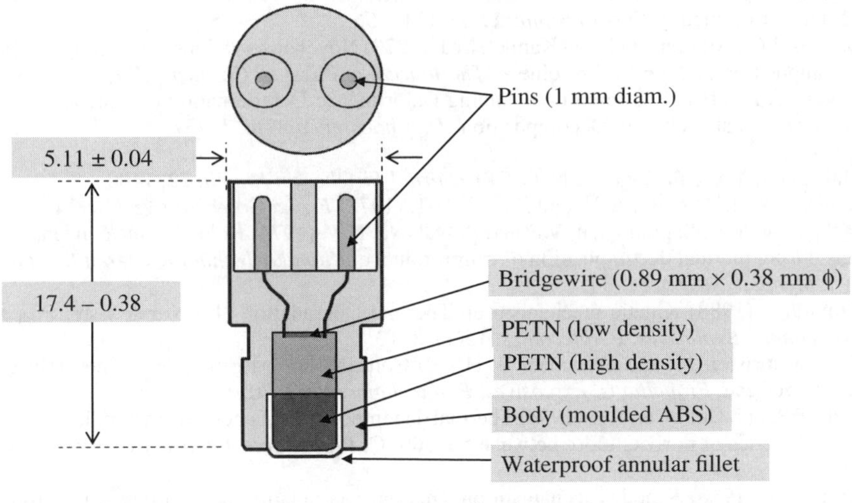

Figure 7.27 Schematic of exploding bridgewire detonator.

because of the danger of igniting flammable gases or suspended coal dust. For the same reason, most of the electric detonators used underground are made of copper, which gives a less incendiary effect than the aluminium ones used above ground. Copper detonators, however, have the inherent disadvantage of forming the unseen, but dangerously sensitive, copper azide, if stored for too long, particularly in moist/wet conditions often found in underground mines.

7.7.1.1 Alternative Primary Systems

Considerable effort has been expended in the search for alternative primary compounds that exhibit greater safety and lower toxicity than heavy metals and the very toxic azides. Some of the possible materials are discussed in Chapter 8.

Bibliography

Bowden, F.P. and Yoffe, A.D. (1951) *Initiation and Growth of Explosions in Liquids and Solids*, Cambridge Press.

Bailey, A. and Murray, S.G. (1999) *Explosives Propellants and Pyrotechnics*, 2nd edn., Brassey's London ISBN 0-08-036249-4.

Asay, B.W. (ed.) (2010) *Shockwave Science and Technology Reference Library*, vol. **5** Non Shock Initiation of Explosives, Springer Verlag, Berlin.

Dodd, J.W. and Tongue, K.H. (1987) *Thermal Methods. Analytical Chemistry by Open Learning*. HMSO: Wiley, Chichester, ISBN 0-471-91333-2.

Wendlandt, W.W. (1974) *Thermal Methods of Analysis*, Wiley InterScience.

References

(1) Pasman, H.J., Groothuizen, T.M. and Vermeulen, C.M. (1969) Thermal decomposition of TNT[trinitrotoluene]. *Explosivstoffe*, **17**(7), 151–156.

(2) Dacons, J.C., Adolph, H.G. and Kamlet, M.J. (1970) Novel observations concerning the thermal decomposition of 2,4,6-trinitrotoluene. *The Journal of Physical Chemistry*, **74**(16), 3035–3040.

(3) Rogers, R.N. (1972) Differential Scanning Calorimetric Determination of Kinetics Constants of Systems that Melt with Decomposition. *Thermochimica Acta*, **3**, 437.

(4) Ref. 4.

(5) Maksimov, Y.Y. and Kogut, E.N. (1978) *Zhurnal Fizicheskoi Khimii*, **52**, 1400.

(6) Balinets, Y.M., Dremin, A-N. and Kanel', G-1. (1978) *Fizika Goreniya i Vzryva*, **14**(3), 111.

(7) Maksimov, Y.Y., Sapranovich, V.F. and Polyakova, N.V. (1974) Role of some condensed products during thermal decomposition of trinitrotoluene. *Khimiko-Tekhnologicheskii Institut imeni D. I. Mendeleeva*, **83**, 51–54.

(8) Kimura, J. (1988) Kinetic mechanism on Thermal Degradation of a Nitrate Ester Propellant. *Propellants, Explosives, Pyrotechnics*, **13**(1), 8–12.

(9) Caire-Maurisier, M. and Tranchant, J. (1979) Contribution à l'Etude de la Thermolyse de la Nitroglycérine. *Propellants, Explosives, Pyrotechnics*, **4**, 67–70.

(10) Brill, T.B. and Gongwer, P.E. (1997) Thermal decomposition of energetic materials: 69 Analysis of the kinetics of nitrocellulose between 50–500 °C. *Propellants, Explosives, Pyrotechnics*, **22**, 38–44.

(11) Kimura, J. (1988) Kinetic mechanism on Thermal Degradation of a Nitrate Ester Propellant. *Propellants, Explosives, Pyrotechnics*, **13**(1), 8–12.

(12) Pollard, F.H. *et al.* (1950) *Nature*, 165.

(13) Garn, P.D. (1975) An examination of the kinetic compensation effect. *The Journal of Thermal Analysis*, **7**, 475–485.
(14) Batten, J.J. (1985) The agent of the autocatalytic thermal decomposition of aliphatic nitrate ester explosives. *International Journal of Chemical Kinetics*, **17**, 1085–1090.
(15) Brill, T.B. (1993) Burning surfaces. *Chemistry in Britain*, 34.
(16) Choi, C.S. (1972) Structure of cyclo trimethylene trinitramine RDX. *Acta Crystallographica Section B*, **28**(9), 2857–2862.
(17) Walder, C.L. (2001) Correlation of Structure and Sensitivity of Energetic Materials. M.Sc. thesis, 11EOE course, Cranfield.
(18) Hrischlaff, E. and Norrish, R.G.W. (1936) A Preliminary Study of the Decomposition of Nitromethane and Nitroethane. *Journal of the Chemical Society*, **1**, 1580.
(19) Schoen, P.E., Marrone, M.J. and Schnur, J.M. (1982) Picosecond UV photolysis and laser-induced fluorescence probing of gas-phase nitromethane. *Chemical Physics Letters*, **90**, 272–276.
(20) Pimental, G.C. and Rollefson, G. (1960) Ch 4 *Formation and Trapping of Free Radicals*, Academic Press.
(21) Guo, Y.Q., Bvhattacharya, A. and Bernstein, E.R. (2009) Photodissociation dynamics of nitromethane at 226 and 271 nm at both nanosecond and femtosecond time scales. *The Journal of Physical Chemistry A*, **113**, 85–96.
(22) Mcdonald, J.W., Schenkel, T., Newman, M.W. *et al.* (2001) The effects of radiation on TATB studied by time-of-flight secondary ion mass spectrometry. *Journal of Energetic Materials*, **19**, 101–118.
(23) Glascoe, E.A., Zaug, J.M., Armstrong, M.R. *et al.* (2009) Nanosecond time resolved and steady state IR studies of photoinduced decomposition of TATB at ambient and elevated pressures. *The Journal of Physical Chemistry A*, **113**, 5881–5887.
(24) Stephenson, J.C. and Mailocq, J.C. (1986) Picosecond laser study of the collisionless photodissociation of dimethylnitramine at 266 nm. *Chemical Physics Letters*, **123**, 390–393.
(25) McQuaid, M.J., Miziolek, A.W., Sausa, R.C. *et al.* (1991) Photodissociation of dimethylnitramine at 248 nm. *The Journal of Physical Chemistry*, **95**, 2713–2718.
(26) Pace, M.D. and Kalyanaraman, B. (1993) Spin trapping of nitrogendioxide radical from the photolytic decomposition of nitramines. *Free Radical Biology and Medicine*, **15**, 337.
(27) Pace, M.D. (1988) Electron paramagnetic resonance of UV irradiated HMX single crystals. *Molecular Crystals and Liquid Crystals*, **156**, 167.
(28) Capellois, C., Papagiannakopoulos and Liang, Y. (1989) The 248 nm photodecomposition of HexaHydro-1,3,5,-Trinitro-1,3,5-Triazine. *Chemical Physics Letters*, **164**, 533.
(29) Greenfield, M., Bernnstein, E.R. and Guo, Y.Q. (2006) Ultrafast photodissociation dynamics of RDX and HMX from their excited electronic statesvia femtosecond Laser pump probe techniques. *Chemical Physics Letters*, **430**, 277.
(30) Guo, Y.Q., Greenfield, M. and Bernnstein, E.R. (2005) Decomposition of nitramine energetic materials in excited electronic states: RDX and HMX. *Journal of Chemical Physics*, **122**, 244310.
(31) Yang, L.C. and Menichelli, V.J. (1971) Detonation of Insensitive High Explosives by a Q-Switched Ruby Laser. *Applied Physics Letters*, **19**(11), 473–476.

8

Developments in Alternative Primary Explosives

8.1 Safe Handling of Novel Primers

The compounds discussed in the following sections are generally very sensitive to some form of initiation, either by impact, friction or electrostatic discharge. This means that their preparation and handling must be performed under very strict safety procedures. Some of the reactions are easy to perform, but the products are very unreliable in their behaviour. Therefore, they should be treated as extremely hazardous until proven otherwise. Preparation should involve initially only milligram-scale quantities until their explosive parameters have been determined. Once these details are known, then safe operating procedures can be ascertained and then gram-scale preparations may be undertaken. Invariably, preparations should be undertaken in secure, explosively safe environments. Since the prime object of this document is to examine laser initiation, because some materials are very sensitive to light, they need special production facilities.

8.2 Introduction

Considerable effort has been expended in the search for alternative primary compounds which exhibit greater safety and lower toxicity than heavy metals and the very toxic azides. In the search for alternatives to lead azide, a number of other considerations were applied.

Researchers at Los Alamos National Laboratory (LANL) elaborated and published in 2006 requirements for 'green' replacements of lead azide (LA), lead styphnate (LS), and lead dinitroresorcinate (LDNR) in primers and blasting caps [1]. They suggested that lead

Laser Ignition of Energetic Materials, First Edition. S Rafi Ahmad and Michael Cartwright.

replacements must retain energetic properties upon exposure to the atmosphere and must conform to the six 'green' primary criteria, namely:

1. Insensitive to moisture and light.
2. Sensitive to initiation but not too sensitive to handle and transport.
3. Thermally stable to at least 473 K (200 °C).
4. Chemically stable for extended periods. Devoid of toxic metals such as lead, mercury and some others.
5. Free of perchlorate, which may act as a teratogen and has adverse effects on the human thyroid gland function.

These materials can be classified under a number of different headings, but the three chosen for this section are Totally Organic, Salts of Organics and Organo-metallic/transition metal complexes. Some materials fall into more than one category, and some of the organic materials readily form salts; these are discussed under the separate headings. We start with the purely organic compounds, since these are closest to the normal high explosives in common use, and were some of the first to be investigated.

8.3 Totally Organic

This group of primers covers a wide range of materials and they are typified by materials such as tetrazene, dinitrobenzofuroxane, (DNBF), diazodinitrophenol, (DDNP) and (1,1-diamino-3, 3-5,5,7,7-hexaazidocyclotetraphospazene) (DAHA). Other materials, such as 5-Azido-1-tetrazole and DAATO3.5 (mixed N-oxides of 3,3-azo-bis(6-amino-1,2,4,5-tetrazine)), will ignite from just the energy supplied by a common camera flash. They have fast ignition times and very fast, low-pressure burning rates, due to decomposition reactions dominated by condensed phase mechanisms.

The long-established commercial primary explosive **tetrazene** {**1**}(1-(5-tetrazolyl)-3-guanyltetracene hydrate is an environmentally friendly energetic material because it does not contain any heavy metal or perchlorate ions. While its preferred structure is still under debate between simple organic and a zwitterion structure, its energetic properties are well established. The compound has a crystal density of ≈1.63 g cm^{-3} and detonation velocity ≈5300 m sec^{-1}. Sensitivity to impact and stab is higher than the sensitivity of LA or LS to these stimuli. The temperature of ignition of tetrazene is ≈430 K (160 °C), and onset temperature for decomposition is about 410 K (140 °C). However, tetrazene slowly decomposes above 333 K (60 °C) and decomposes completely at 363 K (90 °C) in six days.

The minimum charge quantity to initiate PETN in a No 8 detonator amounts to 0.25 g. The compound loses its initiation ability when compressed under pressure exceeding 15 MPa (dead pressing). It is destroyed by wet CO_2 [2], therefore tetrazene does not meet the first and third requirements of Los Alamos National Laboratory's criteria for 'green' energetic materials. In practice, it is used as an additive to LS and LA to improve the impact and stab sensitivity of these materials. The stab sensitivity of LA is reduced from 50 mJ to 5 mJ by the incorporation of 2% tetrazene in the mixture. Substitution of the side chain amine group with methyl reduces the impact sensitivity, so that it is no longer a primer material.

Reaction of tetrazene with sodium nitrite in acidic aqueous medium at 345K (70 °C) affords a new energetic compound **iso-DTET** {**2**}. The most probable structure of this compound corresponds to 5-tetrazolo-azo-1′-tetrazole-5′-amine. The onset temperature for thermal decomposition of DTET is about 487 K (214 °C). This material possesses a high sensitivity to mechanical stimuli. It is more sensitive than tetrazene to impact [3], friction and spark; therefore, like tetrazene, it may have limited uses as a sensitizer for addition to LA and LS. No literature exists on methods of desensitizing this compound, or on trials using the material directly in devices.

A primary explosive, 2-diazo-4,6-dinitrophenol (**DDNP** {**3**}) may also be considered as an environmentally friendly energetic material. Although its structure has been the subject of some discussion [4], its properties are well established. DDNP has a crystal density ≈1.71 g cm^{-3} and detonation velocity ≈6900 m s^{-1} (at a density of 1.6 g cm^{-3}), almost matching the brisance of TNT [5]. The temperature of ignition is ≈445 K (172 °C). DDNP is not stable upon exposure to light, forming a nitroso compound, eliminating nitrogen and losing its explosive performance. The initiation ability of compound 3 is less than that of LA. DDNP is prepared from picramic acid by a diazotisation reaction with sodium nitrite in an acidic medium. Picramic acid is more toxic than LS, and thus this is not a preferred route. DDNP does not meet requirements 1 and 3 of Los Alamos National Laboratory's criteria for 'green' energetic materials.

Triazidotrinitrobenzene (**TATNB** {**4**}), prepared by the reaction of 2,4,6 trichloro, 1,3,5 trinitrobenzene with an alkali azide, is a powerful primary explosive with an impact sensitivity of 5 J but, unfortunately, it has a melting point of only 404 K (131 °C). A further disadvantage is that it also undergoes a slow conversion to hexanitroso benzene and, evolving nitrogen, losing its initiation power; ≈10% of the sample in a sealed system reacts in a month at 323 K (50 °C). It is also degraded by exposure to UV light. Like a number of other primers, the solid can be dead pressed. The starting material is not an environmentally friendly material and researchers have sought to use the less damaging trinitrotrihydroxy benzene as an alternative starting material, since it is used in the preparation of TATB, an insensitive high explosive.

Cyanuric triazide, or 1,3,5-Triazido-2,4,6-triazine (**TAT** {**5**}) has been proposed as a low toxicity, environmentally friendly organic primary explosive [6]. One preparation route is by the reaction of cyanuric trichloride, a readily available sterilizing/bleach product, with sodium azide. The melting point is ≈327 K (94 °C). TAT has crystal density ≈1.73 g cm^{-3}, detonation velocity ≈7300 m s^{-1} (at a density of 1.5 g cm^{-3}), and the gaseous products of detonation of this explosive are non-toxic. Comparison of their relative performance in the dent test shows that TAT is a more effective primary explosive than LA. The minimum charge of TAT required to prime tetryl in a number 8 detonator is 0.02 g. Compatibility tests showed that TAT is compatible with aluminium and stainless steel, as well as with energetic nitramines, including RDX, HMX and CL-20. TAT can be dead pressed at pressures >20 MPa and its sensitivity to spark is 1.2 mJ. Although TAT exhibits a temperature of ignition of 478 K (≈205 °C), it is a very volatile organic substance. It begins to sublime above 303 K (30 °C) and decomposes on heating above 373 K (100 °C) in the solid state. TAT is decomposed by hot water to produce toxic azide ions. Therefore, TAT does not meet the requirements for 'green' energetic materials.

DNBTM {**6**}, 5,5′-dinitrimino-33′methylene-1H-124-triazole, has a thermal stability greater than the 473 K (200 °C) threshold limit for environmentally friendly primer requirements. It is easily synthesized from readily available chemicals in high yields [7]. Its melting point is >513 K (240 °C), meeting another of the 'green' criteria. Its performance is much better than LA, with a VoD of ~8500 m s^{-1}. It is less sensitive to friction and electrostatic discharge than LA but, unfortunately, it is more sensitive to impact (1 J), than LA (2.5–4 J). As such, it is too sensitive for use without some form of phlegmatiser. Like tetrazene, it can be used as a sensitizer in LA systems but the green advantage is lost. Fortunately, the compound readily forms salts with organic bases, and these will be discussed later in the salt section. An organic salt could be used as the phlegmatiser, and these will also be discussed later.

5-nitrotetrazole (**HNT**) {**7**}, is an endothermic compound with a high VoD of ≈8.9 km s^{-1} (at a density of 1.73 g cm^{-3}) [8]. A related compound is 5-azido 1H tetrazole, or **HAT** {**8**}. Both of these compounds are extremely sensitive to accidental initiation and, in particular, the azido compound should be handled with extreme care. Both of these materials are too sensitive to use in a pure form, but considerable research on mixtures has been undertaken. The hydrogen in the 1 position of the tetrazole ring is very acidic (acid dissociation constant (pKa) = −0.82) and, hence, a series of salts are possible. Some of these are discussed in the section on simple salts.

Other tetrazole compounds of interest are those in which two tetrazole rings are linked through nitrogen species. **H2BTA** 5,5′bis (1Hterazoylamine) {**9**}), **BTH** 5,5′bis (1Hterazoylhydrazine{**10**}) and **BTT** 5,5′ bis (1Htetrazolyltriazine {**11**}) differ only in the linkage between the two tetrazolyl rings [9]. All three of these compounds are incompatible with acids because of their free amine groups, and have unpredictable sensitivities. Impact sensitivities are too high to measure by most standard tests. Substitution of the hydrogens on the terazolyl rings with methyl groups lowers their sensitivity and incompatibility but, even so, they are still too sensitive.

The anhydrous 2 Me substituted tetrazole ring compound, **MeBTT** {**12**} has an impact sensitivity of 2.5 J. VoDs are in the range 7.8–7.9 km s^{-1}, but they do not burn to detonation except under forcing conditions. There is an option to use these materials as components in rocket propellant systems, where their high nitrogen and lack of chlorine reduces the acid rain contamination. As a gun propellant, their high nitrogen content reduces gun erosion. Again, salts of some of these molecules will be discussed in later sections.

Nitroguanidine (**NQ**) has been widely used in triple base gun propellants since it has a high nitrogen content and reduces gun flash. It can also be used as a HE but only in combinations with other energetic materials. Further nitration of nitroguanidine produces dinitroguanidine, **DNG** {**13**}, which does have explosive properties in its own right. Unfortunately, it is extremely sensitive to accidental initiation by impact, friction and electrostatic discharge. The hydrogen on the nitramine group is very acidic ($pK_a = 1.1$) and, hence, it can readily form salts, which are much less sensitive to accidental initiation and can easily be prepared in solution without isolating the sensitive DNG. Some of these salts will be considered in the section examining salts.

A related compound in this series is 1,7 diamino,1,7 dinitrimino 246 trinitro 2,4,6, triazaheptane, **APX** {**14**}. This compound has improved performance as a primary explosive, with a VoD of 9650 m s^{-1} at a density of 1.911 g cm^{-3}, which is better than RDX and on a par with some HMX materials [10]. It is sensitive to impact (≈3 J, friction ≈80 N) and very sensitive to electrostatic discharge (≈0.1 J). The sensitivity is dependent on the exact production process, but its performance exceeds LA, while it is less friction-sensitive than LA. Further trials of this material, involving large-scale consistent production, are under way, but it is an improvement on LA in that it requires no intermediary explosive and can initiate normal RDX. Indeed, it is almost possible to consider APX as the perfect self-initiating HE with no other materials present. This would eliminate compatibility problems in the explosive train normally present in LA systems.

Other potential primers are the peroxide-containing compounds diactoneperoxide, **DADP** {**15**} and triacetonetriperoxide, **TATP** {**16**}. Both of these are readily prepared by the low-temperature, trace acid-catalyzed reaction, between hydrogen peroxide solution with acetone [11]. DADP is preferentially formed at ambient temperatures and is the more thermodynamically favoured reaction product. TATP has a higher performance (VoD of 5500 m s^{-1} at 1.2 g cm^{-3}), compared to DADP (3300 m s^{-1} at a similar density). DADP can be pressed to a higher density without initiation. It also has a higher melting point ($\approx$413 K (140 °C)) compared to the low-melting TATP (364 K (91 °C)). The main drawback of TATP is it readily sublimes at temperatures below its melting point. Although it has a better performance than LA, its sensitivity to friction, (0.1 N, impact 0.3 Nm) and static renders it unusable except in improvised explosive devices, where it can be prepared and used *in situ* without any separation and mechanical handling. DADP is much less sensitive to impact and friction than TATP but is not as effective as LA.

Two other peroxide-based explosives can be produced by the reactions of hexamine, a readily available fuel burning block used in camping for heating and cooking, with hydrogen peroxide in the presence of trace acid catalysts. The low temperature reaction of hydrogen peroxide with hexamine produces **DABCO** {**17**}, diazobicylooctane di hydrogen peroxide. This is a simple adduct with two hydrogen peroxide molecules acting as solvate molecules. This material has poor stability and unpredictable performance, such that its properties have not been fully investigated, but it has been included because it could be the first stage in the reaction to form a much more important explosive material, **HMTD** {**18**}, Hexamethylenetriperoxydiamine, or 3,4,8,9,12,13-hexaoxa-1,6-diaza-bicyclo-[4,4,4]-tetradecane. HMTD is a 'green' non-regular primary explosive, readily prepared by another trace acid-catalyzed reaction between hydrogen peroxide and hexamine. The same starting material can also be used for the manufacture of improvised RDX [12]. HMTD crystallizes from solution as white orthorhombic crystals, with a bulk density near 0.66 g cm^{-3}. HMTD is non-hygroscopic, almost insoluble in water and organic solvents. It slowly decomposes in the presence of moisture. HMTD is hydrolyzed by acids and bases at room temperature and by moisture alone above 313 K (40 °C). HMTD is stable for a sufficiently long time in the absence of acid traces but, in the presence of moisture, it easily corrodes metals. Like most peroxide-based explosives, HMTD is readily reduced by solutions of tin (II) chloride. The heat of explosion of HMTD is 5080 kJ kg^{-1}. The ignition temperature of the peroxide is $\approx$473 K (200 °C) but, at elevated temperatures, the product sublimes. Fine powders of HMTD have lower sensitivity to mechanical stimuli, compared to LA, and can be pressed up to the density of 1.3 g cm^{-3}, at which density the measured VoD is $\approx$5600 m s^{-1}, but coarse crystals of HMTD explode when being pressed, probably due to crystal shear and fragmentation under load. The powder is extremely friction-sensitive, being below the threshold values for most conventional friction machines. The static sensitivity is very similar. The threshold for initiation of HMTD charges with the beam from a Q-switch neodymium glass laser ($\lambda = 1.06$ µm, $\tau = 30$ ns) is near 40 mJ cm^{-2}. However, it is difficult to use pure HMTD in optical detonators, because it possesses high sensitivity to friction and static external stimuli. The material is also highly corrosive. HMTD is an effective primer/booster in improvised explosive devices capable of initiating ANFO, ammonium nitrate (fertilizer) /fuel oil, charges. Both HMTD and TATP can be initiated by contact with a low-voltage electric light bulb – typically a flashlight or an indicator bulb.

The reactions of the polymeric cyclic phosphonitrilic halides with nitrogen-containing species are very interesting primer materials. The cyclic trimer $(PNCl_2)_3$ reacts with sodium azide in acetone, to produce the low-melting trimeric phosphonitrilic azide [13] **TPNA**, $(PN(N_3)_2)_3$ {**19**}. This material, like NG, is very impact- and friction-sensitive, even in the liquid state, readily detonating. It is far too sensitive for uses other than chemical research. Similar results are obtained with the tetrameric $(PN(N_3)_2)_4$. However, reactions of the readily available tetrameric, phosphonitrilicchloride $(PNCl_2)_4$, with ammonia and sodium azide under the correct experimental conditions, yields **DAHA** {**20**}, or 1,1-diamino-3, 3-5,5,7,7-hexaazidocyclotetraphospazene. Note that the inclusion of the amine groups provides increased stability to make the material stable for handling as a primer. DAHA is readily initiated by an electric bridgewire, has a rapid deflagration to detonation (DDT) transition, and its output is greater than LA. Limited quantities can easily initiate detonation in RDX or CL20 without a booster intermediate explosive. DAHA is also reported to be laser sensitive, but the level required appears to be sensitive to the wavelength of the laser radiation, indicating that a particular chromophore in the molecule is responsible for the initiation. Further work on this aspect is required.

DAATO 3.5 {**21**} is the name given to a mixture of two nitrogen-oxide species from 3,3-azo-bis(6-amino-1,2,4,5- tetrazine). Some of the molecules have three nitrogen-oxide groups and others have four nitrogen-oxide groups, as shown in the structure detailed in the figure at the end of this chapter. The dashed donor bond in the structure shows where the fourth nitrogen-oxide is attached [14]. This material is extremely sensitive to illumination and can be initiated by a camera flash. It is also sensitive to light from a CO_2 laser. Attempts to record Raman spectra of the compound to elucidate the structure proved difficult, due to initiation before the spectrum was complete. Further work on this compound is required before viable systems can be produced. It has potential for IED systems since it does not require handling but can be used *in situ*.

Another potential primer using nitrogen-oxides is **DNBTDO**{**22**}, 5,7 dinitrobenzo-, 1,2,3,4,triazine 1,3 dioxide. This material is synthesized by a multi-stage reaction from 2-nitroaniline [15]. It is unusual because 1234 triazines are normally unstable, but the presence of the aromatic ring and formation of the nitrogen-oxides increases the materials stability. Its impact sensitivity ranges from 6–4 J, depending on the particle size of the material. Its friction sensitivity is independent of p.s. and, at 360 N, it is not particularly friction-sensitive. Electrostatic discharge sensitivity of 0.15 J means that it must be handled with care. The materials laser sensitivity has not been measured, but the Raman spectrum has been successfully measured using a laser Raman spectrometer. This indicates that it may not be too sensitive to laser initiation, since some of the other compounds of interest cannot survive irradiation in the laser Raman spectrometer. The material shows a slightly higher density than RDX, but its measured VoD is slightly lower at 8400 m s^{-1}, compared to the 8750 m s^{-1} value for RDX.

8.4 Simple Salts of Organics

An unusual salt in this category is **DPDP** {**23**}, 2,4-Dinitrophenyldiazonium perchlorate, which contains an organic energetic cation. DPDP has been used as an environmentally friendly primary explosive for application in commercial detonators [16, 17]. DPDP is a more effective primary explosive than LA, with a minimum charge of 0.007 g to initiate tetryl in a Number 8 detonator. The temperature of ignition is ≈491 K (218 °C). Unfortunately, compound {**5**} is a hygroscopic substance which loses its initiation ability in Number 8 blasting caps after being exposed to a humid atmosphere. Hydrolysis liberates the toxic perchlorate anion. Therefore, DPDP does not meet the requirements for 'green' energetic materials, but has applications in specialist situations.

Other perchlorate {**24**} and nitrate {**25**} salts of 5-aminotetrazoles are primary explosives, but with high sensitivities, with the nitrates showing lower sensitivity than the perchlorates. These compounds are simple salts formed by treating 5 aminoterazole with the corresponding acids when the acid proton goes on the tetrazole ring rather then the external amine base. Substitution of the hydrogen in the tetrazole ring with a methyl group lowers their sensitivity. An interesting mixed salt is formed from aminotetrazoles and silver [18]. This is essentially a mixture of the metal tetrazolate and the metal nitrate or perchlorate. Again, the perchlorates are the more sensitive, and the impact and friction sensitivities are less than for standard LA. Interestingly, the position of methyl substituent has a dramatic effect on the sensitivity, the 1-methyl aminotetrazole mixed nitrate salt being less sensitive than the 2-methyl substituted compound, which is as sensitive as the unsubstituted salt. The x-ray structure shows that the silver is coordinated to the tetrazole species. Formation of salts with other transition metals will be discussed later, since they involve the coordination of the aminotetrazole around the metal cation.

The other potential compounds in this class can be divided into two groups depending on the cation. They all contain an anion derived from an energetic organic molecule but the cation can be either derived from an **organic** molecule or an **inorganic** metal cation. The salts containing organic derived cations and anions can be simply considered as a transfer of protons between two species. There may be no distinct coordination spheres around the ions and there may be considerable covalent contribution to the molecular bonding. Electron density distribution would be an interesting insight into the bonding areas. The most interesting ones are the totally organic since the metal pollution problem is reduced but the materials themselves could be toxic. A number of these have not had their toxic properties completely assessed. There is a bridge between the two types of salts when the cation is the ammonium ion. This is an inorganic cation but contains no metal.

Since the first section involved organic molecules, the discussion will continue with salts containing organic cations. **DNBTM** {**6**}, 5,5′-dinitrimino-33′methylene-1H-124 triazole, contains two potential replaceable hydrogen ions and can therefore form two salts – those with a single charge on the anion and those with a double charge on the anion. The level of charge produced depends on the strength of the organic base to which the proton is transferred. With guanidine, the doubly charged anion is produced {**26**}, whereas with diamino guanidine, only the singly charged anion is produced {**27**}. With the formation of both salts, the sensitivity to friction is decreased by a factor of five and impact sensitivity is decreased by a factor of 40, compared to the original organic molecule [19]. The salts are less sensitive to static as well. Velocities of detonation are also reduced from 8500 m s^{-1} for DNBTM to 7500 m s^{-1} for the bisguanidinium salt. Both of these salts are under further investigation.

Bridging the gap between organic and inorganic salts is the promising inorganic salt ammonium dinitroguanidine, **ADNQ** {**28**}, which has a crystal density of 1.735 g s^{-1}, with a measured VoD of 9060 m s^{-1} at this density [20]. Its decomposition temperature is 473 K (197 °C). It is moderately sensitive to impact (10 J), friction (250 N) and electrostatic discharge (0.4 J), all of which are less sensitive than LA and LS. There is an indication that these values may be dependent on preparation conditions, and this may be attributable to the hygroscopicity of ammonium salts. For this reason, ADNQ is probably better suited to the booster explosive role rather than the primer role.

The simplest salts in the inorganic cation category are potassium and barium salts of perchlorylamides, K_2NClO_3 {**29**}, and $BaNClO_3$ {**30**}. Both of these salts can be precipitated from the acid solutions produced by the reaction of perchloryl fluoride with aqueous ammonia. The solids produced are light-sensitive primary explosives possessing high thermal stability [21,22]. Although they are both non-hygroscopic, the potassium salt is soluble in water, whereas the barium salt is insoluble in water and most organic solvents. The toxic properties of these perchlorylamides are still unknown, but their sensitivities to accidental initiation is extremely high. Typical sensitivities to laser beam irradiation (λ = 1060 nm, 25 ns pulses of 40 mJ intensity) indicate a threshold for the potassium salt of 0.1 J cm^{-2}, and the barium salt of 5–7 mJ cm^{-2}. Both of these salts require further study before viable systems may be possible.

In the purely organic section, mention was made of substituted tetrazoles, with the 5 nitrotetrazole HNT and 5 azotetrazole HAT substituents identified as primary explosives. The acidic hydrogen on the 5-nitrotetrazole (HNT) is easily replaced by a metal cation, giving salts of the tetrazolate anion with better energetic properties and higher thermal stability. The simplest salts have the alkali metal counter-ions. Indeed, sodium nitrotetrazolate is often used as a starting material for the compounds of interest.

Lithium and sodium nitrotetrazolates are not good primary explosives, since their initiation sensitivity is poor, but higher atomic mass alkali metals salts demonstrate increased sensitivity and have been investigated as low-toxicity, environmentally friendly primary explosives. Potassium {**31**}, rubidium {**32**} and caesium {**33**} salts of 5-nitrotetrazole are potentially 'green' primary explosives that have been prepared and thoroughly studied [23]. These salts show primary explosives properties and have sensitivity both to impact and friction similar to a technical grade LA (impact = 3.0–6.5 J and friction = 0.1–1.0 N). The initiation ability of these salts has been demonstrated in blasting caps and they can be qualified as 'green' primary explosives. Unfortunately, none of these salts has decomposition temperatures above 368 K ($\approx$195 °C), and they all melt below 333 K (160 °C).

One of the earliest inorganic nitrotetrazolates to have undergone trial as a primary explosive was the mercury salt $Hg(NT)_2$ {**34**}. Studies [24] have shown that mercury nitrotetrazolate, is a primary explosive which is capable of rapid DDT. The crystal density is 3.32 g cm^3 and its VoD is 6600 m s^{-1} at a pressed density of 3.15 g cm^{-3}. Impact sensitivity is $\approx$2 N and it is sensitive to an electric bridgewire initiation. It is therefore capable of replacing LA and PETN in detonators, thus reducing the number of components in the explosive train. Earlier interest also examined the mercurous salt HgNT {**35**}, which was part of a patent [25]. It is sensitive to impact friction and electric discharge. Its decomposition temperature is 423 K (250 °C), but it is reported as quite corrosive towards metals and is very toxic. Use of these two salts has declined, since most countries operate a ban on toxic mercury compounds.

Copper (II) 5-nitrotetrazole, $Cu(NT)_2$ {**36**}, is another primary explosive with high initiation ability, approximately the same as LA. It has a crystal density of ≈2.11 g cm^{-3}, and the onset temperature for rapid decomposition is about 553 K (280 °C) [26, 27]. The material possesses high sensitivity to the usual initiatory stimuli (impact <1 J, friction <5 N and electrostatic discharge, very sensitive). It is often isolated as the addition salt with an additional hydrogen tetrazolate molecule {**37**} and with two water molecules completing the coordination sphere around the copper ion. This reduces its impact sensitivity to <3 J, but it appears to have little effect on the friction and electrostatic discharge sensitivities. Even so, it is still too sensitive for normal use and is really an interesting coordination complex. The silver salt of 5 nitrotetrazolate {**38**}, is marginally less sensitive than the copper salt and therefore is easier to handle. It is still too sensitive for large-scale usage, is sensitive to laser ignition and also degrades on exposure to light. Both the copper and silver salts can be desensitized to friction by crystallization in the presence of ethylene diamine as a coordinating ligand {**39**}, {**40**}, but the impact sensitivity of the copper salt is still comparable to LA. In this chelated form, it could be more acceptable for bulk handling.

The copper (I) salt of 5-nitrotetrazole $Cu_2(NT)_2$, **DBX-1** {**41**} has a crystal density of ≈2.59 g cm^{-3}, detonation velocity ≈7000 m s^{-1} and sensitivity to spark 3.1 mJ. The heat of explosion of DBX-1 is comparatively low at 3816.6 J g^{-1}, but its sensitivity to impact, friction and spark are similar to those of LA and LS. The onset temperature for rapid decomposition of DBX-1 is about 600 K (≈333 °C), with ignition temperature of 625 K (≈350 °C). The weight loss of DBX-1 after 24 hours of exposure at 181 °C was 0% (for LA, the weight loss was 14.57% under the same conditions). DBX-1, like LA, can not be dead pressed. The minimum charge of DBX-1 to initiate RDX in a standard detonator is 0.025 g. DBX-1 shows good compatibility with RDX, HMX, CL-20, HNS and some other HEs, as well as with construction materials. Several trials have demonstrated its application in detonators.

The salts of 5-azido tetrazole are also very sensitive to accidental initiation, with high sensitivities to impact friction and electrostatic discharge. The alkali metal salts are far too sensitive to use, but the ammonium salt {**42**} and the silver salt {**43**}, show some interesting properties. Those with organic cations such as guanidinium are more amenable to handling, but the guanidinium salt is not a primary explosive. A method of desensitizing these salts is to form the nitrogen oxides on the tetrazole ring. Again, the alkali metal salts are very sensitive, but the ammonium salt, **AATO** {**44**}, is less sensitive, as is the silver salt, **AgATO** {**45**}. Unfortunately, the thermal stability of the azidotetrazolates and the even more thermally unstable nitrogen oxide derivatives do not meet the thermal stability criteria and, thus, can not be used in green primer compositions.

The copper and silver salts of bis tetrazolyl amine CuBTA {**46**} and Ag_2BTA {**47**} have been trialled as primary explosives. Note the requirement for two silver cations, as opposed to the single copper cation. The performance of these materials is a little unpredictable, and they should be treated with extreme care – particularly the silver bistetrazolamine. There is insufficient data in the literature to fully characterize their properties as potential primer compositions.

The copper compound Bis (1-methyl-5-nitraminotetrazolate) copper(II) {**48**} was suggested as a low-toxicity explosive for initiation devices [28,29], and its sensitivity to impact and friction is in the range for primary explosives. The compound exhibits good thermal characteristics, with an onset temperature for extensive decomposition of the salt of about 525 K (252 °C), and its explosive properties are unchanged after 48-hour exposure at 460 K (190 °C). Because it meets the requirements of Los Alamos National Laboratory's criteria for 'green' energetic materials, it could be used in 'green' primary devices, and further work is in progress.

Copper (II) salts of other substituted tetrazolates are primary explosives. Copper 5-chlorotetrazolate {**49**} and 5- bromotetrazolate {**50**} are both very sensitive to laser initiation, with milligram quantities of both compounds exploding instantly when exposed to 200 mW red laser (658 nm) irradiation [30]. These salts also exhibit high thermal stability, but unfortunately they exhibit extreme sensitivity to mechanical stimuli such as impact and friction. They are also extremely spark-sensitive. Further investigation may result in practical laser initiation devices but, because of their extreme sensitivity, their use is restricted to well-protected devices able to resist mechanical and electrical stimuli. The results of performance testing demonstrated an acceptable dent depth. Copper is a bio-metal which has an optimum concentration in bio-organisms and thus is a viable alternative to lead and mercury.

Dinitrobenzfuroxane, **DBNF**, is an explosive produced by the oxidation of orthonitroaniline with sodium hypochlorite, followed by nitric acid nitration. DNBF is not particularly sensitive, but the hydrogen on the benzene ring between the two nitro groups is very acidic and reacts with alkali metal carbonates to give salts. The sodium salt is almost a primary explosive, but the important potassium salt **KDNBF** {**51**} is a primer. A related primer compound is the potassium salt of 4,6-dinitro-7-hydroxybenzofuroxan, **KDNP** {**52**}. One of the synthetic routes to KDNP starts with KDNBF. Both materials have similar crystal structures and similar densities ($\approx$2.21 g cm^{-3}) and have been trialled [31, 32] as environmentally friendly, low-toxicity, energetic compounds. The sensitivity of KDNBF to impact and friction stimuli is the similar to that of LS. The salt has a decomposition temperature of 463 K ($>$190 °C) and the ignition temperature is 483 K (210 °C). KDNBF has an initiating ability less than that of mercury fulminate. It has been used in low-toxicity mixtures with oxidizer KNO_3 and sensitivity additives for primer production in the USA. The onset temperature for extensive decomposition of this compound mixture is about 543 K (270 °C) [33]. KDNP is a fast-deflagrating material with good thermal stability and safe handling characteristics. KDNP was approved as a safe compound suitable for service use, and was recommended for the US weapons improvement program in February 2009.

8.5 Transition Metal Complexes and Salts

This group of compounds covers a wide range of materials and allows a flexible chemical design of energetic complex salts with a wide variation in their physico-chemical properties and differing explosive properties/performance. Many energetic complex salts of d-transition metals have been synthesized and investigated as potentially safe and environmentally friendly replacements for the traditional primary explosives LA and LS. A group of these have the general formula $M_a(L)_b(An)_c$, where M is a d-metal cation,

L is a coordinating ligand, An is an acidic anion, and a, b, c are molecular stoichiometric coefficients.

Complex perchlorates of cobalt (III) ammines with tetrazole derivatives as ligands do not contain toxic heavy metals in their structure [1]. These complexes are safer than classic primary explosives such as LA or LS. They are not hydroscopic, have sufficiently high thermal stability, and their initiation ability is sufficient for application of these complexes as primary explosives in safe blasting caps. The well-characterized cobalt (III) complex perchlorates, such as pentaammine (5-cyanotetrazolato-N2) cobalt (III) perchlorate (**CP**) {**53**}, pentaammine (5- nitrotetrazolato-N2) cobalt (III) perchlorate (**NCP** {**54**} and tetraammine-bis-(5-nitrotetrazolato-N2) cobalt (III) perchlorate (**BNCP**) {**55**}, have been used in safe commercial detonators [34–36].

Complex CP was suggested as a lead-free explosive for safe low-voltage electric blasting caps. However, toxic properties of cobalt complex CP, arising from potential liberation and hydrolysis of the cyanotetrazolate ion and the presence of perchlorate, forced the USA to cease the commercial manufacturing of this compound [37]. Complex NCP was tested as an intermediate/booster energetic component in explosive trains, and still functioned after several hours at high temperature and pressure (423 K (150 °C) and $\approx$80 MPa). The data in the table in Appendix 8.A demonstrates that complex BNCP has the highest initiating ability of the three. The complex has a time of deflagration-to-detonation transition (DDT) of $\approx$10 μsec. BNCP is more sensitive to mechanical stimuli than the other two complexes, but it is less impact sensitive than PETN. BNCP has been used as a primary or secondary explosive charge in blasting caps [27], and also in pyrotechnics.

Pentammine [3-nitrofurazan-4-(5′-tetrazolato-N2′)]cobalt (III) perchlorate {**56**} has improved performance compared to BNCP and, as such, it can be used as a primary explosive in its own right. The minimum charge required for this compound to initiate secondary explosives is higher (0.2 g) than that of BNCP (=0.05 g). The presence of toxic perchlorate anions in these complexes rules against their green credentials. Replacing the toxic perchlorate with non-toxic anions reduces the toxic properties of metal complexes, indicating that the cobalt (III) ammine coordination complex cations are not inherently toxic. The replacement reactions are simple addition of the sodium salt of the anions, such as tetrazole, dinitroguanidine, DNG {**57**} and azide [38, 39] {**58**} to a solution of the relevant cobalt (III) ammine perchlorate complex when the desired product crystallizes out. The N_3^- species replaces one toxic anion with another toxic material. DNG (**17**) has an enthalpy of formation of $\approx$0 kJ/mol, and oxygen balance (Ω) is $+5\%$ for the neutral molecule and $+11\%$ for the anion [40, 41], making the anion a useful oxidizer. The studies of these complexes show that substitution of perchlorate anions with DNG and azide anions decreases the complexes decomposition temperatures by 80–100 K. The initiating ability of these complexes is lower than that of the perchlorate salts. They are insensitive to impact, according to the drop-hammer test, so they may be regarded as more safe energetic compounds than PETN. However, they are not effective primary explosives. Metallic cobalt, the product of the explosion, does have some toxic effects [5].

Complex tris-amminecopper (II) 5-nitrotetrazolate {**59**} is a laser-ignitable primary explosive [42] and, indeed, is ignited by the low-power laser beam during Raman spectroscopy measurements before the spectral accumulation is complete. This is in an analogous manner to a number of other coordination complexes. Special care must be taken while handling this compound, since it is an extremely sensitive material. This complex is essentially insoluble in most polar solvents, and it dissolves to a low extent only in coordinating solvents such as concentrated ammonia or pyridine. Possible practical applications of complex salt {**59**} in impact or laser primers will require more investigation. Complexes of copper nitrate with ligands 5,5′-Bis-(1H-tetrazolyl)amine {**60**} or 5,5′-Bis-(2-methyl-tetrazolyl)amine {**61**} have low sensitivity to friction [43]. Complex {**60**} is more sensitive to impact than complex {**61**}. Evidently, the presence of the methyl groups in the amine ligand molecule phlegmatises the complex towards external stimuli. Both salts were initiated by the low power laser beam of the Raman spectrometer, preventing the collection of their Raman spectrum.

Lead-free perchlorate copper complex of bis-3(5)-hydrazino-4-amino-1,2,4-triazole as a ligand (**62**) has a low threshold of initiation ($\sim 1.1 \cdot 10^{-5}$ J) by Q-switch Nd:YAG-laser beam ($\lambda = 1064$ nm, $\tau = 30$ ns, d = 0.48 mm) [44,45]. This salt is a hazardous explosive because it has high sensitivity to mechanical stimuli with some preparations being more sensitive than lead azide or lead styphnate. Complex 62 has a short deflagration-to-detonation transition (DDT) distance. A minimum weight of the copper complex as an initiating charge for the pressed charge of RDX in a blasting cap No 8 equals to ~0.025–0.030 g. The perchlorate complex requires phlegmatising by mixing with inert material such as a transparent polymer. This complex is a prospective lead-free, light-sensitive, primary explosive, but toxic perchlorate anion limits the possibility of its application in 'green' primers.

Metal complexes of HNT, where the HNT is incorporated as the NT^- ion in the coordination shell of the transition metal ions, Fe and Cu, have been studied as primary explosives [5]. There is a whole series of these complex salts with the general formula of $Cat^+{}_{1\text{–}4}[M^{II}(NT)_{3\text{–}6}(H_2O)_{3\text{–}0}]$. Water and the NT complete the coordination sphere around the transition metal ion. As the number of nitrotetrazolate ligands increases, the performance increases, but so does the sensitivity to mechanical and electrical stimuli. The main cations investigated were sodium, potassium, ammonium, hydrazinium, triaminotriazolium and amino-nitramino-tetrazolium.

The disodium salts of the iron and copper complexes, with four NT- ligands in the coordination sphere {**63**} and {**64**}, exhibit mechanical sensitivities between those of LA and LS, but they are much less sensitive to electric spark than either LA or LS. Both of these complexes undergo DDT processes when used as a complete replacement for LA in M55 stab detonators. Plate dent tests show their performance, which exceeds that of LA, increases with loading pressure up to a maximum VoD of ≈7500 m s^{-1}. They show some dead pressing properties at extreme pressures [25] but, since their decomposition temperatures are >530 K (250 °C), they would appear to meet all the requirements of LANL's criteria for 'green' energetic materials. The ammonium complexes showed the usual tendency to hygroscopicity but, as the number of NT ligands increases, the sensitivity of the ammonium salts increases in the series, as shown in the table (Appendix 8.A). However, the ammonium salts are less sensitive than the corresponding sodium salts, and this is why only the disodium salts have been investigated. The tetrammonium salt {**65**}, is sensitive to both impact and friction.

Other cations trialled were hydrazinium, triaminotriazolium and aminonitrosoterazolium, but these do not exhibit primary explosive properties.

Complexes involving the 3(5) hydrazino4-amino-1,2,4 triazole, HATr, and an oxidizing anion have been identified as light-sensitive primary explosives [46]. The complexes are easily prepared by the reaction of a metal perchlorate or nitrate solution in propanol, with the ligand in propanol solution at room temperature. The complex precipitates out over the course of a few hours. The metals investigated were Cu {**66**}, Cd, {**67**}, Ni, {**68**} and Co {**69**}. The sensitivity to laser initiation decreased in the order given. The copper complex is the most sensitive, requiring only 10^{-5} J of energy in the laser pulse to initiate ($\lambda = 1064$, $\tau = 45$ ns, d = 0.48 mm). All of the complexes that were trialled exhibited short DDT times and distances, with the minimum charge quantity of 30 mg of copper complex to initiate pressed RDX in a standard Number 8 detonator. Unfortunately, these complexes are very sensitive to accidental initiation by mechanical stimuli. The copper perchlorate complex is more sensitive than LA or LS and, therefore, would require complete isolation from sources of mechanical stimuli. The complexes can be desensitized to mechanical effects by treating with ≈5% of a polymer which is transparent to the laser beam.

Silver aminotetrazole perchlorate {**70**} and nitrate {**71**} are potential primary explosives. The perchlorate is very sensitive and the nitrate less sensitive. Perchlorate impact sensitivity is 2 J and friction sensitivity is <5 N, while for the nitrate, the corresponding values are 15 J and 18 N. Both salts/complexes are spark-sensitive. Substitution of a methyl group on the tretrazole ring {**72**} reduces the friction sensitivity by a factor of 10.

A number of complex hydrazinates with transition metal ions, particularly in the II + oxidation state and, with anion oxidizers, may have the properties of primary explosives. These complexes require a greater minimum charge when compared to LA, and their thermal stability is usually less than that of LA. In the first investigation, the metals chosen were magnesium, manganese, iron, cobalt, nickel, zinc and cadmium [47]. The number of hydrazine molecules coordinated into the metal centre depended on the nature of the anion. With non-coordinating perchlorate and weakly coordinating nitrate, three hydrazine molecules occupied the octahedral positions round the metal, whereas, with coordinating ligands such as azide, only two hydrazine molecules were coordinated around the metal. The exceptions were cadmium, which could coordinate two hydrazines with the nitrate, and the magnesium cation, which only coordinated two hydrazines with all three anions, perchlorate, nitrate and azide.

Complex tris-hydrazinenickel (II) nitrate ($Ni(N_2H_4)_3(NO_3)_2$, {**73**} was able to replace LA in commercial blasting caps [48–51]. Pressed charges of complex {**73**} have a maximum pressed density of ≈1.70 g cm^{-3} at 60 MPa but, above that, can be dead pressed. Maximum VoD is ≈7200 m s^{-1} The temperature of ignition of the complex is quoted in the earlier literature as ≈440 K (167 °C), but other experiments [52] give T_{ign} as 488 K (214 °C) with DDT in enclosed samples. The activation energy of decomposition of nickel complex nitrate is ≈80 kJ mol^{-1}, which is lower than the usual values for primary explosives. Performance is 105% that of TNT [53].

This complex is non-hygroscopic and insoluble in water, alcohol and dry acetone. Concentrated sulfuric acid ignites the nickel complex, but dilute acids dissolve/destroy the complex. A 10% aqueous solution of NaOH fully decomposes the hydrazinate, with evolution of hydrazine when hot. The complex is unaffected by sunlight or X-ray irradiations, and shows long-term compatibility with Al, Cu, Fe and stainless steel at ambient temperature. This compound has an impact sensitivity equal to that of lead azide, but the sensitivity to friction and spark is the same as that of the more sensitive AgN_3 {**31**}. It can also be ignited by the beam of a pulsed CO_2 laser with the energy density equal to ≈12.0 J/cm^2. Even a laser diode used in Raman spectroscopy ignited the complex before the Raman spectrum was complete.

A number of detonators and electric primers containing >200 mg charges of the complex have been tested with reliable results in China. All of the other tris hydrazine complexes showed very similar sensitivities to the nickel complex for all three stimuli. Attempts to use the perchlorate derivative, which has a better oxygen balance than the nitrate, have proved very difficult due to spontaneous decomposition at moderate temperatures, which often results in a rapid DDT reaction. Air drying the precipitated solid at 50 °C destroyed the drying equipment. The only perchlorate isolated was the magnesium, but the composition of the product is in doubt.

All of the azide complexes isolated showed high impact sensitivity. Bis-hydrazinenickel (II) azide $[Ni(N_2H_4)_2](N_3)_2$ {**74**} has also been investigated as a replacement for LA in primers {**32**} [40]. Note that the two azide ions occupy the two positions in the nickel coordination sphere vacated by a hydrazine ligand. The temperature of ignition of the complex is ≈470 K (193 °C). The thermal decomposition curve comprises two stages. The activation energy of the first stage of decomposition is ≈142.6 kJ mol^{-1}, while that of the second stage amounts to ≈109.2 kJ mol^{-1}. The non-hygroscopic complex is insoluble in water, alcohol and ether, and is readily decomposed by both acids and bases to yield toxic azide ions. The complex has an impact sensitivity lower than PETN. Hence, the azide complex is a less sensitive and therefore safer compound than the nitrate complex.

A large number of electric primers containing 110–130 mg of complex NHA were prepared and tested in China. Experiments proved the reliability of these devices and their possible application in mining and metallurgical industry. The complex does not meet requirement 3 of Los Alamos National Laboratory's criteria for 'green' energetic materials, because of the production of free azide ions under environmental conditions.

An alternative nickel complex, based on the carbohydrazide ligand, has been investigated. This has an advantage of lower sensitivity, than the orginial hydrazine complex, but it also possesses lower performance than either of the hydrazine complexes. Hydrazine is an absorber for carbon dioxide, which can then react to form the carboxyhydrazide. Precipitating the hydrazine complex in the presence of carbon dioxide produces the hydraziniumnickel tris carboxyhydrazide hydrate {**75**}, with the oxidizing anion lost from the complex. One suggestion is to produce the normal purple powdered nickel trishydrazine nitate and coat with a sample of the carboxyhydrazide to desensitize the mixture. The carboxyhydrazides have uses as both propellant and detonating materials.

Nickel, like copper, also belongs to the group of biometals which are essential for the native activity of some enzymes. There is a warning that high levels of Ni can exhibit carcinogenic activity. The concentration of nickel in environments is rather low, but it persists there for a long period. So, after a prolonged period of commercial use of complexes, their hazardous effects upon living organisms may last for a considerable period of time [54].

8.6 Enhancement of Laser Sensitivity

Pulsed lasers provide a practical means of electric isolation of energetic materials, thereby eliminating hazards associated with ESD, EMI and RI [55–57] and making them safer for civil industrial uses. All of the energetic materials can be initiated by laser irradiation, but some of the materials, typified by RDX, HMX and PETN, are such poor absorbers of laser radiation that the power density has to be very high. This means that the initiation system becomes cumbersome. One method of overcoming this problem is by the addition of good absorbing materials, such as activated carbon black, very small particles of carbon, and carbon nanotubes. The latter can be ignited with a camera flash [58]. An alternative solution is to add nanoparticles of metals – typically gold, silver and copper. The increased sensitivity is thought to arise from the SERS effect occurring on the highly charged surface layers of the particles, as utilized in Raman Spectroscopy [59]. Other solutions to enhancing the sensitivity problem is to form a suitable surface layer by mixing the energetic material with an absorbing polymer and placing this on the surface of the sample to be initiated.

The materials discussed here require lower incident laser power, some of which can be initiated by low-power laser diodes. Some of the materials discussed above can be initiated by direct laser irradiation of the low-density explosives through a transparent window. The initiation threshold can be further decreased when the side of the window in contact with the explosive charge is coated with a thin layer of metal (such as titanium), which can form a plasma capable of initiating low-density PETN directly. This is analogous to the electrical exploding bridgewire detonators. Using a finely divided low-density material lowers the threshold for initiation, but increases both the run-up times and distances for the DDT transition to occur. Typical examples of this are the data for dispersed PETN having the bulk density, which gives run-up times of ≈200 ns in comparison to the shock initiation run up of ≈20 ns. A laser EBW, with PETN at a density of 0.9 g cm^{-3}, has a run-up time for deflagration-to-detonation transition (DDT) of ≈200 ns, compared to the shock initiation run up time of 2 ns. Detonators filed with PETN at a bulk density of 1.0 g cm^{-3} are less effective [60]. In low-density laser detonators containing BNCP {**55**}, the DDT regime is realized as well. Because only finely dispersed and low-density explosives may be used in laser EBW detonators, meeting these requirements reduces the sphere of application of such detonators.

Although a number of lead-free prospective primary explosives have been prepared and successfully studied in primers during recent years, manufacturing of modern 'green' energetic materials on the industrial scale required for practical applications in civil and military primers and detonators is still some way off. Further testing and hazard evaluation for most of these materials is also required.

References

(1) Huynh, M.H.V., Hiskey, M.A., Meyer, T.J. and Wetzler, M. (2006) Green primaries: environmentally friendly energetic complexes. *Proceedings of the National Academy of Science of the USA*, **103**(14), 5409–5412.

(2) Danilov, Y.N., Ilyushin, M.A. and Tselinsky, I.V. (2004) *Industrial Explosives: Primary Explosives*, 2nd edn, St Petersberg State Technol. Inst. (Russian) [2].

(3) Thom, T., Bichay, M. and Woods, A. (2011) Demonstration of DBX-1 as a Green Lead Azide Replacement. Tentative Program 2011 Safe Symposium, October 24–26 Grand Sierra Resort and Casino Reno, Nevada.

(4) Holl, G., Klapoetke, T.M., Polborn, K. and Reinaeker, C. (2003) Structure and bonding in 2-Diazo-4,6-dinitrophenol (DDNP). *Propellants, Explosives, Pyrotechnics* **28**(3), 156–161.

(5) Ilyushin, M.A., Sudarikov, A.M. and Tselinsky, I.V. (2010) *Metal Complexes in Energetic Formulations* (ed. I.V. Tselinsky), A.S. Pushkin Leningrad State University, St. Petersburg (in Russian).

(6) Mehta, M., Damavarapu, R., Cheng, S. *et al.* (2009) D-3. Alternatives to lead azide and lead based materials. Provisional Book of Abstracts for 36th International Seminar and Symposium, 23–27 August 2009, Rotterdam, Netherlands, p. 22.

(7) Dippold, A.A., Feller, M. and Klapotke, T.M. (2011) DNBTM a metal free primary explosive combining excellent thermal stability and high performance. *Central European Journal of Energetic Materials*, **8**, 261.

(8) Ostrovskiy, V.A. and Koldobskiy, G.I. (1997) Energetic tetrazoles. *Russian Chemical Journal*, **41**(2), 84–98 (in Russian).
(9) Kaplötke, T.M., Minar, N.K. and Stierstorfer, J. (2009) Investigations of bis methyltetrazolyltriazenes as Nitrogen Rich Rocket Propellants. *Polyhedron*, **28**, 13–26.
(10) APX reference
(11) Lefebvre, M.H., Falmagne, B. and Smedys, B. (2004) Sensitivities and Performances of Non-Regular Explosives. Proceedings of the VII-th Seminar 'New Trends in Research of Energetic Materials', Pardubice, Czech Republic, April 20–22, 2004, Part 1, pp. 165–174.
(12) Matyas, R. (2003) Chemical Decomposition of Hexamethylenetriperoxidediamine and Triacetone Triperoxide. Proceedings of the VI-th Seminar 'New Trends in Research of Energetic Materials', Pardubice, Czech Republic, April 22–24, 2003, pp. 241–247.
(13) Gudman, C. and Ratz, R. (1955) *Zeitschrift für Naturforschung*, **10b**, 116–117.
(14) Ali, A.N., Sanderstrom, M.M., Oschwald, D.M. *et al.* (2005) *Propellants, Explosives, Pyrotechnics*, **30**, 351.
(15) Klapötke, T.M., Piercey, D.G., Stierstorfer, J. and Weyrauther, M. (2012) The Synthesis and Energetic Properties of 5,7-Dinitrobenzo-1,2,3,4-tetrazine-1,3-dioxide (DNBTDO), *Propellants, Explosives, Pyrotechnics*, **37**, 527–535.
(16) Philippov, Y.V., Falyakhov, I.F., Gilmanov, R.Z. *et al.* (2006) *Influence of the anion nature on the properties of aromatic diazocompounds.* Proceedings of International Scientific and Technical Conference 'Modern Problems of Technical Chemistry', Kazan, December 6–8, 2006, pp. 173–175 (in Russian). DPDP.
(17) Falyakhov, I.F., Gilmanov, R.Z., Nesterov, A.V. *et al.* (2004) Development of Manufacturing Technologies for the Preparation of Environmentally Friendly Primary Explosives in the Series of Aromatic Diazocompounds. Proceedings of International Scientific and Technical Conference 'Modern Problems of Technical Chemistry', Kazan. December 22–24, 2004, pp. 359–362 (in Russian).
(18) Delalu, H., Karaghiosoff, K., Klapötke, T.M., and Miró Sabaté, C. (2010) 5 Amino tetrazoles and Silver based primary explosives. *Central European Journal of Energetic Materials*, **7**, 197.
(19) Dippold, A.A., Feller, M. and Klapoetke, T.M. (2011) 5,5′,Dintrimino3,3′-methylene 124 bistriazole a metal free primary salt combining excellent thermal stability and high performance. *Central European Journal of Energetic Materials*, **8**, 261.
(20) Altenburg, T., Kaploetke, T.M., Penger, A. and Stierstorfer, J. (2010) Two outstanbding explosives based on 1,2, dinitroguaindine. *Zeitschrift für Anorganische und Allgemeine Chemie*, **636**, 463.
(21) Ilyushin, M.A., Tselinsky, I.V., Petrova, N.A. *et al.* (1999) Salts of perchlorylamide as a novel class of light-sensitive explosive for laser initiation. *Energetic Materials*, **7**(3), 122–123.
(22) Rosolovskiy, V.Y. and Kolesnikov, I.V. (1968) About some properties of metal salts of perchlorylamide. *Russian Journal of Inorganic Chemistry*, **13**(1), 180–184.
(23) Klapoetke, T.M., Miro Sabate, C. and Welth, J.M. (2008) Alkali metal 5-nitrotetrazolate salts: prospective replacements for service lead (II) azide in explosive initiators. *Journal of the Chemical Society, Dalton Transactions*, **10**, 1039–1057.
(24) Von Hertz, E. (1937) US Patent 2066954.
(25) Bates, L.R. and Jenkins, J.M. (1978) Production of 5 nitrotetrazole salts. US Patent 4094879.
(26) Bichay, M., Stern, A., Armstrong, K. *et al.* (2004) Strategic Environmental Research and Development Program. SEPDP Project No PP-1364. 'New Primary Explosives Development for Medium Caliber Stab Detonators. *Final Report,* September 2004', 31 p.
(27) Bichay, M., Hirlinger, J. (2005) Lead Azide Replacement Program NDIA. Fuze Conference, April 2005, Presentation of NAVSEA, 22 p.
(28) Klapoetke, T.M. (2011) *Chemistry of High-Energy Materials*, Walter de Grauyter & Co, KG, Berlin-New York.
(29) Geisberger, G., Klapoetke, T.M. and Stierstorfer, J. (2007) Copper Bis(1-methyl-5-nitriaminotetrazolate): a promising new primary explosive. *European Journal of Inorganic Chemistry*, 4743–4750.
(30) Fisgher, D., Klapoetke, T.M., Piercey, D.J. and Stierstorfer, J. (2012) Copper salts of halo tetrazoles: light -ignitable primary explosives. *Journal of Energetic Materials*, **30**, 40–54.

(31) Mehilal, Sikder, A.K., Pawar, S. and Sikder, N. (2002) Synthesis, characterization, thermal and explosive properties of 4,6-dinitrobenzofuroxan salts. *Journal of Hazardous Materials*, **A90**, 221–227.

(32) Lur'e, B.A., Sinditskii, V.P. and Smirnov, S.P. (2003) Thermal decomposition of 2,4-dinitrofuroxan and some of its compounds with metal hydroxides. *Combustion, Explosion and Shock Waves*, **39**(5), 534–543.

(33) Fronabarger, J., Williams, M. and Bichay, M. (2007) *Safe Journal*, **35**(1), 14.

(34) Cudzilo, S. and Nita, M. (2010) Primary explosives from group of coordination compounds. *Wiadomosci Chemiczne*, **64**(3–4), 198–223 (in Polish).

(35) Hafenrichter, T.J., Tarbell, W.W., Fronabarger, J.M. and Sanborn, W.B. (2004) Exploding Bridgewire Initiation Characteristics of Several High Explosives. Proc. Intern. Pyrotechnics Seminar, 31-th, 2004, pp. 707–719.

(36) Ilyushin, M.A. and Tselinsky, I.V. (1997) Primary explosives: status and prospects. *Russian Chemical Journal*, **41**(4), 3–13 (in Russian), 27–29.

(37) Luebeke, P.E., Dickson, P.M. and Field, J.E. (1995) An experimental study of the deflagration-to-detonation transition in granular secondary explosives. *Proceeding Royal Society, London*, **A448**, 439–448.

(38) Smirnov, A.V., Ilyushin, M.A. and Tselinsky, I.V. (2004) Complex Ammine Cobalt (III) Dinitroguanidinates as Energetic Materials. VIIth Scientific Vishnyakov's Talks, Boksitigorsk Leningrad region, April 3, 2004, pp. 112–119 (in Russian).

(39) Smirnov, A.V., Ilyushin, M.A., Tselinsky, I.V. and Sudarikov, A.M. (2006) Investigation of Tetrazolate Ammine Cobalt (III) Thermal Destruction. IXth Scientific Vishnyakov's Talks, Boksitigorsk Leningrad region – St. Petersburg, March 24, 2006, 1, pp. 196–201 (in Russian).

(40) Astrat'yev, A.A., Dashko, D.V. and Kuznetsov, L.L. (2003) Synthesis and some properties of 1,2-dinitroguanidine. *Russian Journal of Organic Chemistry*, **39**(4), 501–512.

(41) Pepekin, V.I., Kostikova, L.M. and Afanas'ev, G.T. (2004) Explosive Properties of 1,2- Dinitroguanidine. 35-th International Annual Conference of ICT, 2004, p. 95, pp. 1–6.

(42) Klapoetke, T.M. and Miro Sabate, C. (2010) Less sensitive transition metal salts of the 5-nitrotetrazolate anion. *Central European Journal of Energetic Materials*, **7**(2), 161–173.

(43) Klapoetke, T.M., Mayer, P., Polborn, K. *et al.* (2006) 5,5′-Bis-(1Htetrazolyl) amine (H2BTA) and 5,5′-Bis-(2-methyl-tetrazolyl)amine (Me2BTA): Promising Ligands in New Copper Based Priming Charges (PC). 37-th International Annual Conference of ICT, 2006, p. 134.

(44) Cudzilo, S., Szmigielski, R. (2000) Synthesis and Investigations of Some Di-(R-1,2,4triazolato) Copper(II) Perchlorates. *Bul. Wojsk. Acad. Technic.*, **49**, 5–17 (Polish).

(45) Chernay, A.V., Sobolev, V.V., Chernay, A.V., Ilyushin, M.A., Dluugashek, A. (2003) Laser Ignition of Explosive Compositions Based on Di-(3-hydrazino-4-amino-1,2,4-triazole)Copper(II) perchlorate. *Combust. Explos and Shock Waves*, **39**, 335–339.

(46) Illyushin, M.A., Tselinsky, I.V., Smirnov, A.V. and Shugalei, I.V. (2012) Physicochemical properties and laser initiation of a copper perchlorate complex with 3(5)-hydrazino-4-amino-1,2,4 triazole as a ligand. *Central European Journal of Energetic Materials*, **9**, 3.

(47) Patil, K.C., Nesamani, C. and Per Verneker, V.R. (1982) WSynthesis and charcterisation of metal hydrazine nitrate azide and perchlorate compexess. *Synthesis and Reactivity in Inorganic and Metal-Organic Chemistry*, **12**, 383.

(48) [36] Zhu, S., Wu, Y., Zhung, W. and Mu, J. (1997) Evaluation of a new primary explosive: nickel hydrazine nitrate (NHN) complex. *Propellants, Explosives, Pyrotechnics*, **22**, 317–320.

(49) Fronabarger, J., Williams, M. and Bichay, M. (2010) Environmentally Acceptable Alternatives to Existing Primary Explosives. paper presented to Joint Armaments Conference. Dallas, Texas, 20th May 2010.

(50) Chhabra, J.S., Talawar, M.B., Makashir, P.S. *et al.* (2003) Synthesis, characterization and thermal studies of (Ni/Co) metal salts of hydrazine: potential initiatory compounds. *Journal of Hazardous Materials*, **A99**, 225–239, 36–39.

(51) (a) Wojewodka, A. and Belzowski, J. (2011) Hydrazine complexes of transition metals as prospective explosives. *Chemk*, **65**(1), 24–27; (b) Talavar, M.B., Agraval, A.P., Anniyappan, K. *et al.* (2006) Primary explosives: electrostatic discharge initiation, additive effect and its relation to thermal and explosive characteristics. *Journal of Hazardous Materials*, **B137**, 1074–1078.

(52) Grant Wing M.Sc thesis EOE Course Cranfield University, Shrivenham, U.K. 2007.

(53) Wojewodka, A., Belzowski, J., Zenon, W. and Justyna, S. (2009) Energetic characteristics of transition metal complexes. *Journal of Hazardous Materials*, **171**, 1175–1177.
(54) Bandman, A.L., Volkova, N.V., Grekhova, T.D. *et al.* (1989) *Harmful Chemical Substances. Inorganic Compounds of Elements of V–VIII Groups* (ed. V.A. Filov), Chemistry, Leningrad (in Russian).
(55) Ilyushin, M.A., Tselinsky, I.V., Zhilin, A.Y. *et al.* (2004) Coordination complexes as inorganic explosives for initiation systems. *Hunneng Cailiao=Energetic Materials*, **12**(1), 15–19.
(56) Ilyushin, M.A., Tselinsky, I.V., Ugryumov, I.A. *et al.* (2005) Study of submicron structured energetic coordination metal complexes for laser initiation systems. *Central European Journal of Energetic Materials*, **2**(1), 21–33.
(57) Chernay, A.V., Sobolev, V.V., Chernay, V.A. *et al.* (2003) Chapter 11, Ignition of explosives by pulse lasers, in *Physics of Impulse Treatment of Materials* (ed. V.V. Sobolev) Dnepropetrovsk, Art-Press, pp. 267–314 (in Russian).
(58) Ajayan, P.M. Terrones, M., de la Guardia, A., *et al.* (2002) Nanotubes in a Flash – Ignition and Reconstruction. *Science*, **296**, 705.
(59) Campion, A. and Khambhampati, P. (1998) Surface-enhanced Raman scattering. *Chemical Society Reviews*, **27**, 241.
(60) Kennedy, J.E., Thomas, K.A., Early, J.W. *et al.* (2002) Mechanisms of Exploding Bridgewire and Direct Laser Initiation of Low Density PETN. Proceedings of the 29th International Pyrotechnics Seminar, IPSUSA, Inc., July 2002, pp. 781–785.

Appendix 8.A: Properties of Novel Primer Explosives

				Sensitivity						
Explosive	No	Density g cm^{-3}	Onset T K	Impact	Friction	Spark	Laser	VoD km/s	Min Charge	Ref
				J	N	mJ				
LA		4.7	315	2.5-4	6	4.7		5.3	0.025	
LS		3.1	282	2.5-5	40	3.1		5.2	>0.5	
Tetracene	**1**	1.63	140	<1	3	7.3		5.3		
DTET	**2**		≈215	<1	<2	3.3				
DADNP	**3**	1.71	180	100 mm	2		vs	6.9		
TATNB	**4**			5			*			
TAT	**5**					1.2	s	7.3		
DNBTM	**6**	1.86	240	1	60	200		8.50		
HNT	**7**			1	5	<5		8.9		
HAT	**8**	1.72		<1	<5	<2		9.0		
H2BTA	**9**			<1						
BTH	**10**			<1						
BTT	**11**			<1						
MeBTT	**12**			2.5				7.8		
DNG	**13**									
APX	**14**	1.91		3	80	100		9.7		
DADP	**15**	1.2						3.3		
TATP	**16**	1.2		0.3	0.1			5.5		
DABCO	**17**									
HMTD	**18**	1.57	395*					5.1	0.05	
TPNA	**19**									
DAHA	**20**						S			
DAATO 3.5	**21**						vs			
DNBTDO	**22**		483	4–6	360	150	nd	8.4		
DPDP	**23**		495						0.07	
5ATP	**24**									
5ATN	**25**									
DNBTG	**26**			10				7.5		
DNBTDAG	**27**							7.6		
ADNG	**28**	1.73	197	10	240	400		9.10		K2
K_2NClO_3	**29**			vs	vs					
Ba $NClO_3$	**30**			vs	vs					
CN_5O_2K	**31**	2.027	195	6.5	1					19
CN_5O_2Rb	**32**	2.489	192	5	0.5					
CN_50_2Cs	**33**	2.986	194	3	<0.5					
$Hg(NT)_2$	**34**	3.32		2				6.6		
Hg(NT)	**35**	3.5	525 K							
$Cu(NT)_2$	**36**	2.11	550	<1	<5	vs				
$Cu(NT)_2$ $HNT.H_2O$	**37**		400*	<3						
AgNT	**38**						s			
$Cu(EDA)_2(NT)_2$	**39**									

(continued)

Explosive	No	Density $g\ cm^{-3}$	Onset T K	Sensitivity Impact	Friction	Spark	Laser	VoD km/s	Min Charge	Ref
				J	N	mJ				
Ag(EDA)NT	**40**									
DBX1	**41**	2.58	333	10 cm	0.1	3.1		7.0		20,21
AAT	**42**									
AgAT	**43**									
AATO	**44**									
AgATO	**45**									
CuBTA	**46**									
AgBTA	**47**									
CuMeNAT	**48**									
27CuCltz	**49**	2.04	300	<1	<5	25	s		0.1	
28CuBrtz	**50**		292	<1	<5	20	s		>0.5	
KDNBF	**51**									14
KDNP	**52**									15
CP	**53**									
NCP	**54**	2.03	265					6.30		
BNCP	**55**	2.03	269					7.12		
tetrazCo	**56**									
DNG	**57**	1.97	280					7.76	0.20	
Azido	**58**	1.83	185					6.42	>0.5	
CuTANT	**59**	1.98	200					6.50	0.5	
CuBisNT	**60**	1.95	165						>0.5	
CuBisMeNT	**61**	2.2	250	5	20	>0.36 0				
CuBishydtet	**62**	2.1	259	5	40	>0.36 J				
NaFeNT4	**63**	2.2								
NaCuNT4	**64**	2.3		10						
ACuNT6	**65**	2.2		2	0.8	350				
CuHATr	**66**						<0.01 mJ			
CdHATr	**67**						0.5 mJ			
NiHATr	**68**						0.6 mJ			
CoHATr	**69**						1.0mJ			
AgATP	**70**									
AgATN	**71**									
AgMeATN	**72**									
NiHN	**73**	2.13	220	<2	15	vs	vs	7.0	0.15	
NiHA	**74**	2.12	186	5	20	s	s	5.42	0.45	
HNiCH	**75**									

Appendix 8.B: Molecular Structures of Some New Primer Compounds

Purely Organic Primers

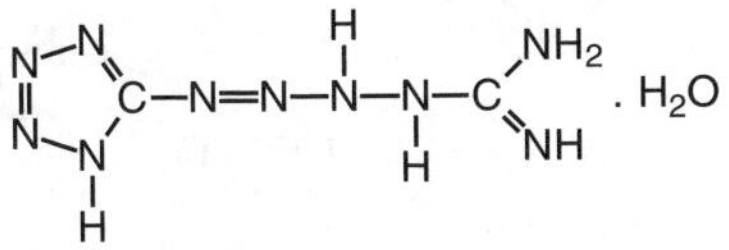

Tetrazene {**1**}

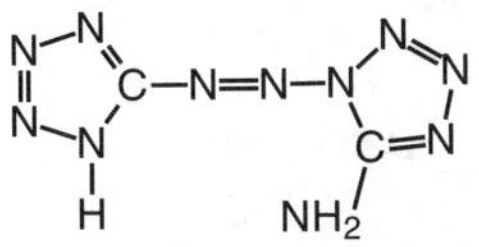

DTET {**2**}

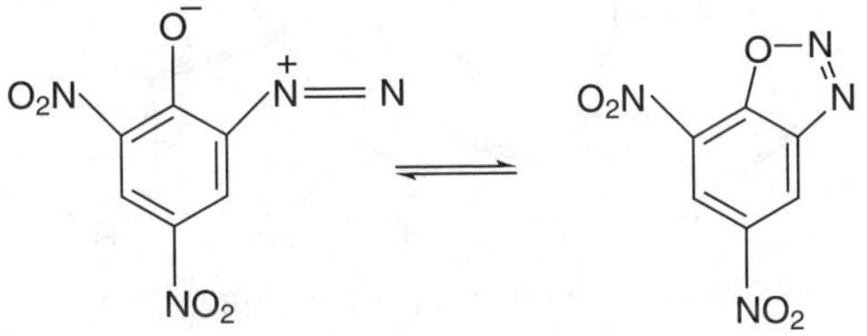

DADNP {**3**}

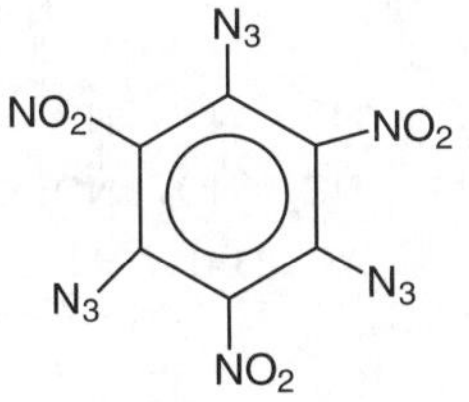

TATNB {**4**}

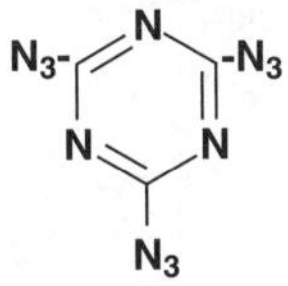

TAT {**5**}

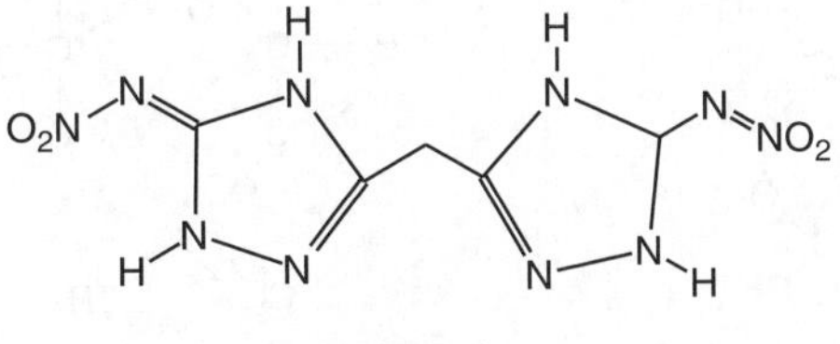

DNBTM {**6**}

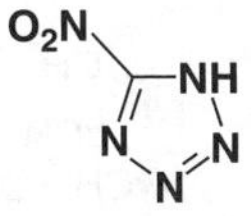

HNT {**7**}

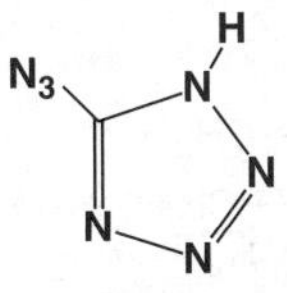

HAT {**8**}

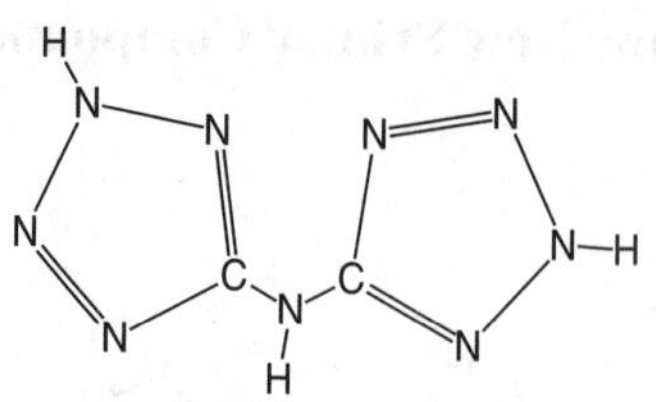

H2BTA {9}

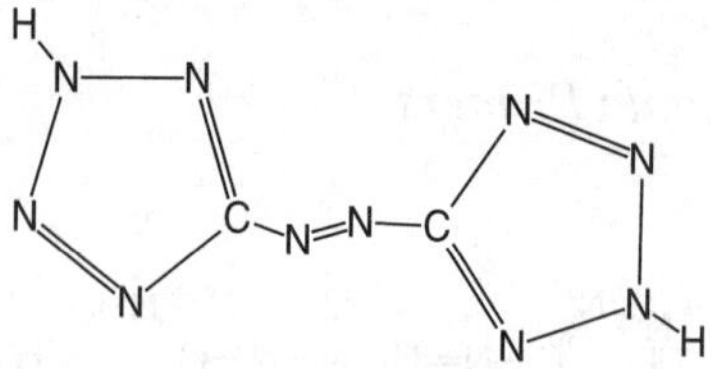

BTH {10}

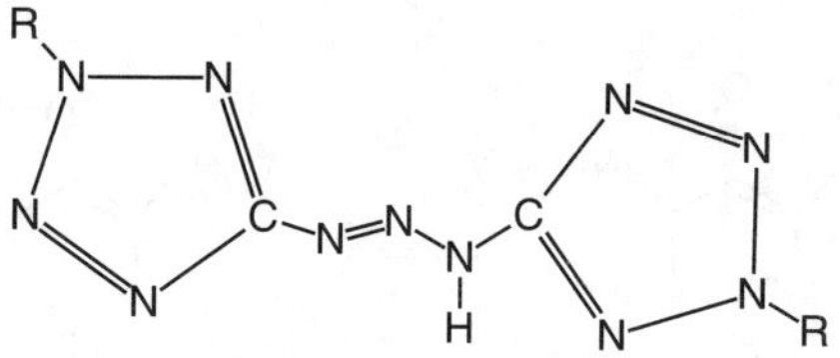

R=H **BTT** {11} R = CH_3**MeBTT**{12}

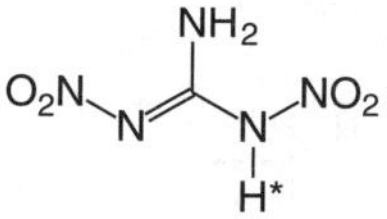

DNG {13}

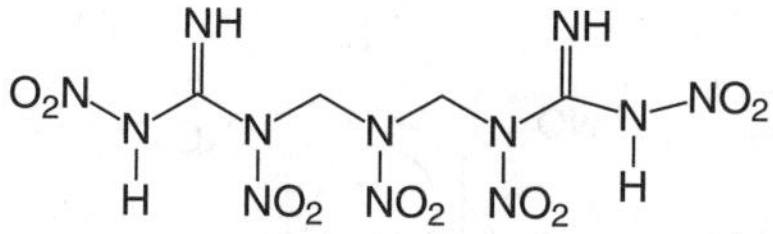

APX {14}

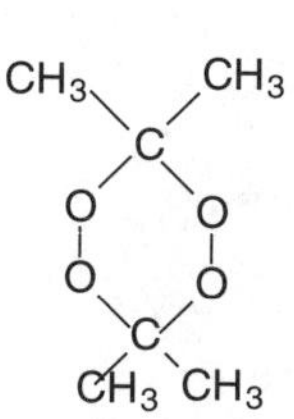

DADP {15}

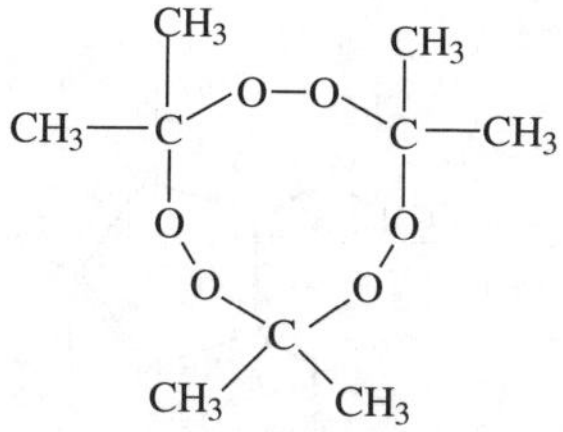

TATP {16}

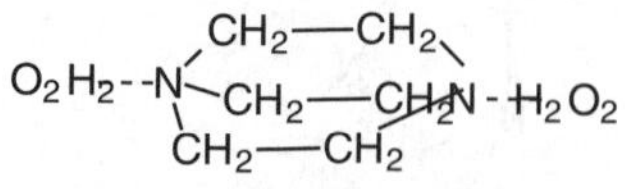

DABCO {17}

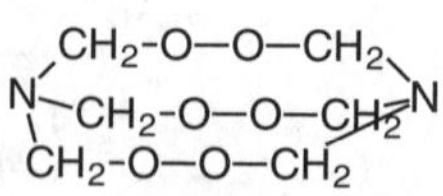

HMTD {18}

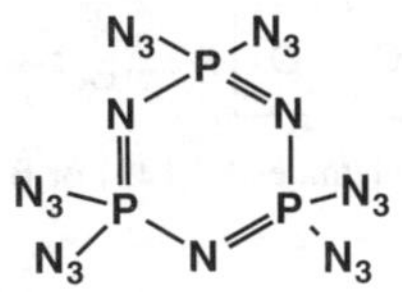

TPNA {**19**}

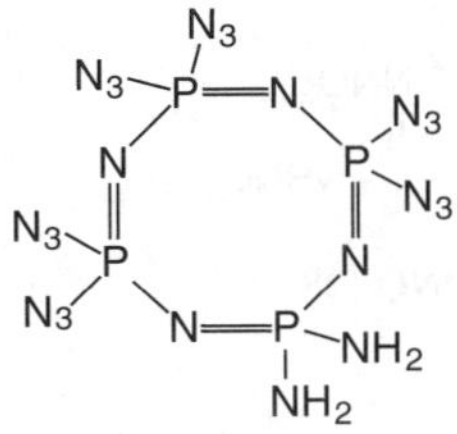

DAHA {**20**}

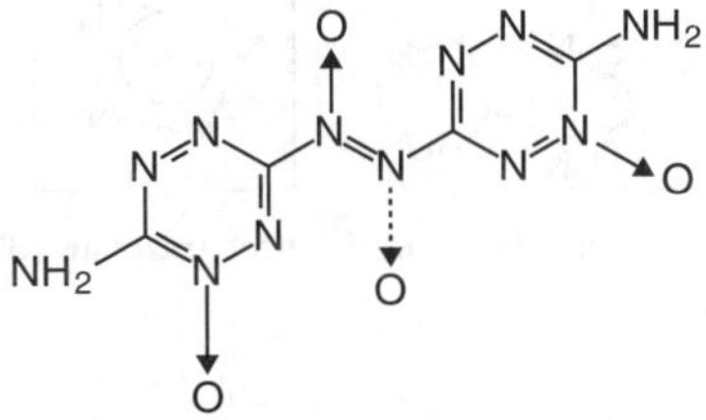

DAATO 3.5 {**21**}

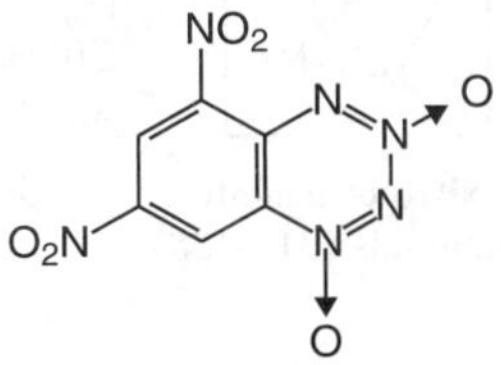

DNBTDO {**22**}

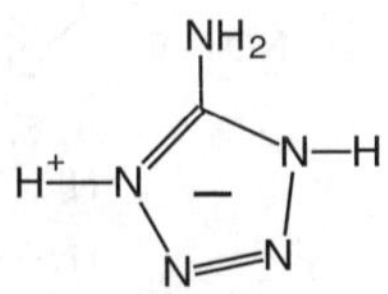

DPDP {**23**}

ClO_4^- (**22**)
Or
NO_3^- (**23**)

5AT salts

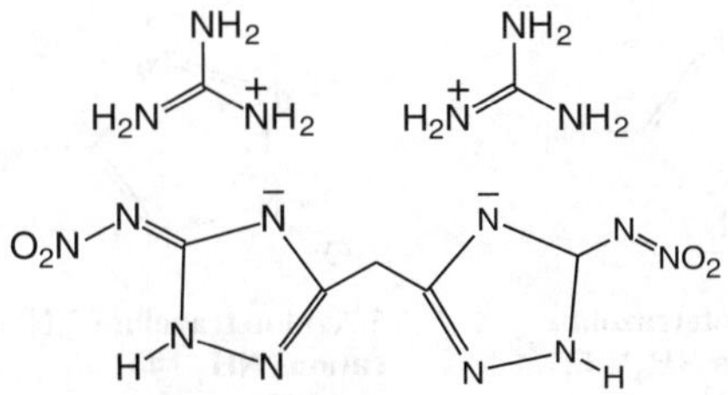

DNBTG {**26**}

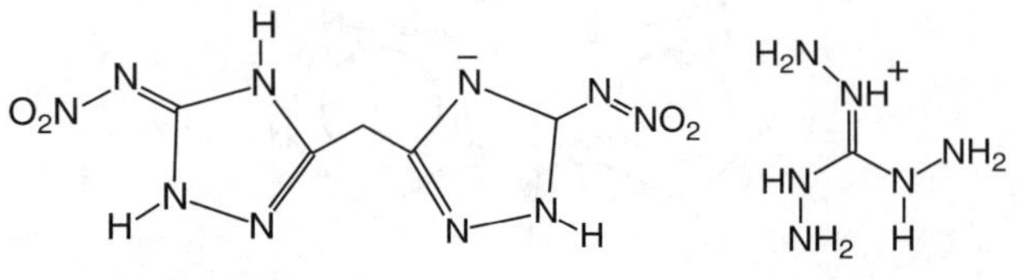

DNBTMDAG {**27**}

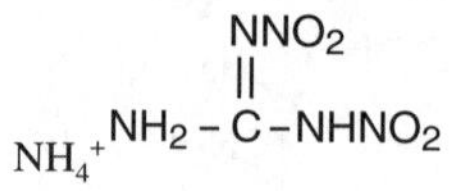

ADNQ {**28**}

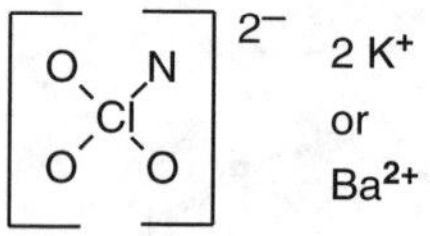

Perchlorylamides K_2^+ {**29**} or Ba_2^+ {**30**}

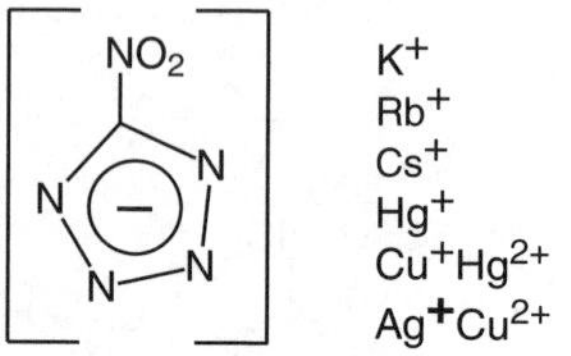

5-Nitrotetrazolate Compounds {**31**}–{**38**}

Cations

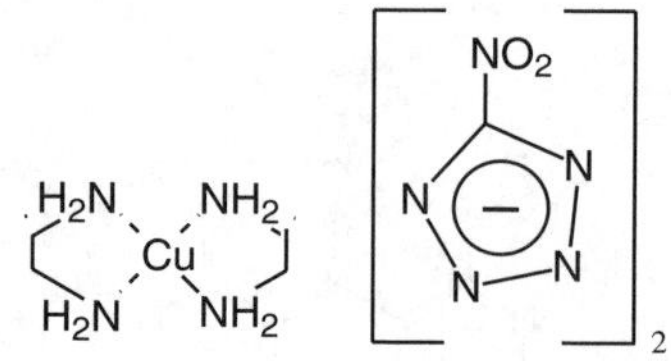

Copper bis EDA 5-Nitrotetrazolate {**39**}

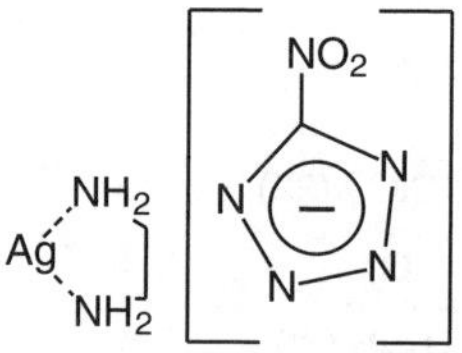

Silver EDA 5-Nitrotetrazolate {**40**}

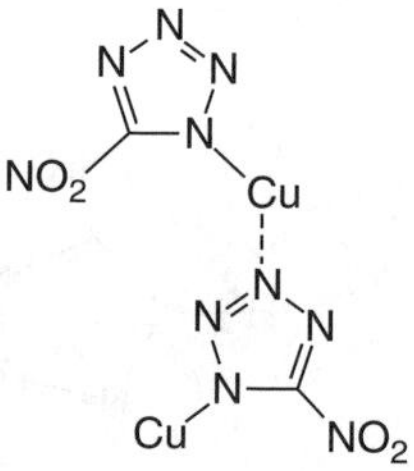

DBX-1 {**41**} $Cu_2(NT)_2$

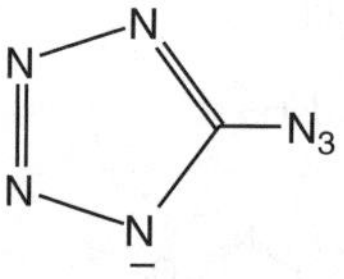

5 Azidotetrazolate Cations NH_4^+{42} Ag^+{43}

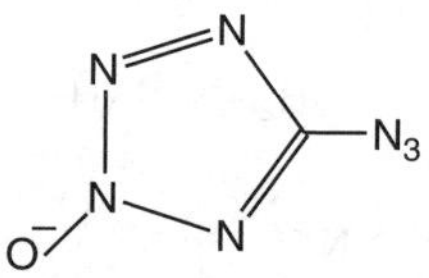

5 Azidotetrazolate 2(N)oxide cations NH_4^+{44} Ag^+{45}

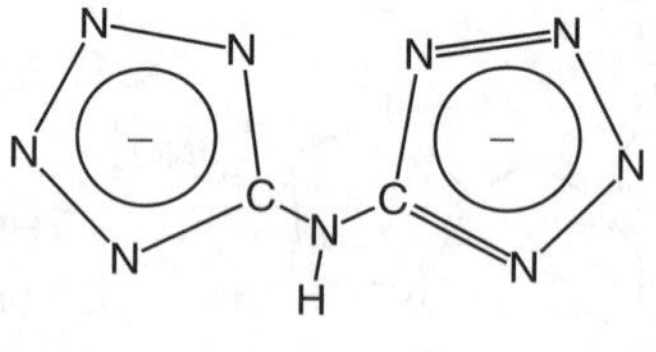

Bis tetrazolylamine with Cu^{2+}{46}. 2 Ag^+{47}

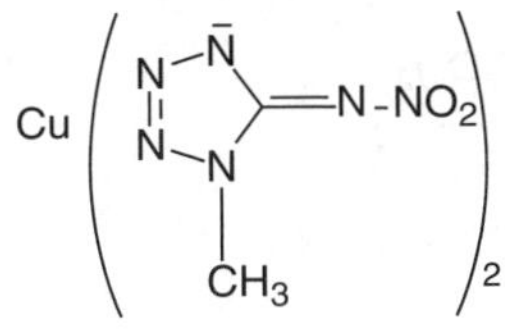

CuMeNAT {48}

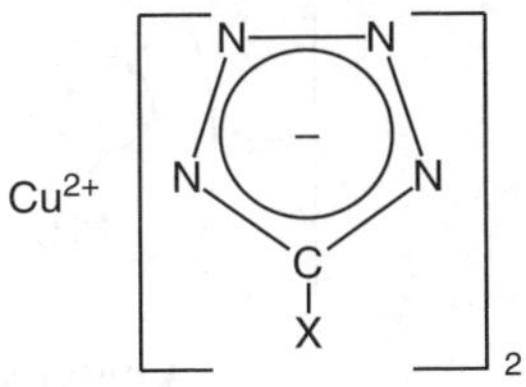

X = Cl CuCltz {49} X=Br CuBrtz {50}

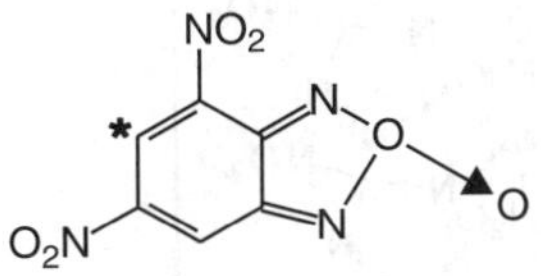

***= K^{+} KDNBF {51}**

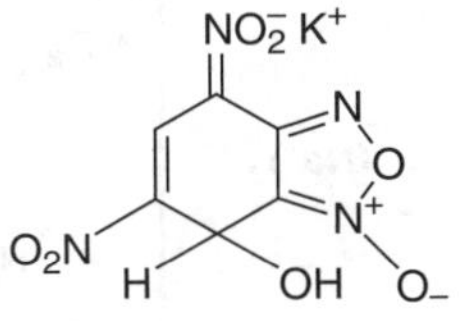

KDNP {52}

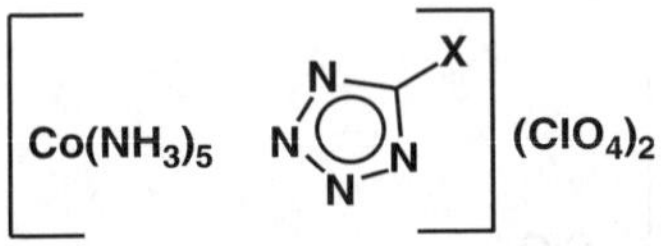

X = CN CP {53}, X = NO_2 NCP {54}

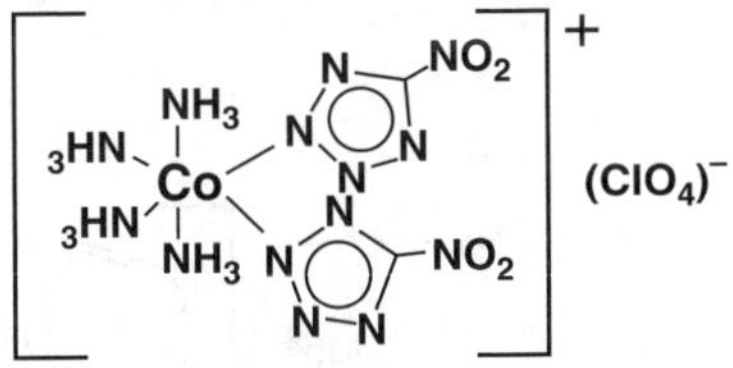

BNCP {55}

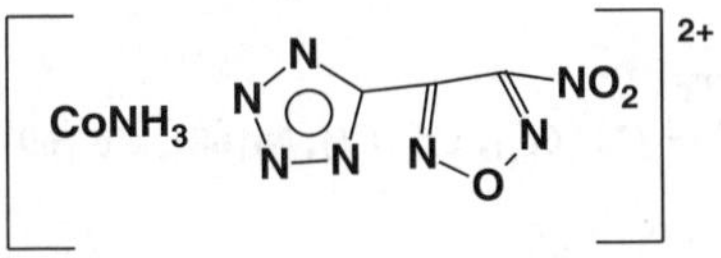

Anions $(ClO_4^-)_2$ {56} or $(DNG^-)_2$ {57}or $(N_3^-)_2$ {58}

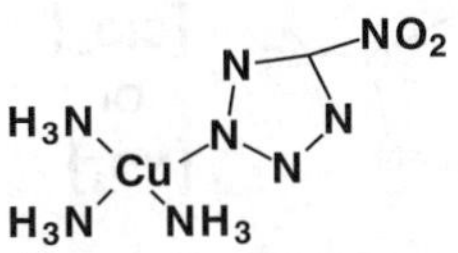

CuTANT {59}

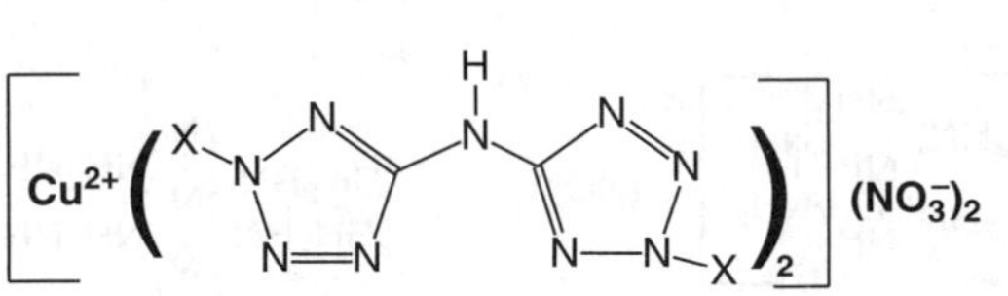

X = H CUBisNT {60}, X = CH_3 CuBisMNT {61}

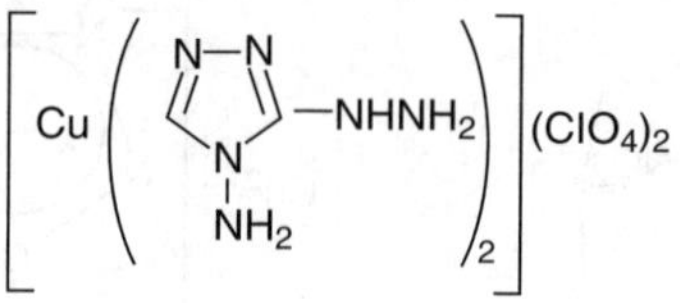

CuBHTAP {62}

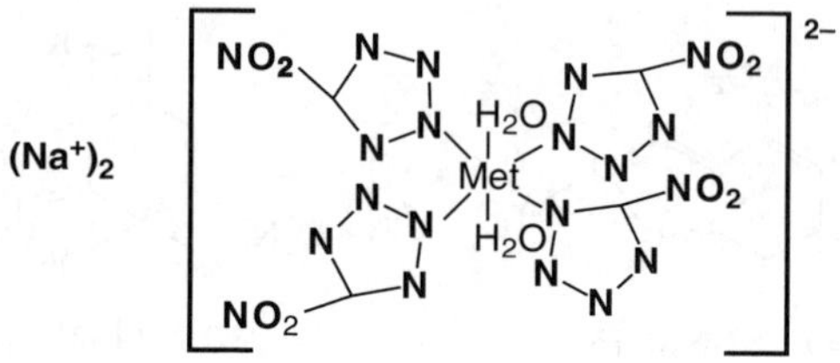

Met = Fe {63} Met = Cu {64}
$(NH_4)_4[Fe(NT)_6]$ = {65}

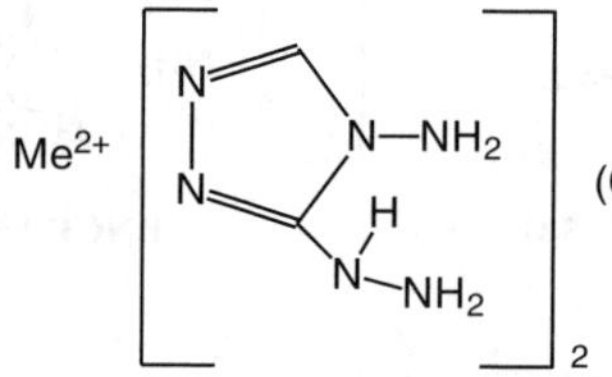

HATr
Me^{2+} = **Cu {66 }, Cd {67}, Ni{68}, Co {69}**

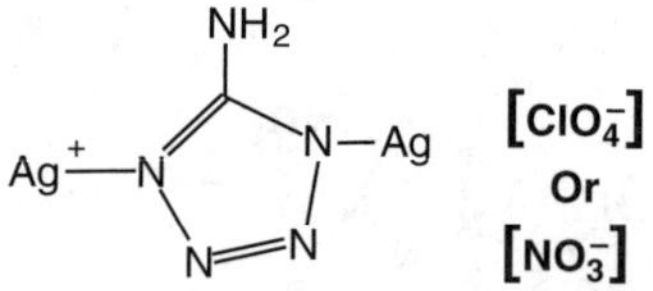

AgATP {70} and AgATN {71} Me substitute for 4Ag {72}

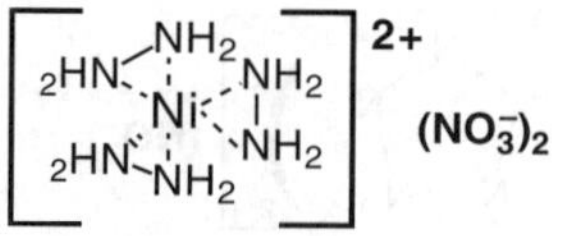

NiHN {73}

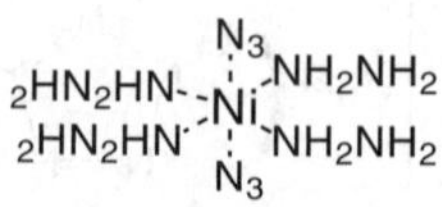

NiHA {74}

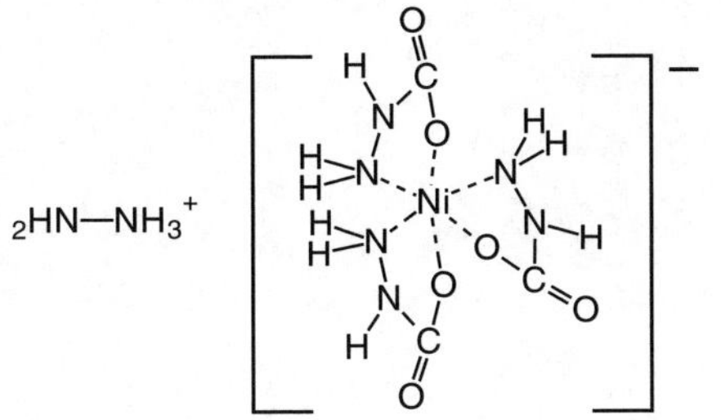

HNiCH {**75**}

9

Optical and Thermal Properties of Energetic Materials

9.1 Optical Properties

9.1.1 Introduction

The absorption of laser light by a target material is a pre-requisite for the initiation of any photochemical or photo-physical event. The other parameters that will determine the extent and efficiency of such events will be governed by the type and constituent of the material, its surface condition, its morphology and, of course, its thermal properties [1]. The dynamics and the kinetics of energy transfer and propagation, and the subsequent effects in energetic material, are discussed in a separate chapter. This chapter will describe the optical properties, particularly the spectral absorption characteristics which are required for the evaluation of the initial stage of the laser interaction process and for the subsequent achievement of effective laser ignition in energetic materials.

Both reflection and scattering of light dictate the amount of optical energy being absorbed by a material, as these cause loss of incident optical energy to the surroundings. These parameters, in turn, depend on the properties of both the optical source and the material. In the case of a laser source, these are wavelength, power (beyond the linear range), angle of incidence and angle of observation. The material characteristics that govern reflection and scattering are: packing density; surface roughness; crystal form; complex refractive index; particle diameter; and the absorbance (the fraction of incident light absorbed by the material) of the component materials. The spatial distribution of specific hot spots, both on the surface and within the bulk, may also play an important role in defining the optical absorbance of samples.

Reflection from practical energetic materials invariably has contributions from both the specular and the diffuse components. The specular component is a surface effect, and it is the dominant feature for polished and shiny surfaces. In this, the reflected beam emerges from the surface at an angle (with respect to the surface normal) equal to that of the of incident

Laser Ignition of Energetic Materials, First Edition. S Rafi Ahmad and Michael Cartwright.

beam. The diffuse reflection arises from both the surface and the bulk of the material. The incident beam (photons) bounces back in all directions, due to multiple reflections by the microscopic irregularities inside the material (e.g. the grain boundaries of a polycrystalline material or the fibre boundaries of an organic material, and by its surface if it is rough).

In this situation, an image of the source, like that in the case of Fresnel's reflection, is not formed. Light is reflected in all directions, with radiance as defined by Lambert's cosine law. This says that the radiant intensity observed from a Lambertian surface is directly proportional to the cosine of the angle, θ, between the observer's line of sight and the surface normal. A Lambertian surface is also known as an ideal diffusely reflecting surface. An important consequence of Lambert's cosine law is that when a Lambertian surface is viewed from any angle, it has the same radiance. This means, for example, that to the human eye it has the same apparent brightness (or luminance). It has the same radiance because, although the emitted power from a given area element is reduced by the cosine of the emission angle, the apparent size (solid angle) of the observed area, as seen by a viewer, is decreased by a corresponding amount. Therefore, its radiance (power per unit solid angle per unit projected source area) is the same.

The proportion of contribution to the loss of optical energy from the two reflection processes is dictated by a variety of parameters, such as: quality of the pressed surface; packing density; particle size; absorbance of the particulates; and so on. For example, pressing a powdered material with two flat-ended metal bars that have smooth surfaces often increases the specular component of the reflectance. It has been reported that the proportion of this component increases with increasing pressure, due to the increase in uniformity of particle orientations. Roughening a smooth surface or pressing the powder with a paper will, in general, increase the diffused component of the reflectance. The scattering component may be considered as a part of the diffuse reflection process, as it is also influenced by the inhomogeneities and discontinuities within the bulk of the material. A high proportion of scattering and diffused reflection from many solid energetic materials does not allow line-of-sight measurement (Fresnel reflection analysis) for the estimation of reflection loss.

9.1.2 Theoretical Considerations

From a phenomenological consideration, the absorbed intensity (power per unit area) at a specific wavelength, I_a, may be considered, for most practical purposes, to be the difference between the overall reflected intensity, I_r, and the incident intensity, I_i. The intensity of the reflected light will have contributions from both the specular and diffuse components, I_s and I_d, which is expressed as:

$$I_a = I_i - (I_s + I_d) = I_i - I_t \tag{9.1}$$

Absorption of optical radiation may be quantified by a parameter, A, absorption efficiency or 'absorbance', defined as the fraction of the optical intensity, I_i, which has been absorbed by the material. Dividing Equation 9.1 above by the incident intensity, we obtain:

$$A = 1 - R_t \tag{9.2}$$

where R_t is the fractional loss of light intensity due to both the reflection events. In similar way. the loss due to both specular and diffuse (including scattering) reflection may be defined in terms of specular reflectance, R_s, and diffused reflectance, R_d.

For efficient and effective use of optical energy from a laser for the initiation of ignition or combustion processes in energetic materials, data on the spectral absorbance properties of the materials are required. The wavelength-dependent absorption parameter (absorbance) can be calculated using reflectance data measured using some commercial absorption spectrometers with specially adapted devices, and assuming that the fraction of light intensity that has not been reflected has been absorbed to raise the temperature of the irradiated surface, with the propagation of the resulting heat as a dynamical process. While the absorption data thus obtained may be adequate for matching the spectral absorption maxima of materials with the wavelength of the laser beam, in some cases, where theoretical evaluation of the ignition parameters are conducted, a parameter called the absorption index, k, is required, as is elaborated below.

For the measurement of k, using commercial absorption spectrometers, the substance needs to be semitransparent. Such materials may be produced from opaque solid substances by dissolving or dispersing them (grinding or making powders) in suitable solvents and making thin films. Insoluble solids can be dispersed (by dry mixing in the powder forms) within the bulk of a semitransparent solid such as KBr crystals and making thin slices of pallets. Such pallets can then be used to obtain spectral absorption data, using a commercial spectrometer, and the values of absorption index can be obtained applying Beer-Lambert's law. In this, the first order reaction dynamic dictates that the fractional quantities, absorbance (A) and transmittance (T) are defined as:

$$\mathrm{T} = (I_t/I_i) \quad \text{and} \quad \mathrm{A} = \ln(1/\mathrm{T}) \approx K_a \times l$$

where l is the thickness of the slab (or the cuvette in the case of a fluid sample), I_t is the transmitted intensity and K_a is the absorption coefficient. The refractive index of a semi transparent substance can be written as, $\mathrm{N} = k\,(\text{real}) + i\,n\,(\text{imaginary})$. The imaginary part of the refractive index essentially is the contribution from the absorption loss, and the real part provides contribution from reflection losses. Fresnel interface reflectance, $\mathrm{R_t}$, may be calculated from the formula below:

$$\mathrm{R_t} = \{(n-1)^2 + k^2\}/\{(n+1)^2 + k^2\} \tag{9.3}$$

The absorption coefficient, K_a (cm^{-1}), is related to the absorption index, k through the wavelength of the radiation field as follows:

$$k = K_a \lambda_i / 4\pi \tag{9.4}$$

It is apparent from the above formula that the absorption index is a dimensionless parameter.

The refractive index (n) and the absorption index (k) parameters may be calculated from the measured spectra for absorption index at a specific wavelength, $k(\lambda)$, and adopting an optimization-based curve fitting routine and the established dispersion theory relations [1]:

$$n^2 - k^2 = n_\mathrm{e}^2 + \sum\nolimits_\mathrm{j} \left\{ \omega_\mathrm{p,j}^2 \left(\omega_\mathrm{0,j}^2 - \omega^2 \right) \right\} / \left\{ \left(\omega_\mathrm{0,j}^2 - \omega^2 \right) + \gamma_j^2 \omega^2 \right\} \tag{9.5}$$

and,

$$2nk = \sum\nolimits_\mathrm{j} (\omega_\mathrm{p,j})^2 \gamma_\mathrm{j} \omega / \left\{ \left(\omega_\mathrm{0j}^2 - \omega^2 \right) + \gamma_\mathrm{j}^2 \omega^2 \right\} \tag{9.6}$$

The parameters $\omega_{p,j}$, and ω_0 are the plasma and oscillator frequencies of the jth oscillator respectively, ω is the frequency of the radiation (laser) field and, γ_j is the line width of the oscillator. The residual value of the refractive index, n_e, occurs at the visible/near-infrared wavelength region. This procedure allows determination of the oscillator and line width parameters and, finally, the values of n and k at any wavelength. Note that the application of the above procedure is based on the assumption that the sample is dispersed in a crystalline medium (e.g. potassium bromide), which is pressed into a slab of very small thickness, d, so that the criterion, $(K_a \times d) < 1$, and the Rayleigh-Gans scattering [2] conditions are satisfied. The procedure has been applied successfully by some authors to calculate the optical constants of some energetic materials, such as energetic binders, high explosives and oxidizers [3]. The evaluations have so far been carried out in the IR wavelength regions, as isolated and strong absorption lines occur mostly at such wavelength bands. However, optical properties of energetic materials at visible and near-IR wavelengths are of particular interest for the practical relevance of laser ignition mechanisms. This is primarily because of the commercial availability of a wide variety of cheap, efficient and lightweight diode lasers and light-emitting diodes operating at this wavelength band.

At UV-Vis-Near-IR wavelength bands, a conventional commercial absorption spectrometer may be used to obtain data on the absorption spectra of samples. Various procedures may be applied to obtain absorption results, using conventional spectrometers based on the transmission of light through semi-transparent samples. For this, the sample needs to be dissolved in an appropriate solvent, and the effect of solvent on the absorption properties of the material needs to assessed and taken into consideration. To avoid introducing solvent effect, the samples may be ground into sub-micron sizes and dispersed within a practically non-absorbing substrate such as KBr, to form a sub-millimetre-thickness pallet. Caution must be exercised to ensure that the results from such measurements represent absorption data of the practical sample that is under consideration for laser ignition studies. In general, however, the depth to which the optical field of a laser beam will penetrate a sample may be estimated using a standard absorption spectrometer, and applying the Beer-Lambert law given as follows:

$$I_t = I_i \exp - \{K_a(\lambda) \times d\} \tag{9.7}$$

where:

I_t and I_i are the transmitted and incident intensities
d is the thickness of the sample
$K_a(\lambda)$ is the wavelength dependent absorption (attenuation) coefficient.

It is evident from Equation 9.4 that the absorption index is linearly dependent on the wavelength of the incident beam. The penetration depth is defined as the length within the sample at which the incident intensity has dropped down to its $(1/e)$th value and, from Equation 9.7 above, this will happen when, $d = 1/K_a(\lambda)$, showing that the penetration depth will be larger for shorter wavelengths. This means that the volume of sample activated before any photochemical or photo-physical activities take place will be larger at UV-Vis wavelengths than that for the IR wavelength band. For small penetration depths, the reactions take place mainly at the surface and are prone to being affected by external pressure or surface confinement (such as cover plate or external pressure). For large absorption depths, bulk heating takes place and, at the initial stage of heating, the explosive becomes inertially confined. This reduces the ignition threshold laser power and renders the process independent of external conditions [4].

9.1.3 Practical Considerations

Optical absorption depends not only on the wavelength of the light beam and the optical properties of the material at that wavelength, but also on the beam qualities at the focused spot at the target. Additionally, optical absorption depends upon various physical conditions of the material – for example, surface roughness, presence of voids, compaction (particulate density) and presence of impurities and, of course, in the case of samples tailored for high absorption at a specific wavelength band, on the optical properties of the deliberately incorporated sensitizer. Therefore, the spectral absorption properties of individual energetic material samples belonging to the same category or type may vary from one sample to another, depending upon their physical conditions and temporal histories. The latter is dictated by the extent of degradation due to weathering. It is, therefore, important to experimentally evaluate the spectral absorption properties of samples to be tested for laser ignition, matching the wavelength of the light to that of the absorption maximum of the material, and not to the generic data for single crystals or an arbitrarily pressed inhomogeneous material.

For laser ignition studies, the beam has to be focused onto the surface of the target in order to achieve the high power density required for the ignition initiation process. It is noted that the high degree of coherence of the laser beam renders lenses to focus the beam to a very small area. As has been elaborated in Chapter 3, the laser beam may be focused down to a diffraction-limited spot area more than four orders of magnitude smaller than the area at the exit point, with proportional increase in power density at the target. This is possible for a Gaussian intensity profile of the beam. In practice, most common laser sources will have inhomogeneities or hot spots across the beam cross-section, which may cause higher beam area at the focus of a lens than that expected from the diffraction limited case. In any case, at high power densities, absorption may be governed by a multi-photon process, causing photo-ionization or photo-fragmentation prior to initiation of ignition, detonation or combustion.

In the case of pulsed excitation, the delivery of full laser power may not be considered instantaneous, particularly for pulses of milliseconds durations. In such scenarios, the dynamic of the process dictates that the laser beam will encounter heated plasma or gas emanating from the irradiated target area towards the laser as the pulse develops up to its peak value. Such scenarios will be dealt with later, in the chapter on laser interaction processes.

Absorption spectroscopy can be conducted with commercial spectrometers, using the material in a suitable solvent and comparing results with that of a standard (1 cm path length) cuvette. While data from such measurements may give a pointer for a matching wavelength, this may not provide a reliable quantitative value for the sample in solid state. For solid-state measurement of spectral absorbance, the so-called FTIR/KBr – pallet method is gaining popularity for applications of diode lasers operating in the near-IR wavelength band. This can be carried out for crystalline solids which lend themselves to be crushed into fine powders. In this, the sub-micron size powder of the target material is mixed with a similar-sized KBr powder. The mixture is then pressed into pellets of small thickness so that, in the normal mode of operation, the transmitted beam is detectable above the noise level. For non-crystalline materials such as some propellants, the samples may be prepared by slicing a thin sliver using a microtome. For some energetic binders, such as N5, poly GLYN, etc., the fluid may be sandwiched between two Plexiglas slides, and curing them by heat or UV radiation, before using them for the measurement of spectral absorption spectra using commercial spectrometers.

9.1.4 Examples of Absorption Spectra

Absorption spectra of energetic materials, including propellants, pyrotechnics, explosives and some energetic binders, have been recorded at the author's laboratories. Some examples are presented below.

9.1.4.1 Pyrotechnics

Pyrotechnics are widely used for fireworks, and also as initiators for 'difficult to ignite' energetic materials. A large number of pyrotechnic compositions can be produced by the combination of different oxidizers, fuels and required binders, plasticizers, stabilizers, curing/cross-linking agents and so on. Details of these materials, relevant to this book, have been given elsewhere. Commonly used pyrotechnic materials include G20 (also known as gunpowder), SR371a and SR44. All of these contain potassium nitrate as the oxidant; SR44 uses boron powder as the fuel, and SR371a uses granules of magnesium as the fuel. SR44 and G20 contain some carbon black and hence appear black, while SR371a contain different additives and has a wood-ash colour. Their absorption spectra, recorded at the authors' laboratories and produced from the recorded diffused reflectance spectra, are shown in Figure 9.1.

It is noted from the spectra that all three materials exhibit very high absorption (over ≈88%) across the UV-Vis and near-IR bands, extending from ≈350 nm up to 850 nm (limit of range for the instrument). While the absorption decreases with the increasing

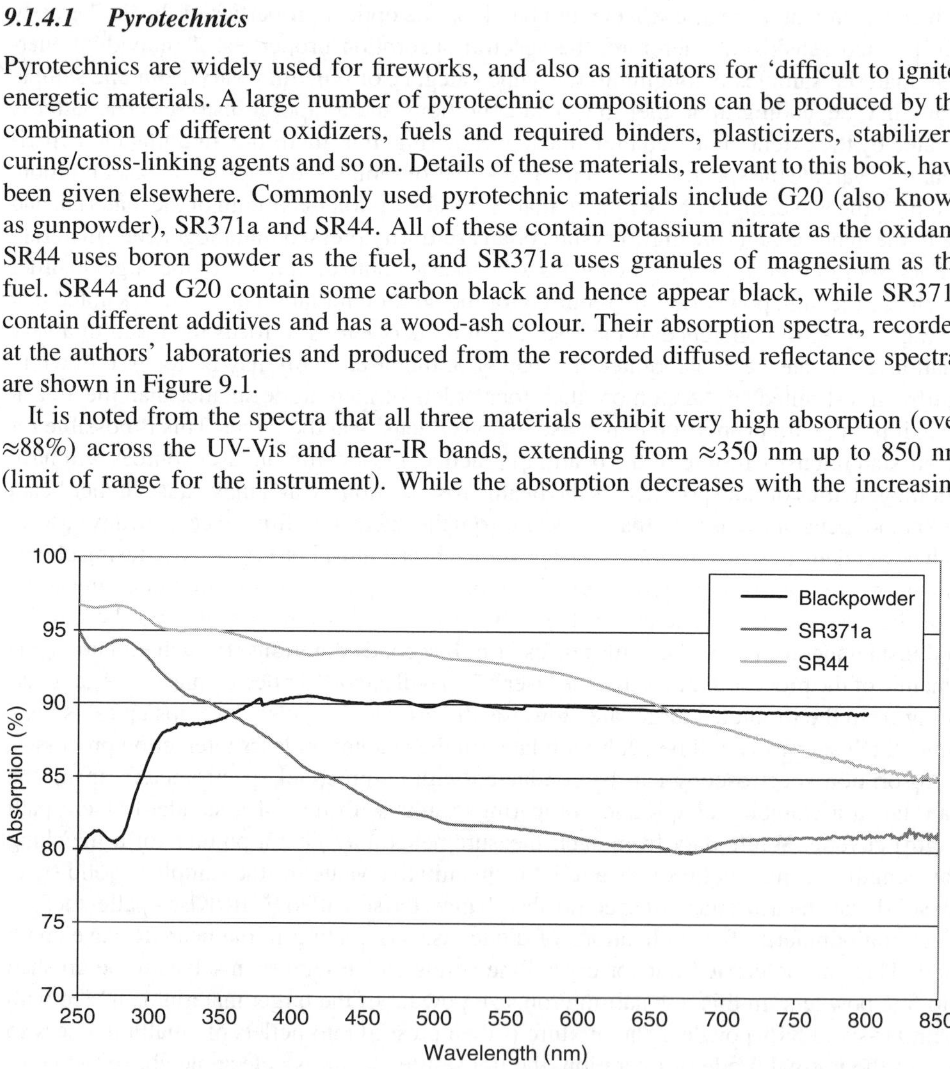

Figure 9.1 *Absorption spectra of some typical pyrotechnic materials.*

wavelength for SR 44 and SR371a (more rapidly for SR371a than for SR44), the value remains practically constant over the whole spectral range and up to the near IR band for the G20 pyrotechnic.

9.1.4.2 Propellants

Propellants are used mostly as gas generators and to provide energy to propel or push in a variety of applications, including the launching of rocket motors. There are three main categories of propellants that have found wide spread commercial and military applications. Propellants are divided into four categories, depending on the types of the main energetic constituent. Single-base propellant has Nitroglycerine (NG) as the main energetic material, in addition to usual plasticiser (binder), stabilizers and so on. The double-base propellants have both NG and NC (nitrocellulose) as the energetic materials and are extruded using a machine (EDB). Composite double-base propellants (CDP) do not contain any NC or NG. They normally contain metallic aluminium and some oxidizer, and are chemically synthesized.

Another commonly used propellant is that based on ammonium perchlorate (AP), which includes hydroxyl-terminated polybutadiene (HTPB) and a small amount of nitramine. Absorption spectra of two double-base propellants, both containing some carbon black and oxidizers in addition to nitrocellulose (energetic resin) and nitro-glycerine (explosive plasticiser), were recorded. Also recorded were the absorption spectra of a composite propellant. These samples were chosen for their common use (in service). The samples were practically opaque, and the use of conventional absorption spectroscopy, based on transmission measurement, was not possible. Therefore, absorption spectra were evaluated from the reflectance data recorded by using a 150 mm internal diameter integrating sphere (Lambda 9B013-8277) attachment in a UV-VIS absorption spectrometer (Perkin Elmer, Model-Lambda 9). In this arrangement, a small sliver of approximately 2 mm thick sample was cut with a scalpel from the solid material, and the diffused reflectance from its surface was recorded over the limited spectral range of the spectrometer. Incident light was diffusely reflected from the surface over the solid angle of 2π sr. The spectra were compared with that of a standard (compressed magnesium powder) recorded under the same setting. A simplifying assumption was made, in which the absorbance as a percentage of the intensity (flux density) is expressed as:

$$\mathrm{A}(\lambda) = 100 - R(\lambda)$$

where $\mathrm{R}(\lambda)$ is the reflectance in percentage. The absorption spectra for the three propellant materials are shown in Figure 9.2.

It is interesting to note that both the CDB- and AP-based propellants exhibit high absorption, practically at a constant level, whereas for the extruded double-base, although it shows a similar wavelength dependency for wavelengths up to ≈550 nm, the absorption decreases drastically with the increase of wavelength. The results indicate that, in the visible laser wavelengths, all the propellant samples are likely to be easily ignitable. However, although the CDB- and AP-based samples continue to show high levels of absorption, the absorption drops down drastically toward the near-IR wavelengths for the EDB sample. This indicates that this type of propellants may not be a suitable candidate for laser initiation using a near-IR laser which is suitable for practical implementation of laser ignition.

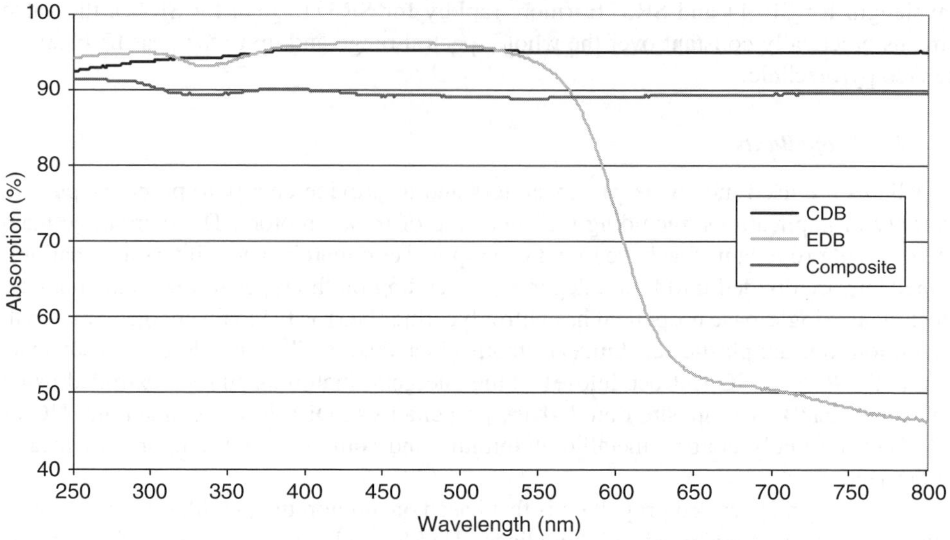

Figure 9.2 *Absorption spectra of some propellant samples.*

9.1.4.3 Explosives

Energetic materials in general, and explosives in particular, exhibit absorption peaks in the deep UV wavelength band (109–360 nm). These are a consequence of $n \rightarrow \pi^*$ transitions in nitrate (313 nm), carbonate (217 nm), nitrite (360 and 280 nm) and azide (230 nm). An example of a typical absorption spectrum of a high explosive material in this particular case, HNS (Hexanitrostilbene), having the chemical formula $C_{14}H_6N_6O_{12}$, is shown as exemplar in Figure 9.3. In the near-infrared band, around 800 nm (output from most efficient diode lasers), the absorption by high explosives is rather poor. For almost all reported tests on diode laser ignition of energetic materials, use of sensitizer dye, or inert materials such as carbon black, were found necessary.

9.1.4.4 Low Vulnerable Ammunition (LOVA)

One of the main requirements for the practical application of energetic materials is, of course, maximum explosive performance. Because of increasingly stringent safety requirements in recent years, two other factors have gained prominence in the formulation of a new generation of energetic materials. One is the increasing emphasis on safety during storage, use and transportation. The other is the impact of the manufacture and disposal of explosive materials on the environment. Consequently, the emphasis now is on the requirement of high-performance munitions to be balanced and complemented with the achievement of through-life safety, whole-life cost-effectiveness and end-life environmentally friendly disposal. The development of new energetic materials having such properties and criteria is a complicated process, in which many candidate molecules have to be considered, a few synthesized, even fewer formulated, and only a small handful adopted by the military or industry. The laborious process involves computer modelling, plenty of laboratory work, and thorough testing.

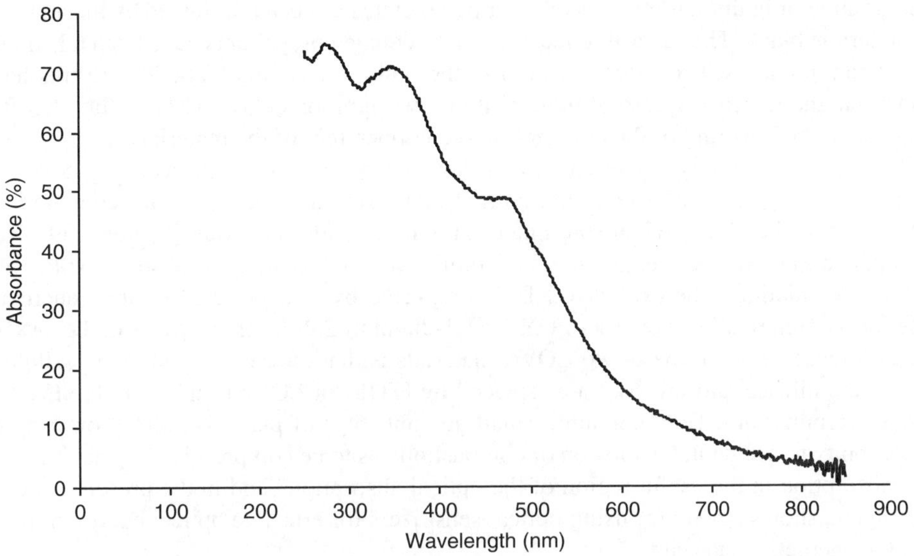

Figure 9.3 *Typical absorption spectrum of the explosive material, HNS.*

The search for LOVA materials which will satisfy the above requirements has given rise to nitramine-based propellants (e.g. RDX, HMX), incorporating both energetic and non-energetic plasticizers and binders, to replace conventional nitrocellulose-based propellants. Additionally, new insensitive high explosives such as FOX-7, PolyNIMO, HTPB, etc. are also formulated to replace relatively more sensitive nitramines as energetic fillers in the formulation of propellants. The chemical structures of these formulations are presented in Figure 9.4 as exemplars.

As a consequence of their reduced response to accidental stimuli, the LOVA propellants and explosives are difficult to ignite by conventional electric stimuli. 'Non-ideal ignition property', resulting in loss of accuracy, misfire, hang-fire and catastrophic failure of gun systems are the major drawbacks of such LOVA systems, and they have contributed to the paucity in the development of commercially available practical systems based on insensitive ammunitions (IM).

Direct laser ignition of propellants and explosives, particularly LOVA ones, has not been reported as widely as that for pyrotechnics. As for other propellants, for LOVA materials, the ignition threshold and ignition delay have also been reported to be dependent on laser parameters, chemical composition of the charge and the pressure within the confinement.

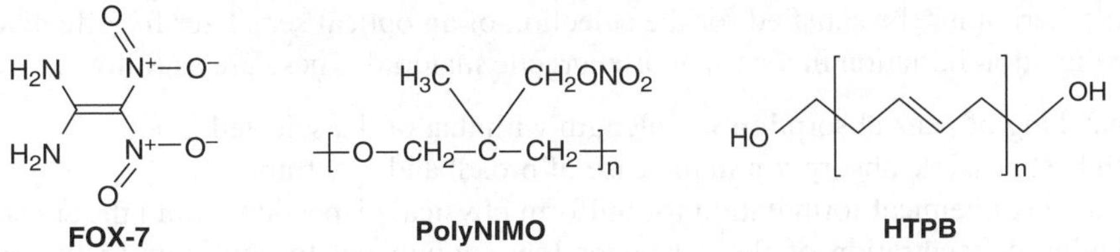

Figure 9.4 *Chemical structures of some potential LOVA energetic fillers.*

The variation of ignition properties on laser parameters has been minimized by the addition [5] of carbon black. The surface reflectance of carbonized propellants have been taken into consideration and used for normalization of the ignition threshold data. The results have shown that the increasing ease of ignitability (lower ignition delay and laser flux density) is related to the lowering of the burning surface temperature of the materials.

There has been a spate of research and development activities in recent years on the prospect of using insensitive ammunition (IM). Absorption spectra of two such materials, coded as LOVA1 and LOVA2, are presented in Figure 9.5. Such materials [6] are mainly used as propellants and can be used, with some modification, in formulations of some explosives. In both formulations, the explosive RDX is replaced by a much less sensitive energetic material, commercially known as FOX 7 (1,1-diamino-2,2-dinitroethylene or DADNE). The difference between these two LOVA materials is that the sensitive binders, cellulose acetate and nitrocellulose (16%) are replaced by HTPB in LOVA1 and by PolyNIMO in LOVA2 formulations. Both contain a small amount (6%) of plasticizer. The lowering of vulnerability to accidental explosion or combustion has to be compromised by having very poor absorption in the useful region of the optical absorption band in the present context. This, again, shows a need for using optical sensitizers for effective optical energy transfer in to the energetic materials.

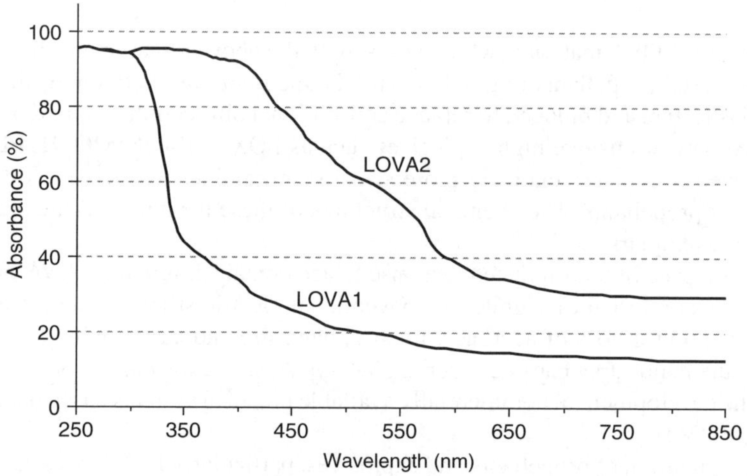

Figure 9.5 *Absorption spectra of two different experimental LOVA propellants.*

9.1.4.5 Optical Sensitizers

Several criteria must be satisfied for the selection of an optical sensitizer for effective and efficient ignition initiation in a particular energetic material. These are as follows:

i. Matching of peak absorption wavelength with that of the selected laser.
ii. Highest possible absorption in the case of broadband spectrum.
iii. Allowing chemical formulation for uniform physical dispersion within the substance.
iv. Having concentration of the sensitizer low enough not to compromise the energy content (%) of the material.
v. Using material which is practically inert for the host substance.

Table 9.1 *Properties of some commercial near-IR absorbing dyes as potential sensitizers.*

Product codes	Product suppliers	Solvent	Max. abs. wavelength ($\pm 1/2$ band at 80%)	Absorptivity ($L\ g^{-1}\ cm^{-1}$)
Epolight™5768	Epolin, Inc. USA	Acetone	791 nm (≈15nm)	292
IR dye 9798	Adam Gates Company LLC, USA	—	798 nm (≈15 nm)	157
NIR800A	QCR Solutions Corp, USA	Methanol	800 nm (≈25 nm)	297
NIR805B	QCR Solutions Corp, USA	Methanol	805 nm (≈15 nm)	400
IR Dye 9807	Adam Gates Company LLC, USA	—	807 nm (≈15 nm)	409
NIR811A	QCR Solutions Corp, USA	Dimethyl formamide	811 nm (≈15 nm)	342
ADS800AT	American Dye Source, INC. USA	Methanol	811 nm (≈20 nm)	254
ADS815EI	American Dye Source, INC. USA	Methanol	815 nm (≈20 nm)	352

The use of IR-absorbing dyes as sensitizers have the advantage that such dyes are often commercially available at reasonable cost, or may be tailored to have an isolated absorption line (narrow absorption band) to match with the wavelength of the initiating laser line (wavelength). Because of usually high absorption of some dyes at specific near-IR wavelengths, these may be used at trace concentrations. However, chemical dyes are often chemically interactive with the host matrices of energetic materials, and may not be resistant to exposure to weathering or temperature during the storage or processing of such materials.

The advantage of wavelength selectivity offered by chemical dye sensitizer outweighs the high absorption (70–80%) and wide band absorption afforded by carbon black (CB). Commercial CB powders, having different grain sizes (size distributions), and conforming to the above criteria for a potential optical sensitizer, are readily available. However, high concentration levels, needed for effective and efficient laser ignition initiation in energetic materials, may compromise the energy content of the materials. Reports on the use of chemical dyes as a sensitizer for energetic materials in the near-IR wavelength band are scarce. Examples of some commercial infra-red absorbing dyes are included in Table 9.1. Further research is needed to evaluate the compatibility, reactivity and stability of such dyes in the matrices of energetic materials of interest.

9.2 Thermal Properties

9.2.1 Introduction

For the analysis of laser ignition in energetic materials, knowledge of thermal properties, along with their optical properties, is needed. Thermal properties are usually quantified by three parameters: heat capacity; thermal conductivity; and thermal diffusivity. These are properties of the bulk of the sample, and they can be measured using various commercial instruments or evaluated by laboratory base apparatus. However, caution must be exercised when using data on these parameters from the literature, as the sample under consideration may be far removed from an ideal sample of the same category for which data are quoted. The definition and other relevant aspects of these parameters are briefly described below. These parameters will be needed to estimate ignition temperatures, ignition delay time and minimum laser energy for ignition initiation in energetic materials.

9.2.2 Heat Capacity

When an object is heated, its temperature rises and, conversely, its temperature decreases when it is cooled. The relationship between the heat (q) that is either delivered or removed, and the change in temperature (ΔT) is simply related as:

$$q = C\Delta T = C(T_f - T_i) \tag{9.8}$$

The constant of proportionality in the above equation is called the heat capacity (C). The heat capacity, therefore, is the amount of heat required to raise the temperature of a material by one degree. The temperature change is the difference between the final temperature (T_f) and the initial temperature (T_i). In SI (International System) units, it is expressed as Joules per degree Kelvin of temperature (J K^{-1}). For many experimental and theoretical purposes, it is more convenient to report heat capacity as the bulk property, which is independent of the size or quantity of the material. This is most often done by expressing the property in relation to a unit of mass. For practical considerations, the international standard requires that the heat capacity is designated as specific heat capacity, or 'specific heat', and expressed as per unit of mass, i.e. J kg^{-1} K^{-1}. In some applications, the mass is expressed in molar concentrations (moles), and the specific heat is then expressed as J mol^{-1} K^{-1}. In some circumstances, and particularly for solids and liquids, the specific heat may need to be expressed as volume heat capacity, and in SI unit that must be expressed as: J cm^{-3} K^{-1}, the specific heat is simply divided by the density of the material.

9.2.3 Thermal Conductivity

Thermal conductivity (k) relates the ability of a material to conduct heat. Conduction takes place when a temperature gradient exists in a solid (or stationary fluid) medium. Conductive heat flow occurs in the direction of decreasing temperature because higher temperature equates to higher molecular energy or more molecular movement. This involves transfer of energy within a material without any mass motion as a whole. Thermal conductivity (k) is defined as the quantity of heat (Q), transmitted through a unit thickness (L), in a direction normal to a surface of unit area (A), due to a unit temperature gradient (ΔT), under steady state conditions and when the heat transfer is dependent only on the temperature gradient. This is expressed as:

$$k = Q \times L/(A \times \Delta T) \tag{9.9}$$

Thermal conductivity, k, is expressed in unit of Joules per unit cm per second and per degree Celsius (J cm^{-1} s^{-1} $°C^{-1}$)). Since J s^{-1} is the unit of power of the source, the unit of conductivity in SI unit may also be expressed as watts per metre, per Kelvin (w m^{-1} K^{-1}).

The law of heat conduction, also known as Fourier's law, states that the time rate of heat transfer through a material is proportional to the negative gradient in the temperature and to the area, at right angles to that gradient, through which the heat is flowing. In differential form, the flow rates or flux of energy is considered to take place locally. For a homogeneous material, and for the consideration of one-dimensional geometry between two endpoints at constant temperature, the heat flow rate, which forms the basis for the derivation of the heat equation, is given as:

$$\Delta Q/\Delta t = -k\,A\,(\Delta T/\Delta x) \quad (9.10)$$

where:

A = the effective cross-sectional surface area of a thin slab of the material.
ΔT = the temperature difference between the ends
Δx = the distance between the ends
k = the thermal conductivity, as defined earlier.

Several commercial systems for measuring the conductivity parameter of energetic materials are available in the market. These are either based on micro-calorimeter or photo-acoustic spectroscopy. Thermal properties of some energetic materials, derived from published literatures, are presented in Table 9.2.

Table 9.2 *Thermal properties of some energetic materials and sensitizers.*

Materials		Temperature of Ignition (°C)	Heat capacity (J g^{-1} °C^{-1})	Thermal conductivity (J cm^{-1} s^{-1} °C^{-1})
Secondary explosives	HMX	335 [7]	1.05 (at 37 °C)	5.02×10^{-3} (at 25 °C) 4.05×10^{-2} (at 160 °C)
	HNS	325 [7]	1.02 (at 37 °C)	—
	RDX	260 [7]	0.97 (at 37 °C)	1.08×10^{-3} (at 41 °C)
Optical sensitizers	Carbon	400 [8]	0.71 (at 27 °C)	1.19–1.65 (at 27 °C)
	Chemical dye D	222 [9]	—	—
Propellants	Double-base (DB)	≈200 [10]	1.38	—
	Composite modified DB	≈200 [9]	1.3	—

9.2.4 *Thermal Diffusivity*

Thermal diffusivity is, for all intent and purposes, the measure of thermal inertia. Heat moves rapidly through a substance with high thermal diffusivity because the substance conducts heat quickly relative to its volumetric heat capacity or 'thermal bulk'. The substance generally does not require much energy transfer to or from its surroundings to reach thermal equilibrium. In heat transfer analysis, the thermal diffusivity, α, is the thermal conductivity divided by density and specific heat capacity and, in SI units, it is given as:

$$a = k/(\rho C) \quad (9.11)$$

where:

k is thermal conductivity (Wm^{-1} K^{-1})
ρ is density (kg m^{-3})
C is specific heat capacity (J kg^{-1} K^{-1}).

Thermal diffusivity is often measured with the so-called flash or laser flash method. It is based upon the measurement of the temperature rise at the rear face of the thin-disc specimen produced by a short energy pulse on the front face. With comparison to a reference sample specific heat of any unknown sample can be measured. The energy pulse heats one side of a plane-parallel sample, and the time-dependent temperature rise on the back side due to the energy input is detected. The higher the thermal diffusivity of the sample, the faster the energy reaches the back side. Commercial equipment for measuring thermal diffusivity is widely available. In the one-dimensional, adiabatic case, the thermal diffusivity, α, is calculated from this temperature rise as follows:

$$a = 0.1338\left(d^2/t_{1/2}\right) \tag{9.12}$$

where d is the thickness of the sample and $t_{1/2}$ is the time to reach the half maximum.

References

(1) Brewer, M.Q. (1992) *Thermal Radiative Transfer and Properties*, John Willey and Sons, NY, p. 162.
(2) Bohren, C.F. and Huffmann, D.R. (2010). *Absorption and Scattering of Light by Small Particles*, Wiley-Interscience, New York, ISBN 3-527-40664-6.
(3) Isbell, R.A. and Brewer, M.Q. (1998) Optical properties of energetic materials: RDX, HMX, AP, NC/NG and HTPB. *Propellants, Explosives, Pyrotechnics*, **25**, 218–224.
(4) Östmark, H., Carlson, M. and Ekvall, K. (1994) Laser ignition of explosives: effect of laser wavelength on the threshold ignition energy. *Journal of Energetic Materials*, **12**, 63–83.
(5) de Yong, L., Nguyen, T. and Washi, J. (1995) Laser ignition of explosives, propellants and pyrotechnics. Technical report, Aus. Gov. Of Def. Report # DSTO – TR-60068.
(6) Sanghvi, R.R., Sundaram, S.G., Kulkarni, M.W. *et al.* (2005) Studies on ignition of TPE based RDX propellants by laser impulse. *Journal of Scientific and Industrial Research*, **64**, 175–180.
(7) Jacqueline, A. (1998) *The Chemistry of Explosives*, The Royal Society of Chemistry, Athenaeum Press Ltd., UK.
(8) http://www.taftan.com/thermodynamics/IGNITION.HTM.
(9) Crawford, B.L., Huggett, C. and McBrady, J.J. (1950) The mechanism of the burning of the double-base propellants. *The Journal of Physical Chemistry*, **54**(6), 854–862.
(10) Polo, M.J. (2010) Low flux radiative ignition studies of ammonium perchlorate composite propellants. Thesis, University of Illinois, USA, p. 21.

10

Theoretical Aspects of Laser Interaction with Energetic Materials

10.1 Introduction

Energetic materials come with many diverse chemical and physical characteristics. Lasers also come in a plethora of sizes, shapes and, of course, with a wide variety of output power, operating at different discrete wavelengths. For efficient and effective ignition initiation commensurate with practical applications, there must be a match between the properties of the selected material and those of the selected laser system. For a specific material, and for a specific application, an appropriate laser system needs to be located in the market or custom-built for optimum optical performance. The performance may be characterized either by the ignition delay time or by the minimum laser power density (power per unit area at the target) requirements.

For a commercially available suitable laser system, the sample may be modified by the use of sensitizers to achieve the desired matching. It is, however, desirable to achieve ignition in such materials using minimum power from a technically feasible, economically viable and commercially available laser source, and with the need for minimum chemical modification of the materials. In the UV wavelength band, for example, the fourth harmonic of Nd:YAG laser at 215 nm (corresponding to the photon energy of. 4.4 eV), direct photo-dissociation or ionization leading to exothermic bond-breaking is possible. In this case, a low level of laser power density at the target may be sufficient to produce an ignition event. However, such laser sources, albeit operating at low power level, are usually large, expensive and cumbersome to operate, and are deemed to be unsuitable for practical applications in laser ignition initiation systems.

Alternatively, infrared lasers such as CO_2 lasers, operating at around 10.6 μm (corresponding to photon energy of $\approx$ 0.11 eV) and Nd:Yag laser at 1.06 nm (corresponding to photon energy of $\approx$ 1.1 eV), operating in the range of GWatts cm^{-2} peak power densities, may cause direct photo ionization, leading to ignition events by multi-photon photo effect.

Laser Ignition of Energetic Materials, First Edition. S Rafi Ahmad and Michael Cartwright.

These types of lasers are also big and bulky and are deemed unsuitable for most laser ignition applications. It is envisaged that commercially and techno-economically viable laser sources, operating in the UV and in the IR wavelength bands, may be available in the future. For the present, however, the best prospect for lasers for this application is afforded by cheap and cheerful diode lasers. In the present analysis, therefore, emphasis is put on the use of diode lasers operating in the near-IR wavelengths within the band 750–850 nm. This is because small, efficient and cheap diode lasers operating at this wavelength band and providing high output power are commercially available for a wide variety of possible practical uses for initiating ignition in explosives, propellants or pyrotechnics.

10.2 Parameters Relevant to Laser Interaction

Ignition initiated in energetic materials may lead to different kinds of events in different types of materials. In pyrotechnics ignition at a point at the surface or within the bulk of the material may lead to sustained burning, commonly known as 'combustion'. This process may propagate through the depth and breadth of the material and ultimately consume it totally to produce a mass of ash. In solid propellants ignition, under confined or semi-confined condition, may lead to the propagation of burning at a fast but still sub-sonic speed within the material, commonly known as 'deflagration'. This may lead to the production of gasses, giving rise to a propellant force into a confined system. In solid high explosives, ignition may, under some circumstances, lead to deflagration, propagation within the material at a supersonic speed, and lead to detonation. Parameters affecting laser initiation of ignition in energetic materials, which will be introduced in the theoretical analyses of the optical interaction process, are summarized as follows:

10.2.1 Laser Parameters

- Photon energy (wavelength of the laser light; nm/μm).
- Photon flux (rate of photon/average power; J. sec^{-1})
- Peak power (maximum power within the laser pulse; watts).
- Power density (peak value) (intensity/fluence; watts cm^{-2})
- Beam uniformity (defining hot spots or Gaussian profile).
- Duration of irradiation (Pulse length; ns/μs/ms).
- Dimension of focused spot (defining power density and governing heat dissipation).

10.2.2 Material Parameters

- Spectral absorptivity of the target surface.
- Properties of optical sensitizers within the material.
- Thermal conductivity of the bulk material.
- Uniformity within the bulk of the material.
- Chemical composition and morphology of the material.
- Ignition temperature of the material.

Definitions, symbols and dimensions of the laser parameters used in the analyses that follow are given in Table 10.1.

Table 10.1 *Material and laser parameters used in the analyses of LI processes.*

Notation	Description	Dimension (SI unit)
α_a	Optical absorption coefficient*	(1/length), m^{-1}
R_{rel}	Optical reflection coefficient	No dimension
P_0	Laser power (at the centre of the beam).	(watt), W
$I(r)$	Laser power density on the surface of a target at a distance *r* from the centre.	(watt/area), $W.m^{-2}$
I_0	Laser power density at the focused area of the beam	(watt/area), $W.m^{-2}$
τ_l	Duration of laser irradiation (pulse width)	(second), s
ω	Beam radius ($1/e^{th}$ value)	(distance from the centre), m
ε_l	Laser energy deposited over the irradiation time on to the target area.	(joule/area), $J.m^{-2}$
E_{ign}	Minimum laser energy for the initiation of ignition, (ignition energy)	(joule/area), $J.m^{-2}$
E_a	Activation energy (for a photochemical reaction event).	(joules/mole), $J.mol^{-1}$
Q_{las}	Heat energy generated per unit volume (absorption length × effective beam area) of the sample due to absorbed laser energy	(joules/volume), $J.m^{-3}$
α_d	Thermal diffusivity, i.e. rate of temperature flow ($= k/(\rho \times C_p)$)	($meter^2$/time), $m^2.s^{-1}$
k	Thermal conductivity	watt/(distance × temp.), $W.m^{-1}\ K^{-1}$
ρ	Density of the material	(mass/volume), $kg.m^{-3}$
C	Specific heat (heat capacity)	(joule/(mol.K)), $J.\ mol^{-1} K^{-1}$
T_0	Ambient temperature of the sample	(kelvin), K
T_{ign}	Ignition temperature	(kelvin), K

*Note that the symbol used for absorption coefficient in Chapter 9 is, K_a.

Laser power density on the surface of a target may be varied by two conventional means. One is by varying the power by controlling the voltage in the electronic power pack regulating the current in a diode laser or by using neutral density filters in the path of the beam. Another alternative means is to vary the beam diameter at the target using a focussing lens.

10.3 Mathematical Formalism

10.3.1 Basic Concept

Theory governing the interaction of a laser beam with energetic materials is based on the conventional heat transfer equation in solid materials [1]. Ignition initiation process is a direct consequence of the instability of energetic materials on heating. This instability makes energetic materials, particularly high explosives, quite unsafe to process, transport, handle and store. However, this vulnerability to heat makes these materials release their internal energy either through combustion or explosion. For effective performance, the material used in any application needs to be heated throughout the bulk. Such volumetric heating must start at the surface or at a point on the surface and be then conducted away throughout the volume. In order to effect combustion or explosion, the material needs to be ignited. The ignition (burning), depending upon the circumstances, may sustain and propagate along the depth and breadth of the material to consume it, or propagate at a supersonic speed within the volume, culminating in a detonation event.

There are various methods for the initiation of ignition, and commonly applied methods are percussion, stabbing and electrical signal (vide afro). For secondary explosives, the initiation method is almost always by using a shock wave which, in turn, is generated by the explosion of a primary explosive, known as a donor charge. The use of directed energy from a laser for ignition initiation for affecting combustion or detonation had been considered in earnest a couple of decades ago, for safety considerations. The topic is now high on the political and economic agenda of agencies who manufacture, process, distribute and, most importantly, use such materials, particularly high explosives.

When a laser beam is incident on the surface of a solid material, the optical energy is partly reflected, partly scattered and partly absorbed within the skin depth of the material. The fraction of the energy absorbed will depend upon the optical, physical and thermal properties of the material in relation to the intensity and wavelength of the laser radiation. Under a steady state condition, laser heating will proceed consecutively through phases of inert heating, pre-ignition chemical reaction and self-sustained combustion, while, at the same time, the combustion process is being transported away from the point or region of the laser impact into the bulk of the sample. Depending on the competing temporal histories of the laser pulse and the material emission dynamic, attenuation of laser power due to self-absorption will ensue. The scenario depicting the temporal history and the phases that this induce in the material are sketched in Figure 10.1, which is a hypothetical and idealized plot of the temporal history of the development of sustainable ignition following laser irradiation.

The temporal history of laser ignition, so-called 'ignition map', is a plot of laser irradiation time (in milliseconds) against the laser flux (power density). For low laser fluxes, the sample will only undergo inert heating, as the heat loss through conduction and convection is higher than the heat generation by the laser source. Beyond a threshold value of the laser flux, exothermic or endothermic reaction will commence. This region will be attained faster with increasing laser flux values. For still higher flux values, the dynamic will change, and the ignition threshold will be reached at a higher value. Beyond this threshold value, the ignition energy will be strong enough to give rise to sustained combustion or deflagration, allowing the process to consume the bulk of the sample. The final outcome (i.e. the generation of gases in propellant materials or detonation in explosives) will be governed by the rate of the combustion process (subsonic or supersonic), and will be discussed later.

10.3.2 Optical Absorption

The volume that is affected at the initial stage of the laser interaction is defined by the product of the area of the (focused) beam on the surface of the target and the 'skin-depth'.

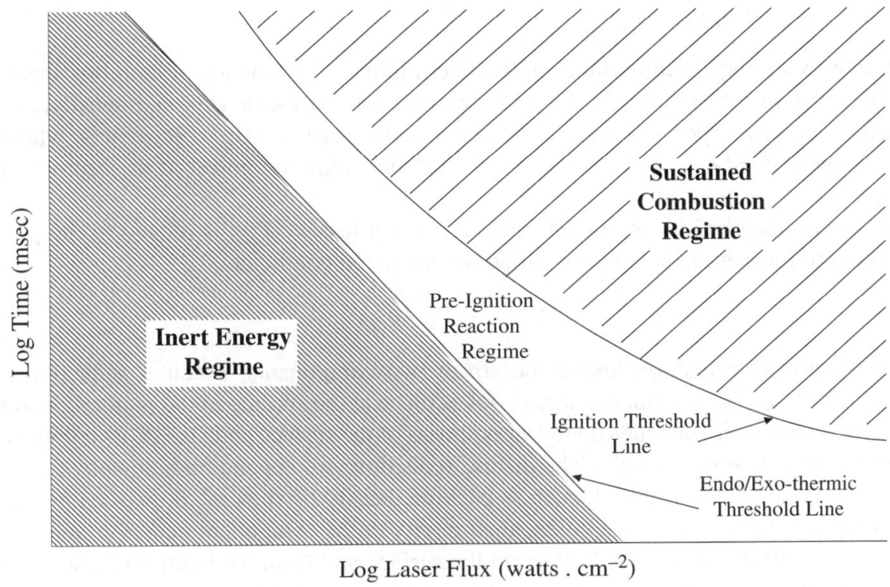

Figure 10.1 *Schematic representation of the temporal history of laser ignition (ignition map).*

The later is defined as the penetration depth of the laser beam at which the beam intensity has dropped to $(1/e)^{th}$ of its peak value, and it may be expressed as:

$$\delta_s = 1/\alpha_a \tag{10.1}$$

where α_a (cm^{-1}) is the wavelength-dependent absorption coefficient, which appears in the standard Beer-Lambert equation:

$$I(z) = I_0 \ \exp(-\alpha_a z) \tag{10.2}$$

where:

I_0 is the power density (P_0/area) incident on to the target
$I(z)$ is the intensity at a depth z, which is the axis along the depth of the sample.

Note that this equation does not contain a factor representing the reflection losses. The absorption coefficient, α_a, is related to a dimensionless parameter, the absorption index, k, which is the imaginary part of the refractive index of the material, as $\alpha_a = (4\pi k)/\lambda$, so that the skin depth, from Equation 10.1 is given as:

$$\delta_s = 1/\alpha_a = \lambda/(4\pi k) \tag{10.3}$$

where λ is the wavelength of the light. For opaque materials, k can be measured or calculated theoretically from other properties of the materials (see Chapter 9).

10.3.3 Optical Reflection

Reflectivity will play an important role in determining the efficiency of coupling incident optical power to the sample. In the analyses that follow, incident laser intensity, I_{in}, is defined as the power per unit area of the beam at the target, i.e. $I_{in} = P_{in}/(\pi w^2)$, where w is the radius of the beam at the target at $1/e^{th}$ of the maximum power at the centre of the beam.

The power absorbed by the target (after reflection losses), P_{ab}, is related to the power incident onto the sample by a factor called the reflection coefficient:

$$R = 1 - (P_{ab}/P_{in})$$

Contribution to reflection loss comes from two components: specular reflection and diffuse reflection. The specular reflection takes place at the surface of the material, and the loss due to this process depends on the smoothness of the surface and also on its colour with respect to the colour of the light. The diffuse reflection will depend on many physical and optical properties of the material, such as packing density (in the case of pressed powder), crystal form, complex refractive index, absorptivity of the component materials and so on. This type of reflection can arise both from the surface and from the interior of the sample; the later is due to scattering processes. The reflection coefficient is most often measured with an integrating sphere attachment in a commercial absorption spectrometer. For this, the use of standards (e.g. pressed $BaSO_4$ powder) is common. The absorbed power is then related to the incident power as:

$$P_{in} = P_{ab}/(1 - R) \tag{10.4}$$

10.4 Heat Transfer Theory

Generation of heat in a material by a laser beam incident on its surface, and the subsequent propagation of this heat across the bulk of the sample, constitute the theoretical basis of laser ignition of energetic materials. Before setting up the mathematical formalism for the process in an energetic material, it is considered prudent to review the established heat transfer theory within a solid material as follows.

The heat transfer equation is deduced from the first principle of energy conservation in any interaction process. When heat is generated on the surface of a solid material (or compact powdered material) by an external source, the conservation of energy demands that:

$$(Q_{in} - Q_{out}) = (Q_{gen} - de/dt),$$

Where:

Q_{in} and Q_{out} are the amount of heat going into and the amount coming out of the system respectively

Q_{gen} is the amount of thermal energy generated within the system

de/dt is the rate of dissipation of the internal energy.

The left hand side represents the heat flux (i.e. the net amount of heat dissipated within the sample), and the right hand side represents the net power per unit volume, generated after taking account of the increase in the internal energy ($e_0 \rightarrow e_t$) per unit volume of the sample within the laser irradiation period, dt ($t_0 \rightarrow t_t$). In mathematical notation, this is expressed as:

$$\nabla Q = (Q_{gen} - de/dt) \tag{10.5}$$

The change in the internal energy is quantified by the change (rise) in the temperature of the sample over that from the ambient, and is given by,

$$de/dt = \rho\, C(dT/dt) \tag{10.6}$$

where ρ and C are the density and specific heat of the material respectively. The phenomenon of heat flow from a cold to a hot region is described by the Fourier heat transfer law (after the name of the French physicist):

$$Q = -k \times \nabla T \tag{10.7}$$

where the – sign represents the fact that heat propagates in the direction of a colder to a warmer region, and k is the thermal conductivity of the material, having a dimension of watts. m^{-1}. K^{-1} (in SI unit). Combining Equations 10.5, 10.6 and 10.7, and rearranging terms, we obtain the generalized heat transfer equation, as follows:

$$\nabla^2 T - (1/\alpha_d)(\partial T/\partial t) = -(1/k) \times Q_{gen} \tag{10.8}$$

The parameter, α_d is known as the thermal diffusivity of the sample and is defined as, $\alpha_d = k/(\rho\, C)$ with the dimension of $m^2\ sec^{-1}$, which is the rate of the spread of heat across the surface.

The above equation represents the relation between time-dependent net rise in temperature with the net heat energy generated in the sample by an external stimulus. For a laser beam as the energy source, and for a homogenous solid energetic material (e.g. an explosive sample), the energy term should be a combination of contributions from the volumetric heating of the sample due to absorbed laser beam and the heat generated due to the exothermic reaction within the explosive material. The above equation (rearranging the terms) can now be written as follows:

$$(1/\alpha_d)(\partial T/\partial t) - \nabla^2 T = 1/k(Q_{ex} + Q_{laser}) \tag{10.9}$$

Where:

Q_{laser} (watts) = power generated by the laser beam per unit (effective) volume
Q_{ex} (watts) = power generated per unit volume due to exothermic reaction.

The heat generated due to exothermic reaction has been assumed (justified by experiments) to follow Arrhenius kinetics. For a zeroth order Arrhenius kinetic, it is expressed as follows:

$$Q_{ex} = E_{re}(\rho \alpha_d C/k)\exp(-E_a/k_B T) \tag{10.10}$$

Where:

E_a is the activation energy (J molecule^{-1})
E_{re} is the energy generated due to reaction (J)
k_B is the Boltzmann constant
T is the ambient temperature of the sample in Kelvin unit.

Note that $k_B = R/N_A$, where R is the Universal Gas constant and N_A is the Avogadro's number. In chemistry, R is used instead of k_B, and the activation energy in that case is expressed as KJ mole^{-1}.

The second term in Equation 10.8 takes account of the power generated due to the absorption of laser light. For a Gaussian intensity distribution across the diameter of the beam on the target material, intensity distribution is expressed as:

$$I(r) = I_0 \exp(-r^2/\omega^2) \quad (10.11)$$

where:

I_0 is the effective peak (central) laser intensity (after reflection losses) from the surface of the target
r is the radius of the effective heated region
ω is the radius of the beam at $1/e^{th}$ of the peak intensity level. In this case, $I(r) = P(r)/(\pi\omega^2)$, P being the total effective optical power (integrated over the beam area of $\pi\omega^2$).

For a laser source, the time and space-dependent power generated per unit effective volume of the sample is expressed as the product of the net incident power (after reflection losses) and the fraction absorbed. Therefore, from Equations 10.2 and 10.10, and expressing the power as intensity, we get:

$$Q_{laser} = (\pi\omega^2)I_o \exp(-\alpha_a z)\exp(-r^2/\omega^2) \quad (10.12)$$

Using Equations 10.9 and 10.11, the heat transfer equation set up in Equation 10.8 can now be expressed as:

$$\{(1/\alpha_d)(\partial T/\partial t) - \nabla^2 T\} = E_{re}(\rho\alpha_d C/k)exp\{-E_a/(k_B T)\} + (\pi\omega^2/k)I_o\exp(-\alpha_a z)\exp(-r^2/\omega^2) \quad (10.13)$$

One of the objectives in the theoretical analyses of laser interaction with energetic materials is to deduce an equation which may be used for the estimation of the threshold optical energy for the initiation of ignition in a sample at the point of interaction. This means the equation needs to provide a relation for the critical or minimum laser energy required for crossing the boundary in the ignition map (Figure 10.1) to achieve sustainable combustion in the material, in terms of laser and material parameters. The laser energy per unit area, E_l, deposited onto the sample is defined as:

$$E_l = (P_l \times t)$$

where t is the laser irradiation time.

The critical energy may be obtained either by increasing the laser intensity (above the lower limit) for a fixed irradiation time, or by increasing the irradiation time for fixed laser intensity. Equation 10.12, relating the laser parameters and the parameters of the material for temperature rise due to an incident laser beam, can not be solved analytically to obtain the critical values of ignition temperature or minimum laser intensity for ignition initiation. However, for special cases, this may be solved with some simplifying assumptions. For this analysis (collected from the cited literature), the following assumptions have been made:

i. The sample is homogenous and effects due to hot spots are negligible.
ii. The sample is chemically inert for temperatures up to that corresponding to the ignition boundary line (see ignition map in Figure 10.1).
iii. The beam area is relatively large for transverse heat losses to be negligible.
iv. Ignition takes place at a temperature T_g, (ignition temperature) and after a time τ_i, from the onset of laser irradiation (ignition time).

Under the above conditions, and for heat transfer consideration only along the thickness (depth, z) of the sample, the laser-initiated heat transfer equation (Equation 10.13) has been expressed as:

$$(1/\alpha_d)(\partial T/\partial t) - \nabla^2 T = (\pi\omega^2/k)I_o \exp(-z)\exp(-r^2/\omega^2) \qquad (10.14)$$

Temperature, $T(z, t)$ is a function of distance, z along the depth of the sample, so that the initial temperature within the volume defined by the laser beam area at the surface and the skin depth (Equation 10.4) is written with the proviso, $T(z, 0) = T_0$, which is the ambient temperature of the sample before laser interaction. As the laser irradiation time increases, the heat diffusion starts, both across (radially, r) and along the depth (z) and, the temperature gradient:

$$\partial T/\partial z \to 0 \text{ as } z \to \infty, \quad \text{for } t > 0 \qquad (10.15)$$

Above criteria define the ignition temperature, T_{ign}, as the temperature at the point when, for all intent and purposes, gradient $\partial T/\partial z = 0$, i.e. the temperature is maximum.

For a laser irradiation time, τ_l (or, in the case of pulsed irradiation for laser pulse width, τ_p) the last factor in Equation 10.14, which takes account of transverse diffusion of heat, can be expressed in terms of the diffusivity parameter, as introduced in Equation 10.18, so that $r = (\alpha_d \times \tau_l)$, so that the Equation 10.14 can now be written as follows:

$$(1/\alpha_d)(\partial T/\partial t) - \nabla^2 T = (\pi\omega^2/k)I_o \exp(-\alpha_a z)\exp(\alpha_d^2\tau_l^2/\omega^2) \qquad (10.16)$$

The above equation can be further simplified with the assumption that the thermal diffusivity is small enough for transverse heating to be negligible. This will be strictly true if the laser beam area is much greater than the absorption depth, i.e.

$$\omega > a_d \times \tau_l.$$

Therefore, for a relatively short laser irradiation (pulse duration), samples with slow thermal diffusion rate, and for large laser beam area commensurate with the required laser power density, the heat transfer equation can be further simplified as below:

$$(1/\alpha_d)(\partial T/\partial t) - \partial^2 T/\partial^2 z = (\pi\omega^2/k)I_o \exp(-\alpha_a) \qquad (10.17)$$

The absorption parameter, α_a in the above equation is now expressed in unit of per unit length (along z direction). For the above scenario, the skin depth needs to be redefined as the ignition depth in terms of the laser irradiation time (pulse duration) and is expressed as follows:

$$\delta_{ign} = (\alpha_d \tau_1)^{-1/2} \tag{10.18}$$

The heat transfer equation can now be solved to find an expression for the threshold ignition intensity, $I_{ign.}$ (corresponding to the ignition energy, $\epsilon_{ign} = (I_{ign} \times \tau_1)$ – that is, the threshold value of laser energy required for achieving sustained combustion (or detonation in the case of explosives under confined condition), relating to some experimentally measureable laser and sample parameters. There have been many reports [2–5] providing analytical solutions to Equation 10.16, under various conditions and simplifying assumptions. However, the solutions and formalism proposed below, for two different special situations by two different authors, have been experimentally verified to provide good estimates of relevant parameters for laser ignition studies. These are as follows:

i. Laser pulse duration is small so that the heat conduction along the depth (z direction) may be neglected. This condition may approximately be met when ignition depth is smaller than the skin depth, $\delta_s > \delta_{ign}$, for this case $\partial^2T/\partial^2z = 0$ and the heat transfer equation can be written as:

$$(1/\alpha_d)(\partial T/\partial t) = (\pi\omega^2/k)I_o\exp(-\alpha_a) \tag{10.19}$$

The above equation for a short burst of laser irradiation (or short pulse) has been solved [6] analytically using the notations below:

$$T = T_0 \text{ at } t = 0; \text{ and } T = T_{ign} \text{ at } t = \tau_{ign}$$

The solution provides an expression for the threshold laser power required for achieving sustained combustion. Taking into consideration the reflection loss in Equation 10.14, the expression for the threshold incident laser power is given as follows:

$$(P_{in})_{ig} = \pi\omega^2/(1 - R) \times (\rho C/\alpha_a\tau_1)(T_{ig} - T_o) \tag{10.20}$$

In terms of total incident laser energy:

$$\varepsilon_{ign} = (P_{in})_{ig} \times \tau_1 = \pi\omega^2/(1 - R) \times (\rho C/\alpha_a)(T_{ig} - T_o) \tag{10.21}$$

$$T(\tau_1) = T_0 + (1 - R)(\pi\omega^2)^{-1} \quad (\alpha_a/\rho C)P_{in}\tau_1^{-1} \tag{10.22}$$

ii. In a situation encountered most often in practical circumstances, where absorption depth is much smaller than the skin depth, i.e. $\delta_{ign} < \delta_s$, the laser energy is mostly absorbed on the surface ($z = 0$) and $k = 0$, so that the heat transfer equation can be simplified to:

$$(1/\alpha_d)(\partial T/\partial t) - \partial^2 T/\partial^2 z = 0 \tag{10.23}$$

The temperature gradient can be related to the incident laser intensity as $\partial T/\partial z = I_0$. Based on this, an analytical solution to the above equation (Equation 10.23) has been found [7]. Assuming that temperature gradient decreases rapidly with the increasing depth, i.e. $\partial T/\partial z \to 0$ as $z \to \infty$ at the surface ($z = 0$) and writing $T(r, 0) = T_0$, an equation relating the laser intensity, I_{ig}, at which the temperature reaches the critical value, T_{ig}, for sustained ignition to occur (see ignition map in Figure 10.1) is written as:

$$(I_{ab})_{ign} = (k/2)[(\pi/\alpha_d)]^{1/2}(T_{ign} - T_0) \tag{10.24}$$

Taking account of reflection loss (Equation 10.4), and expressing in terms of total incident, threshold laser power is expressed as follows:

$$(P_{in})_{ig} = (\varepsilon_{in})_{ign}/\tau_l) = \pi\omega^2 k/2(1 - R) \times [(\pi/\alpha_d)]^{1/2}(\tau_l)^{-1/2}(T_{ign} - T_0) \tag{10.25}$$

Equations 10.20 and 10.25 provide guidance to the dependence of the threshold laser power (or energy) on the ignition delay time, among other parameters of the materials, under two different regimes of relevance for specific experimental conditions. Note that for short laser pulses or irradiation times, the threshold ignition energy is practically independent of laser pulse duration, as given in Equation 10.21. On the other hand, if the thermal diffusivity (i.e. if the ignition depth) is large in comparison to the skin depth, the threshold energy will depend on the laser pulse or irradiation duration, as shown in Equation 10.25.

For long laser pulses, the temperature rise will be dependent on the laser pulse width or the irradiation time, and the expression for this parameter is deduced from Equation 10.23, as given below:

$$T(\tau_l) = T_0 + [2(1 - R)(\pi\omega^2)^{-1}k^{-1}((\pi/\alpha_d)^{-1/2} P_{in} \times (\tau_l)^{-1/2}] \tag{10.26}$$

It is shown that for a specific sample and a laser wavelength, and under a quasi-steady state condition (i.e. P_{in} is constant), the temperature rise will depend inversely on the square root of the pulse duration or the irradiation time of the laser. For the case of a short pulse, the temperature rise depends linearly on τ_l. Note, from Equations 10.21 and 10.25, that there is a trade-off between the minimum laser power and the irradiation time for achieving ignition. Achievement of maximum magnitude of the temperature rise and, for that matter, for minimum the threshold ignition laser power, will depend on various parameters and, the criteria that need to be considered (see Equations 10.21 and 10.25) are as follows:

1. Minimum laser beam area at the target for a fixed power level (i.e. high power density by focusing the beam).
2. Minimum reflection loss.
3. Long laser pulse duration or irradiation time (above the critical power to undergo sustained inert heating and temperature rise).
4. High diffusivity ($\alpha_d = k/(\rho C)$, i.e. low thermal conductivity, high packing density and high specific heat capacity.

References

(1) Frank-Kamenetskii, D.A. (1955) *Diffusion and Heat Exchange in Chemical Kinetics*, Princeton University Press, Princeton.
(2) Abdulazeem, M.S., Alhasan, A.M. and Abdulrahmann, S. (2011) Initiation of solid explosives by laser. *International Journal of Thermal Sciences*, **50**(11), 2117–2121.

(3) Rubenchik, A.M. (2007) On the initiation of high explosive by laser radiation. *Propellant, Explosive, Pyrotechnic*, **34**(4), 296–300.
(4) Binggorg, X. (2006) Experimental study and numerical simulation on laser ignition process. *Acta Armamenteri*, **27**(3), 533–536.
(5) Zhang, H. *et al.* Study on laser ignition model as its analytical solution of solid energetic material. *Chinese Journal of Lasers B* (English edition), **B9**(4), 376–384.
(6) Ostermark, H. (1985) Laser as a tool in sensitivity testing of explosives, in *Proc. 8th. Int. Symp. on Detonation* (ed. J.M. Short), NSWC, pp. 473–484.
(7) Carslow, H.S. and Jaeger, J.C. (1959) *Conduction of Heat in Solids*, 2nd edn, Oxford university press, Oxford.

11

Laser Ignition – Practical Considerations

11.1 Introduction

When the laser was first invented, it was hailed as a 'tool looking for applications' – and, as we know, applications it had, in almost every sphere of science and technology. A spate of research and development activities in recent times, on its applications for safe and remote ignition of energetic materials to replace the conventional electro-thermal ignition mechanism, was a natural consequence of this ethos. Because of the diversity of potential application scenarios and academic interests, there had been a plethora of reports over the last three decades on laser ignition of energetic materials, involving different types of lasers and different types of energetic materials, including the newly developed low vulnerable ammunitions (LOVA), also known as insensitive ammunitions (IM).

Theoretical analyses, reported in the previous chapter, were aimed at understanding the interaction dynamics between optical energy from lasers and energetic materials in order to allow prediction of suitable laser-material combinations for eventual real-world applications in safer and remote ignition initiation in energetic materials. Earlier experimental works mostly involved high-power gas lasers, such as the CO_2 laser ($\approx$10.6 μm), the Ar-ion laser (@500 nm) and so on. Such lasers are bulky and costly and are therefore not likely to find any significant applications in real-world laser ignition initiations. However, initial research using such lasers had allowed validation of laser-interaction formalisms, and provided indications of the limit and scope of applicability of the theory established so far. A futuristic application scenario that may be pictured is the use of a compact gas or solid-state laser for the ignition of explosives, propellants and pyrotechnics for the overall control of performances in an intercontinental ballistic missile, as shown as an artist's sketch in Figure 11.1.

For both military and civilian applications of laser as a safe and efficient igniter of explosive devices, two most important practical considerations are: cost-effective laser

Laser Ignition of Energetic Materials, First Edition. S Rafi Ahmad and Michael Cartwright.

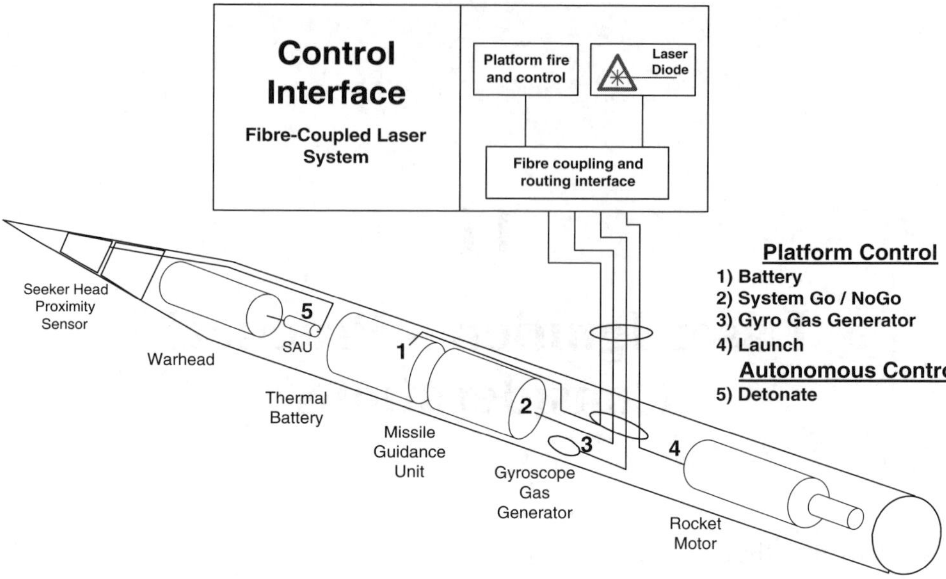

Figure 11.1 *An artist's sketch of a guided weapon system (GWS) operated by different laser beams taken from a single powerful laser.*

source and, suitable beam delivery system. These two aspects for the practicability of replacing conventional igniters with laser based systems are discussed below.

11.1.1 Laser Source

For any practical application, an igniter for energetic materials needs to be compact, small, lightweight and cost-effective. The types of lasers that are most likely to fit the bill are the small and highly efficient semiconductor lasers commonly known as 'diode lasers'. Commercially available diode lasers are capable of delivering a beam with high power, and they conform well to the criteria specified above (i.e. they seem to be quite cost-effective). These lasers operate at wavelengths within the near-infrared wavelength band between ≈780 nm to ≈820 nm and, unfortunately, these wavelengths rarely coincide with the peak of the absorption bands of most energetic materials.

One of the ways to achieve ignition in energetic materials having poor absorptivity at diode laser wavelengths is to use quite high laser power. Use of a high-power laser, however, will defeat the objective of 'cost-effectiveness' and also compromise other criteria such as portability, safety and so on. The cost of diode lasers goes up with increasing maximum output power and better beam quality. The latter is dictated by the spatial homogeneity of the beam, its closeness to a Gaussian profile and the beam divergence. Good beam quality will allow better focusing, i.e. a smaller beam area at the target. This, in turn, will provide higher intensity (power per unit area) from a small laser system to augment ignition efficiency. To fulfil the first requirement for the practical viability of laser ignition, cheap-and-cheerful low-power diode lasers can be used effectively if the target material is sensitized to enhance its absorptivity using an appropriate optical absorber. This will also minimise reflection losses in the interaction process.

Most energetic materials are organic powders or solids. For powders, the ignition process is usually carried out on compressed pallets. Contribution to reflection loss comes mainly from diffuse reflection. For all intents and purposes, such reflection may be considered to follow a Lambertian cosine law distribution (equal radiance in all directions). This arises from multiple reflections from surface irregularities arising from the grain boundaries of crystalline materials or the cell boundaries of roughened organic materials. For powdered materials, the grain size and compactness of the sample will make a significant contribution to the magnitude of the reflectivity of the sample. When optical sensitizers are used for matching the wavelength of the laser with the wavelength of the absorption band of the sensitizer (absorber), the reflection loss is minimized and the absorption is increased. To evaluate the effectiveness of sensitization, overall reflection loss (reflection coefficient) is experimentally determined using a special attachment in a commercial spectrometer, and the absorption spectra are constructed from these data. The practical aspects of optical sensitization in the laser ignition process are discussed later.

11.1.2 Beam Delivery System

An efficient way to deliver the laser beam to the target is an important consideration in the design of a practical laser igniter. The type of beam delivery system will depend on the particular application, and could either be a free-space one (using a lens to focus the beam onto the target) or through a low-loss fibre optic cable. In the latter case, the laser beam is focused onto the well-polished input end of the fibre, so that the beam area at the surface of the cable does not spread significantly beyond the effective area of the cable. The output end of the cable is placed as close to the target as possible. The effective area of the beam at the target will then be governed jointly by the effective area of the fibre and the distance of the tip from the surface if no lens is used for focussing the beam.

Although direct delivery of laser energy to the target is more efficient and less cumbersome, fiber optic delivery system has advantages. A synopsis of an indirect energy delivery scheme, considered by some researchers including one of the authors, is included in the following.

11.2 Laser Driven Flyer Plate

The 'Laser Driven Flyer Plate (LDFP)' technique is an indirect method for ignition initiation in explosives using lasers. Under a confined condition, and for a reasonably high laser power density at the target which is in the form of a thin metal foil, the laser irradiation may lead to detonation, and the setup for such a device is shown schematically in Figure 11.2.

In this technique, a thin (typically 5–10 μm) metal foil or a film (typically of aluminium), coated on to a transparent surface (e.g. a glass disc), is placed on to the groove of a plastic (e.g. PMMA) cylinder with a hole through the middle. The hole's dimension loosely matches that of the explosive stick (e.g. PETN) and, when inserted into the system, it maintains a small air gap (typically, 100 μm) between the surfaces of the metal and the end of the explosive column. This end of the device forms the widow of a confinement chamber. The

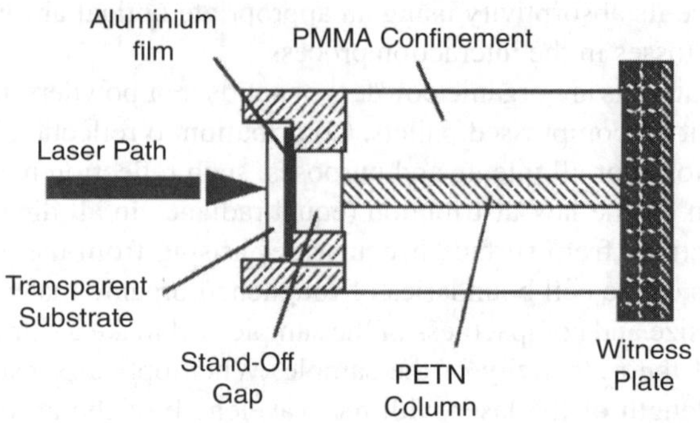

Figure 11.2 Schematic of a laser-driven flyer plate technique for the initiation of detonation in high explosives.

other end of the plastic cylinder may be blocked by a so called 'witness plate' to test the occurrence of the detonation and to estimate impact velocity from the depth and structure of the impact on the plate.

A pulsed laser beam, normally from a Nd-YAG laser, operating at 1.06 μm and typically providing an output of about 100 mJ, is focused on to the metal foil through the transparent window. The irradiated area gets heated rapidly and is ablated as plasma. This causes further heating of the underlying layer of the metal foil and imparts a mechanical force to eject the layer as a molten, fast-moving flat bullet onto the surface of the explosive column. The ensuing shock energy propagates along the bulk of the column as a wave at a supersonic speed (4–5 km s^{-1}), creating a 'shock to detonation transition (SDT)' within the confined space. If the confinement can withstand the shock force of the detonation, an indentation on to the witness plate will be produced.

Despite the fact that, for any practical applications, such a device will be rather complicated, it has the advantage that it can be used for all type of explosives, particularly HNS, which had been found to be very difficult or impossible to ignite directly by low power diode lasers. Although this method of ignition does not impose any restriction on the beam quality of the laser (spatial profile), it requires a high-power pulsed laser (i.e. a laser having a big size and weight), thereby making such a device less cost-effective. Despite the above drawbacks, research is under way [1] to improve its efficiency, and offer the possibility of using small, cheap-and-cheerful lasers by employing multi-layered films to make the metal film more absorbing, easier to vaporize and efficient in providing a thermal barrier, so that the flyer plate is ejected as a solid slab. For this, the use of other materials, such as Ge, Hf and Ti, are also being tried to produce a single layer having all the above properties.

11.3 Direct Laser Ignition

Research and development on the feasibility of laser ignition has been reported on a plethora of expulsive propellants and pyrotechnic materials over the last four decades. Tests and experiments were conducted with various types of lasers, particularly ultraviolet lasers (e.g. excimer lasers), visible and near-infrared lasers (e.g. Ar-Ion, Nd-YAG lasers) and frequency-converted outputs from high-power solid state Q-switched lasers and IR lasers (e.g. CO_2 laser). The latter was used for the near coincidence of the laser lines within the wavelength band of 10.4 μm to 11.2 μm and the wavelength of absorption peaks of most of HE materials.

Lasers operating in the UV-Vis wavelengths provide high peak power and the wavelength often coincide with maximum absorption bands of many explosive materials. While research using these lasers had been useful for the understanding of the basic interaction processes, these lasers are not considered to be practically viable for the present application. Therefore, in the following experimental considerations, outcome of tests and evaluation of performance, state-of-the-art technology, prospects for practical application, and the research need for specific applications, mainly concentrating on the use of diode lasers, will be evaluated under headings specifying the types of energetic materials. A typical experimental layout for laser ignition tests, using a cw laser and free space beam delivery system is presented in Figure 11.3a.

Diode lasers do have high beam divergence. Free space beam delivery from these types of lasers is inefficient and mostly ineffective, unless the laser head is very close to the target. This will certainly destroy the laser every time it is used for ignition test. For diode lasers, the beam delivery is carried out by using fibre optic cables, as shown in Figure 11.3b. Recent developments in high quality optical fibres have allowed very good transmission of light at the near-IR wavelengths and with adequate tolerance for high-power beam transmission, particularly for small distances in laser ignition trials.

11.3.1 Explosives

11.3.1.1 Types and Relevant Characteristics

Explosives are classified as 'low' and 'high', depending upon the rate at which they decompose (see Chapter 4). At a relatively low rate of decomposition, these materials undergo fast burning, known as deflagration, and produce huge amount of gas. These are generally known as propellants, and the laser ignitibility of such materials will be discussed later in this chapter. High explosives are, in turn, classified into two categories depending on their sensitivities to external stimuli: primary and secondary.

Primary explosives, such as mercury fulminate, some metal azides, nitro-glycerine and so on, are generally highly sensitive and their use is restricted due to their vulnerability to accidental combustion during processing, storage and transportation. Besides, these cannot be shaped or processed for controlled ignition initiations. The secondary explosives are relatively insensitive solid materials such as RDX, HMX. PETN, TNT, etc. In some applications, these are used as boosters for initiating detonations in primary explosives. There are other, much less insensitive, explosive materials which are currently undergoing tests and evaluation but which have not yet found in-service or commercial applications.

11.3.1.2 Sensitivity and Sensitiveness

An inevitable consequence of the lower sensitivity of some secondary HE materials to mechanical impact, heat pulse, electrical impulse and so on is their relatively lower sensitivity to laser radiation. To specify the degree of sensitivity to laser radiation impact, a term, 'sensitiveness' has been coined and generally accepted. Explosives having low sensitivity (high vulnerability), in turn, will invariably have lower sensitiveness to laser radiation. For practical applications, the choice of laser will, therefore, be dictated by commercially

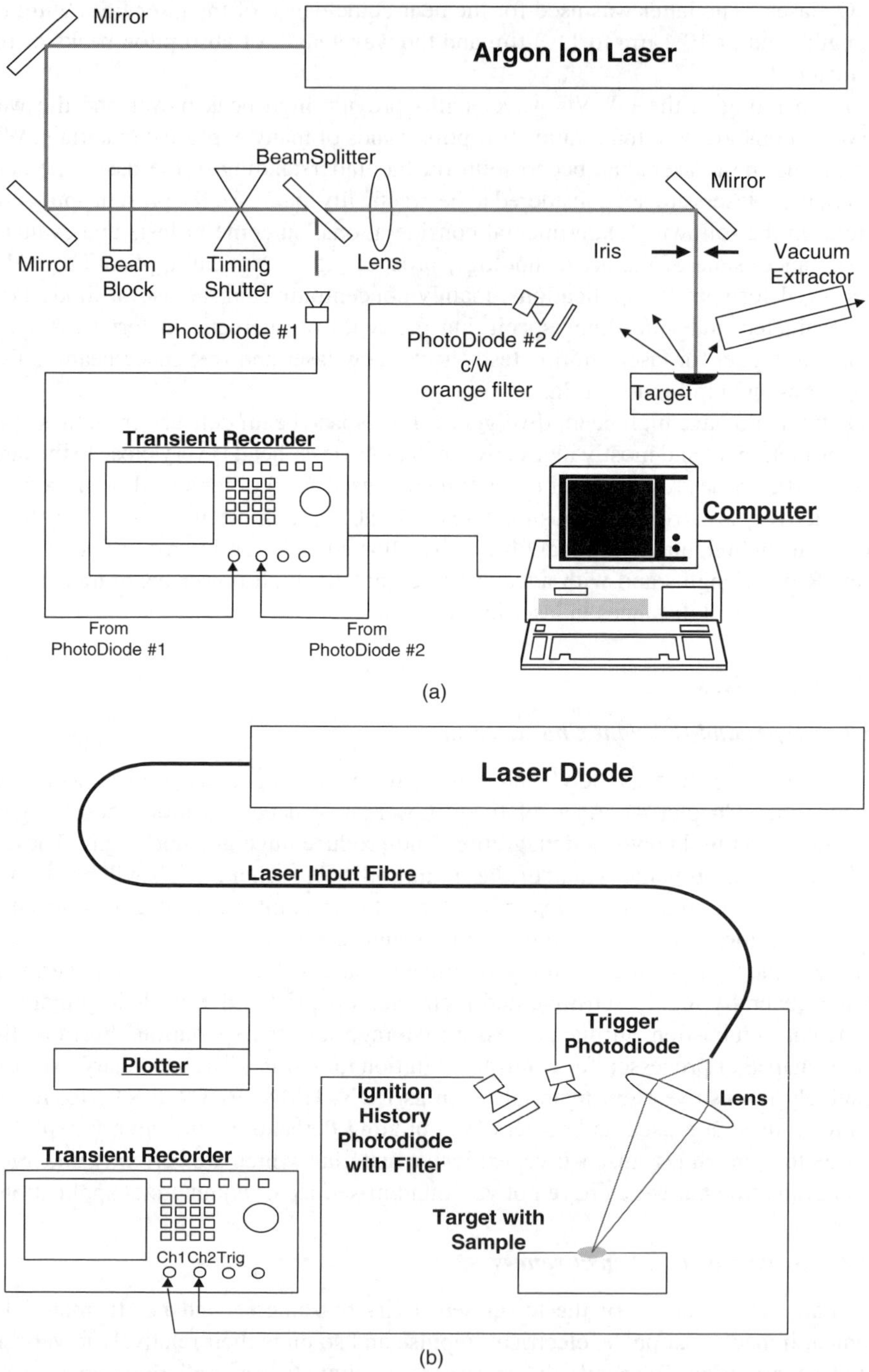

Figure 11.3 *(a) Schematic of a typical experiment layout for laser ignition test in open air using an Ar-ion laser. (b) Schematic of a typical experiment layout for laser ignition test in the open air using a diode laser. Reproduced from [16]. Copyright © 2005 WILEY-VCH Verlag GmbH & Co. KGaA, Weinheim.*

available cost-effective lasers having adequate output power at a suitable wavelength. The wavelength must be in near resonance with the wavelength band of the absorption line of the material. In some cases, materials may have a broad absorption band, making the choice of wavelength easier. Additionally, the material of choice needs to be capable of releasing high explosive force. These criteria are put as follows:

i. Low sensitivity to mechanical, thermal and electrical stimuli. Materials having such properties will allow safe storage, handling and processing.
ii. High sensitiveness to laser impact. Materials with this property will be easily ignitable by laser beams. In other words, they will undergo ignition at relatively low laser power levels and also have a faster ignition delay time.

The first criterion may be met by using so-called 'insensitive ammunition (IM)' such as secondary high explosives. Extensive studies on laser ignitibility of high explosive (HE) have, so far, been reported on only three different materials – HMX, RDX and PETN. Attempts to directly ignite another widely used high explosive, HNS, using diode lasers have so far not been successful.

11.3.1.3 Optical Absorption

Strong absorption of the laser beam by the selected target material is a prerequisite for any effective and efficient photo-thermal activity leading to ignition. Therefore, it is mandatory to find or create a match between the wavelength of the selected laser and the peak wavelength of the absorption band of the material. Absorption of light, invoking its wave nature, takes place through the interaction of the electrons that whirl around the atoms and molecules and the electric field of the electromagnetic radiation (light). This interaction is affected by the way the electrons are associated within the atomic structures of the molecules of the materials and the strength of the atomic bonds within the molecules. The molecular structures of some commonly used explosives are shown in Figure 11.4.

In these examples, all three HEs have NO_2 groups bonded as side branches. While the pattern of such bonds within HMX and RDX are very similar, it is considerably different for HNS. In the latter, the NO_2 group is bonded directly with the benzene ring structure, thereby giving rise to saturation. The absorption of optical energy is primarily governed by the vibrational modes of the various components of the structures. These correspond to the energies in the infrared wavelength band. However, strong electronic transition corresponding to the NO_2 (O–N–O) symmetric stretching mode will give rise to some strong absorption in the UV-Visible bands.

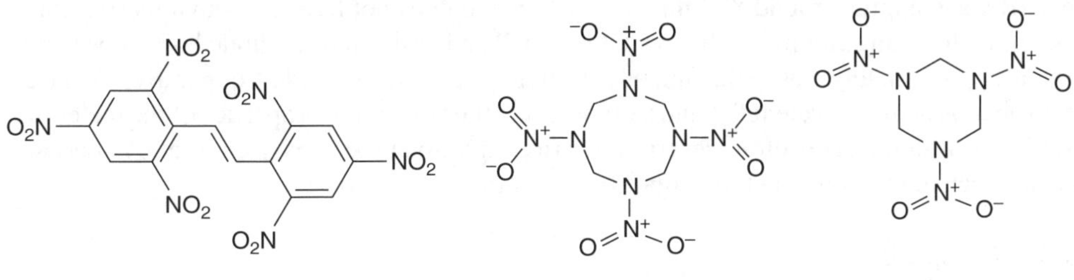

HNS **HMX** **RDX**

Figure 11.4 *Molecular structures of some common high explosives.*

A typical absorption spectrum of solid HNS-IV, covering the UV to near-IR wavelength band (i.e. 250–850 nm), recorded in our laboratory, has already been shown in Figure 9.3 of Chapter 9. Conventionally, the HE sample, as a powder (or as a solution in an appropriate solvent), is thoroughly mixed with a small fraction of powdered KBr crystals (approximately at a 6 : 1 ratio). Discs, typically 1 mm thick, are then made using ether a press or in the case of a solution, by casting and evaporating the solvent. The KBr has very low absorption in the Vis-near-IR wavelength band and, therefore, is an ideal candidate for use in the absorption spectroscopy of some solids. The discs thus produced are used as slides in the absorption spectrometer. The spectral absorbance of KBr is used as a reference standard for the calibration of the system. The spectral absorbance may also be estimated from the diffused reflectance spectra obtained by using a commercial spectrometer with an integrating sphere. Examples of such spectra for some commonly used explosives are shown in Figure 11.5.

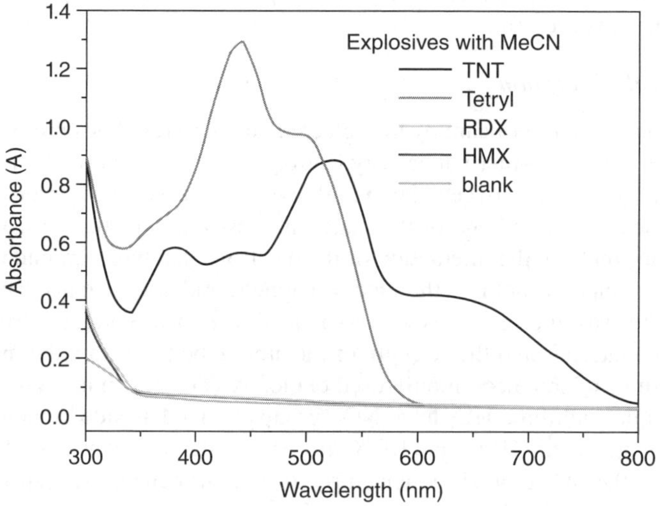

Figure 11.5 *Typical optical absorption spectra of some common explosives. Reproduced from [18]. Copyright © 2008, Royal Society of Chemistry.*

The peaks at the UV-Vis wavelength bands for most explosives are the characteristics of the strong symmetric vibrational stretching modes, typical of NO_2. Absorption at the near-IR wavelengths around 800 nm is very low and does not have any obvious structure. Therefore, for using commercially available, small and cost-effective diode lasers operating within approximately 780–802 nm, HE samples need to be made cooperative to laser interrogation at this wavelength band by modifying the absorption properties of the materials using optical sanitizers. Before describing various ignition processes in energetic materials, it is instructive to define the term 'ignitibility' in the present context.

11.3.1.4 Ignitibility

Ignitibility is defined by the combination of two laser parameters: ignition delay time (also referred to as 'reaction time') and the minimum laser intensity (power per unit area) at which the ignition is achieved for a particular material within an acceptable delay time. Ignition delay time is the time taken for the start of the ignition event following the start of the laser irradiation, as illustrated in Figure 11.6.

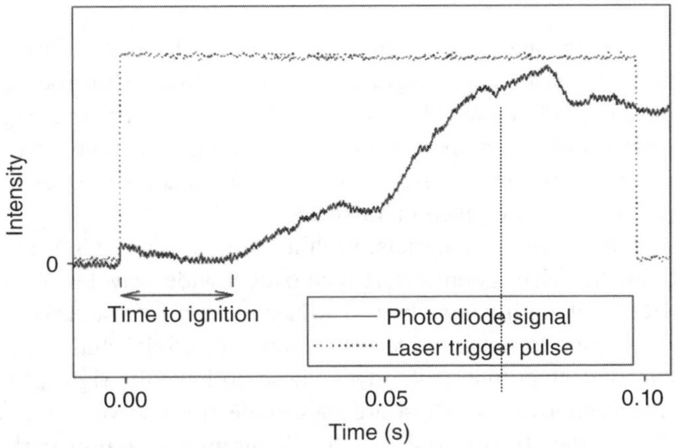

Figure 11.6 *Example of temporal history of a laser-induced ignition event. Reproduced from [10]. Copyright © 2001 WILEY-VCH Verlag GmbH & Co. KGaA, Weinheim.*

For the implementation of laser as the igniter for energetic materials, it is essential to locate lasers with optimum output parameters and, at the same time, select and tailor target materials for the best possible match for maximum laser energy transfer that will allow ignition with minimum laser intensity, to make the device cost-effective. For this, the use of commercially available, compact, ruggedized and cheap diode lasers will be necessary. Such lasers, capable of delivering adequate power output for efficient ignition initiation, operate only at the near-IR wavelength band between 780–820 nm.

Output power of some diode lasers operating in the UV-Vis wavelength bands, at which some HE exhibit strong absorption (see Figure 3.3 in Chapter 3) is normally too weak to be of any practical use for laser ignition application. It now transpires that HE materials need to be tailored to make them absorb strongly at the near-IR wavelength band, using highly absorbing inorganic powders such as carbon black, gunpowder and so on, or chemical dyes, generally known as optical sensitizers. A brief survey on current research on the use of sensitizers to facilitate or enable laser ignition in energetic materials is presented below.

11.3.1.5 Optical Sensitization

For making energetic materials, particularly the HEs, amenable to relatively easy laser ignition, optical sensitizers to be incorporated in to the materials need to fulfil three main criteria:

i. These should be inert, i.e. chemically non-reactive to the molecules of the host matrices. At the least, these should not drastically change the colour and texture, unacceptably reduce the explosive power or lower the thermal conductivity. Finally, these should not enhance the sensitivity of the materials to non-optical stimuli.
ii. These should preferably be soluble in a solvent common to both the explosive material and the sensitizer chemical. This will allow formation of chemical bonding within the

host matrix or, at least, should allow uniform physical dispersion throughout the bulk of a solution of the sample in a non-reactive medium. When the solvent is removed through evaporation, an optically sensitized sample should be obtained.

iii. These should have good stability and resistance to weathering.

It goes without saying that selected sensitizers should have very high absorbance at around the peak of the laser wavelength band and should be inexpensive in order to make the procedure cost-effective. The chemical dyes may offer advantages in terms of wavelength selectivity and the prospect of incorporation into the molecular structure of the host matrices. Examples of some commercially available chemical dyes having potential for use as optical sensitizers are given in Table 11.1.

In addition to chemical dyes, materials, such as carbon black (CB), carbon nano-tube (CN) and so on, having fairly good absorbance over a wide wavelength range, covering UV-Vis to near-IR wavelength bands, have also been tested for use as optical sensitizers. Although these materials do not provide wavelength selectivity and have comparatively much lower absorptivity than that of the dyes, these do have the right properties to fulfil two of the three criteria cited above. These are not soluble in any solvent, but can be dispersed in solutions containing the HE materials, so that the granules are uniformly coated by the CB particles on drying into solid forms. Because of these criteria, and reasonably good absorbance at the diode laser wavelength band at around 800 nm, carbon black has been used in tests and evaluation for laser ignitibility of explosives in many laboratories around the globe over last two decades.

The absorptivity parameter of carbon black powders, unlike those of the dyes, cannot be quantified easily, as different samples from different manufacturers, and of different particle sizes, have been found to have different values for this parameter. Moreover, the sizes of the grains have been found to affect the absorptivity of samples coated with such optical sensitizers. It has been observed that particles with smaller sizes (size distributions) give better coatings of the explosive grains, whereas samples coated with CB having relatively larger particle sizes (size distributions) show much better laser ignitibility. The shape of the absorption spectra of samples of HEs coated with CB are invariably flat over the UV-Vis-Near-IR bands, and could be an order of magnitude lower than that at the peak wavelengths of some strongly absorbing dyes. Despite this drawback, several authors have experimented with CB as an optical sensitizer for diode laser ignition of explosives, taking the above mentioned advantages and with the proviso that such materials will ultimately be easer to procure, safer to use and more cost-effective than dyes.

It is desirable to keep the concentration of the sensitizer to a minimum, so that effects on the physical and chemical properties of host matrices are kept to a minimum. The technology of producing tailored explosive materials by incorporating sensitizers into solid or powdered HEs is still evolving. The procedure adopted for each type of HE sample will depend on its phase (powder, solid, thick gluey, etc.), the solvent type and its solubility. The conventional procedure is to dissolve or disperse a measured quantity of the HE materials in a suitable solvent such as dichloromethane, THF, etc., add an appropriate fraction of CB and dry the mixture while stirring the solution. However, in some cases, it has been found that the target materials are made into granular form through dilution in a solvent and mixed with CB particles. The admixture powder is then pressed into pallets for laser ignition studies.

Table 11.1 *Specifications of some commercial dyes suitable for sensitizing explosives for diode laser ignition.*

Product code	Supplier	Molecular structure	Solvent	Peak abs. WL/Bandwidth	Absorptivity
ADS815EI	American Dye Source, Inc. USA	C42H44IN2Cl	Methanol	815 nm ± 20 nm	352
ADS800AT	American Dye Source, Inc. USA	C54H54N2O4S	Methanol	811 nm ± 20 nm	254
NIR811A	QCR Solutions Corp, USA	Dimethylformamide	**	811 nm ± 15 nm	342
IR Dye 9807	Adam Gates Co. LLC, USA	Polycyanine	**	807 nm ± 15 nm	409
NIR805B	Adam Gates Co. LLC, USA	**	Methanol	807 nm ± 15 nm	409
NIR800A	QCR Solutions Corp, USA	**	Methanol	800 nm ± 25 nm	297
IR dye 9798	Adam Gates Co. LLC, USA	Polycyanine	**	798 nm ± 15nm	157
Epolight™5768	Epolin, Inc. USA	**	Acetone	791 nm ± 15 nm	292

*($L\ g^{-1}\ cm^{-1}$). **Data not available/commercial in confidence.

Two important things need to be considered in the formulation and preparation of samples namely, negligible conglomerations of either the HE granules or the CB particles, and uniformity of the coating of the granules on the surface of the sample. For the latter, the CB particles should be as fine as possible. Note that the ignitibility of HE will go down with decreasing size of the CB granules. CB powders with mean particle sizes of a few tens of nanometres, bimodal size distributions and a concentration of ≈0.5–3% by weight are commonly used in reported [2, 3] test results. The sizes are roughly an order of magnitude smaller than those of granules of HE substances. The results quoted in the references are for unconfined test conditions at normal atmospheric conditions.

For test purposes, diode lasers operating at around 800 nm and giving outputs of a few watts at the exit port are used for laser ignition studies. Such lasers, when focused on to a fibre optic surface having, for example, 100 μm core diameter, will provide (after reflection and coupling losses) a few kW cm^{-2} power densities at the target surface. Conventionally, ignition maps (a plot of ignition delay time versus the laser intensity (flux density)) are created. A typical temporal history of the ignition event for an explosive material is illustrated in Figure 11.7.

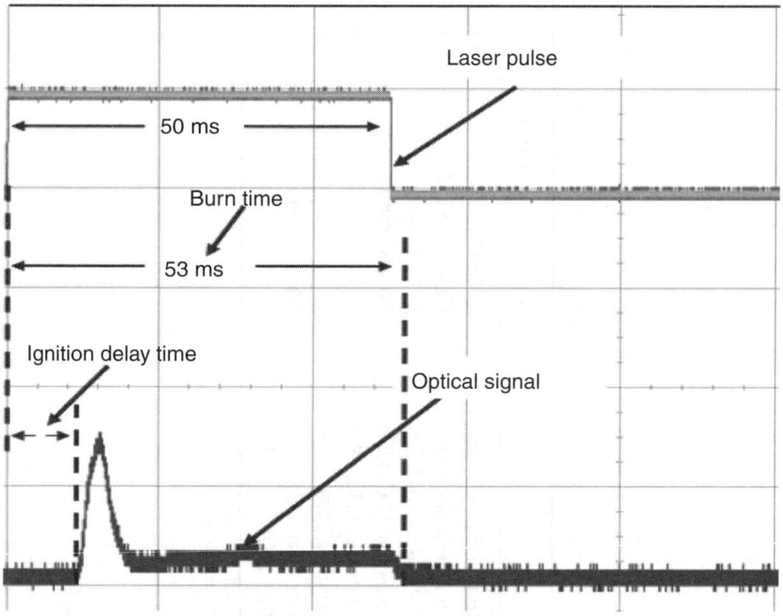

Figure 11.7 *Typical oscilloscope trace of ignition event in an explosive material.*

From the ignition map of an energetic material the threshold (minimum) values of power density for ignition to take place, and the corresponding ignition delay time (reaction time), the parameters which dictate the laser ignition characteristics of particular sensitized HE materials are evaluated. For example, a threshold laser flux density of approximately 2 kW cm^{-2} for HMX (with 1.5–4% CB) has been reported to give ignition delay time in the range of 2–4 ms. In most test cases, the interaction results in fast burning. Note that these values will change drastically with sample type and concentration of the CB, and also with the laser type. However, the same material is reported [4] to have exhibited ignition delay time similar to that of the above for a similar incident power density from a CO_2 laser (without sensitizers). Ignition tests of HE samples under airtight confinement may, under some circumstances, give rise to 'shock to detonation transition' (STD). However, in most laboratory tests, fast deflagration, particularly for slow laser heating, is normally the outcome for laser irradiation at IR wavelengths. The ignition delay time has been found [5] to decrease by a factor of $\approx$2.4 for a decrease in pressure within the confinement chamber by a factor of 4 mega Pascal (1–4 MP).

For prompt ignition initiation leading to detonation of explosives (under confined conditions), the optical energy delivery rate (power density) from a laser source has to be quite high, and the wavelength should preferably be in the UV-Vis. wavelength band. Most early research and some recent studies, has been conducted using UV wavelengths from excimer lasers (e.g. 248 nm, 308 nm etc.) and Q-switched outputs (in the order of GW cm^{-2} and nanoseconds pulse duration) from Nd:YAG lasers operating a near IR wavelength (1.06 μm). It is generally accepted that, at longer wavelengths in the near IR and above, the interaction at low-to-moderate power densities is governed by the process of a slow temperature rise through the formation of local hot spots within the bulk of the material (often requiring optical sensitizers). Under some conditions, this may grow into a fast deflagration event, culminating in to a detonation event if the test is conducted in a confined environment. However, in the case of UV and high-power visible laser wavelengths (the latter might be due to multiphoton interaction at high power densities), the interaction is considered [6, 7] to be primarily due to a fast photo-physical process, resulting in bond-breaking, shockwave formation, and eventually leading to prompt detonation, even without sensitizers or confinement.

The effect of the irradiated laser spot on the ignition energy threshold or the ignition delay time has also been a subject of some in-depth studies [8]. While the incident intensity (power density) can be increased to minimize ignition delay (reaction) time or reduce threshold energy, there exists a minimum spot size below which burning ensues, culminating into creating a crater or hole on to the target but no sustained ignition takes place at all.

The effect of binders on laser ignitibility of secondary explosives has also been a subject of investigation in recent times. In practical applications, HE materials are most commonly formulated with some polymeric materials known as binders, to give the materials mechanical strength and for ease of physical shaping according to the type of application. The use of inert binders reduces the energy content per unit mass of the composition. This has prompted researchers to locate energetic binders and to study the effects of such binders on the ignitibility of HEs. Recent research has suggested [9] that while HMX on its own could not be ignited by $\approx$9 kW cm^{-2} output from a diode laser (801 nm), this becomes amenable to laser initiation, as expected, when sensitized with CB. A considerable reduction of delay time (reaction time) is achieved when it is formulated with some energetic binders such as, Poly-GLEN, HTPB and particularly, polyphosphozene.

11.3.2 Propellants

The constituents of most propellants are stabilizers, oxidants, binders, fuel, curing agents and plasticisers, and of course, the fuel material. Constituents that will affect optical absorption properties and, hence, the laser ignitibility of propellants, are given below:

- PNMA: an amine-based stabilizer. Chemical name is para-nitromethylaniline, chemical formula is $C_7N_2O_2H_8$, and usual concentration in propellant is less than 2%.

- 2NDPA: an amine-based stabilizer. Chemical name is 2 nitro-diphenylamine, chemical formula is $C_{12}N_2O_2H_{11}$, and usual concentration in propellant is less than 2%.
- Plastinox 2246: used as a plasticiser. Chemical name is 2, 2′-methylene-bis (4 methyl-6-tert-butylphenol), chemical formula is $C_{23}O_2H_{32}$, typical concentration in propellant is less than 3%.
- Nitroglycerine: used as the fuel in double-base propellant. Chemical formula is $C_3H_5N_3O_9$; present at a proportion of about 30%.
- Nitrocellulose: used as the oxidant in double-base propellant. Chemical formula is $C_{12}H_{14}N_6O_{22}$; present at a proportion of 50–60%.
- Ammonium perchlorate: used as an oxidizer in propellant. Chemical formula is NH_4ClO_4; typical concentration in propellants is not yet specified.
- AND: used as an oxidizer. Chemical name is ammonium dinitramide, having the chemical formula: $NH_4N(NO_2)_2$. Typical proportion in propellant is not yet quoted.
- GAP: used as an energetic binder; classified as a liquid propellant. Chemical name is glycidyl azide polymer or, more exactly, azido-methyl-ethylene-oxide. Chemical formula is $(C_3N_3O_2H_7)_n$; typical proportion in propellant is not yet quoted.
- DANPE: used as a plasticiser. Chemical name is 1.5-diazido-3-nitrazapentane, chemical formula is $C_4N_8O_2H_8$. Typical proportion in propellant is not yet quoted.
- RDX: an oxidizer. Chemical name is cyclo-1,3,5-trimethylene-2,4,6-trinitramine, chemical formula is $C_3H_6N_6O_6$. Typical concentration in propellant is 40–50%.

Studies on laser ignition of propellants are not as extensive as those on explosives and pyrotechnics. Some ‘in-service’ rocket motor propellants which have been reported to have undergone laser ignition tests are: Cast Double-Base (CDB), Extruded Double-Base (EDB) and Composite (Com). Some experimental propellants, such as a combination of CL-20 and polyGLYN and a combination of ADN and polyNIMMO, along with a new generation of low vulnerable ammunitions (LOVA), otherwise known as insensitive ammunitions (IA), currently under trial, have also been subjected to laser ignition tests. The state of the art of the topic is reviewed in the following.

11.3.2.1 Cast and Extruded Double Base Propellants

Both of these materials normally contain some carbon black (0.2–0.3%) and an oxidizer, in addition to nitrocellulose (40–50%) as the energetic resin (binder) and nitroglycerine (35–45%) as the explosive plasticizer. The CDB is black with yellow granules, and the EDB is reddish-orange. The latter contains approximately 14% Di-methyl Phthalate, which accounts for its reddish-orange colour and the consequent high absorption at this colour (wavelength) band, as shown in Figure 11.8. The spectra show that the CDB has a broadband absorption at around 95% level over the entire UV-Vis-Near IR band, whereas, the EDB sample exhibits similar absorption for UV-Vis wavelengths up to about 550 nm. The absorption plummets to about 40% level for wavelengths above this up to the near-IR band at about 800 nm.

Laser ignition tests on these two double-base propellants, reported in the open literature, were conducted using both an Ar-ion lasers operating in the visible wavelength (@500 nm) and a CO_2 lasers operating in the IR wavelength (@10.6 μm). At the visible excitation wavelength, ignition delay time @100 W cm^{-2} of power density had been estimated [10] to be between one and two seconds.

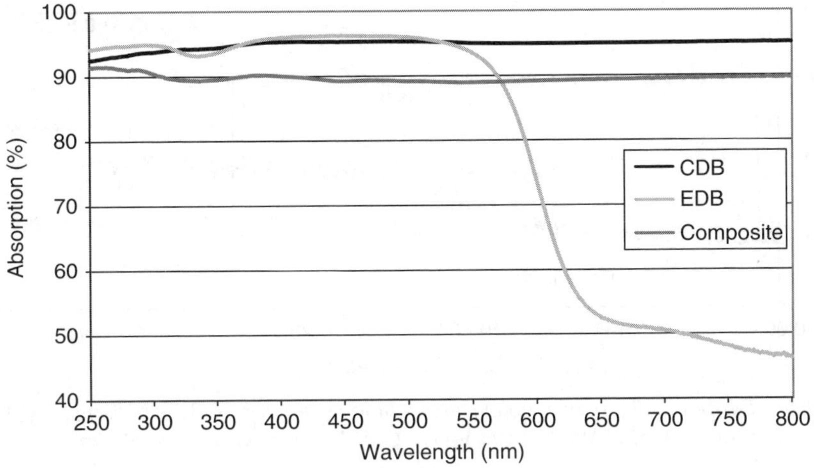

Figure 11.8 *Typical absorption spectra of some in-service propellants. Reproduced from [10].*

11.3.2.2 Composite Propellants

The constituent of this type of propellant is quite different from those of the double-base ones. The primary oxidizer in this material is an ammonium compound, commonly ammonium perchlorate (≈65–70%), a metal powder, e.g. Al (≈15–20%) as the high energy fuel, an energetic polymeric binder such as HTPB or HMX (15–20%) and small amounts of epoxy resin and a few other chemicals. The absorption spectrum of such a propellant material (see Figure 11.8) is practically similar to that of the CDB. It has been reported [10, 11] to have exhibited, for incident power densities of ≈100 W cm^{-2}, ignition delay times of approximately 100 ms and 400 ms for visible (@500 nm) and IR (@10.6 nm) wavelengths respectively. This indicates that the interaction process in the IR wavelength is primarily a slow thermal process and, for the visible wavelength region, this is most likely to be a photo-physical process.

Ignition tests under confined environment [11] seem to have some significant effect on the ignitibility of these propellants, but not as prominent as for the explosives (*vide afro*) or pyrotechnics (see below). The CDB and composite propellants have been found [12] to undergo ignition, albeit not quite reproducibly, when irradiated with diode lasers operating at 780 nm and at power levels at or above ≈120 W cm^{-2}. The EDB sample is not ignitable at this wavelength because of its very poor absorbance at the near-IR wavelength band. It is expected that the ignitability of these materials could be greatly enhanced at the diode laser wavelengths if they are optically sensitized. It has been found that the addition of small quantities of energetic propellants/explosives such as RDX or HMX increases the ignitibility of composite propellants. The effect was found [13] to be considerably higher for HMX than for RDX. There are a host of other propellants, as listed above, some of which have undergone scouting tests, with results similar to those of the widely used CDB, EDB and composite ones. An example of the temporal history of the ignition event for CDB following laser irradiation is presented in Figure 11.9.

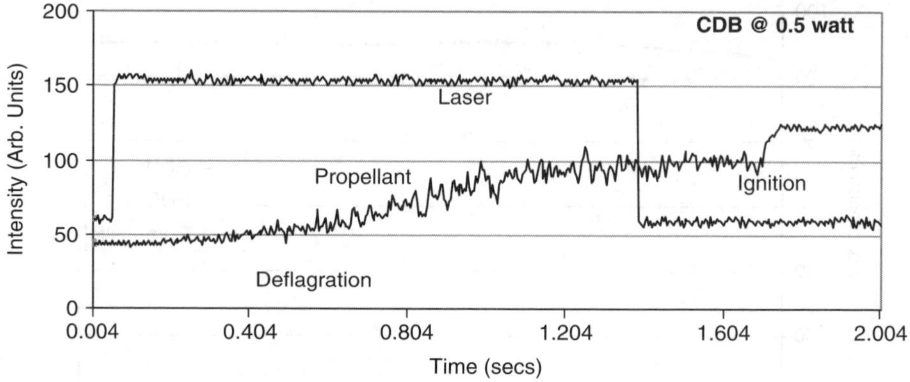

Figure 11.9 *Temporal history of a typical laser ignition event for CDB propellants. Reproduced from [17]. Copyright © 2001 WILEY-VCH Verlag GmbH & Co. KGaA, Weinheim.*

11.3.2.3 Low Vulnerable Ammunition (LOVA) – Propellants

While maximum explosive performance is necessary in all practical applications, two other factors need to be considered when formulating new generation of propellants. Safety during storage, use and transportation, and the impact on the environment during manufacturing, shaping and disposal of such materials, are now topics on the top of the environmental and economic agenda of explosives-producing countries. Consequently, the emphasis now is on the synthesis of high-performance munitions complemented with through-life safety, whole-life cost-effectiveness and end-of-life environmentally friendly disposal protocol. The search for low vulnerable ammunition (LOVA) to satisfy the above requirements has given rise to nitramine (e.g. RDX, HMX)-based propellants incorporating both energetic and non-energetic plasticizers and binders, to replace conventional nitrocellulose-based propellants. Additionally, new insensitive high explosives such as TATB, FOX-7, NTO and so on, are also formulated to replace relatively more sensitive nitramines as energetic fillers in the formulation of propellants. As a consequence of the reduced response to accidental stimuli, the LOVA propellants render themselves difficult to ignite by conventional electric stimuli. 'Non-ideal ignition' resulting in loss of accuracy, misfire, hang-fire and catastrophic failure of gun systems, have been major drawbacks of such LOVA propellants.

Direct laser ignition of LOVA materials has not been reported as widely as that for other energetic materials. Ignition threshold and ignition delay time for propellants (*vide afro*) are dependent on laser parameters, chemical composition of the charge and the pressure if the sample is within a confinement. The variation of ignition properties on laser parameters can be minimized by the addition of carbon black. Interest in the laser ignitibility of the second-generation insensitive munitions (IM) and LOVA propellants is now growing, but the R&D in this area is still in its infancy. LOVA propellants may be formulated by replacing the sensitive binder (e.g. cellulose acetate) with inert binder materials such as HTPB, PolyNIMO and so on, replacing RDX with relatively insensitive FOX7 as filler, and avoiding nitrocellulose altogether.

Tests conducted on such formulations of propellants at the author's laboratory has shown that ignition with diode laser is possible for laser power densities in excess of 2 kW cm^{-2}, with ignition delay times of 1–2 seconds. The formulation with PolyNIMO was found to be insensitive to visible wavelengths, whereas that with HTPB was ignited at power densities in excess of 1 kW cm^{-2} at visible wavelengths, albeit with an unacceptably long (8–10 secs.) delay time. The delay time is reduced to 1–2 seconds for power densities in excess of 10 kW cm^{-2}. The use of insensitive components in the formulation of LOVA propellants makes these materials not only insensitive to mechanical, thermal and electrical stimuli, but also to laser irradiation. Therefore, for the practical applications of laser ignition in such LOVA propellants using diode lasers, further tests and evaluation using different optical sensitizer are needed.

11.3.3 LI of Pyrotechnic Materials

Pyrotechnics are quite sensitive to conventional external stimuli and, therefore, are vulnerable to accidental or hostile ignition initiation, for example, electrical signals from RF sources in electro-explosive devices (EED). Research on the use of laser radiation as a safe igniter, particularly in rocket motor propellant beds, is driven by this need for removing or minimizing the vulnerability that exists in conventional electrical initiation systems. There have been many recent reports on the effects of various parameters on various aspects of interaction of laser radiation with pyrotechnics. However, details on in-service laser ignition systems are somewhat sketchy, except for a patented prototype system [14] based on a match head made of nitrocellulose paste.

Many different compositions of pyrotechnics have been tested for laser ignitibility. These are primarily compositions of magnesium/sodium with potassium nitrates, boron with ferrous oxides/potassium nitrates at different proportions. Sometimes, a small proportion of zinc oxide or charcoal (5–10%) may also be added. Commonly used pyrotechnic compositions on which studies had been undertaken at the author's laboratory [12] are:

- G20 (gunpowder): sulfur (10%)/charcoal (15%)/potassium nitrate (75%) composition,
- SR44: boron (30%)/potassium nitrate (70%) mixture, and,
- SR 371C: magnesium (42%)/potassium nitrate (50%) composition with 8% acaroids resin.

The absorption spectra of the samples (pressed into compact pallets) evaluated from their reflectance spectra are shown in Figure 11.10. The absorption properties vary from sample to sample, as expected. A composition (SR112) containing Tetranitrocarbzole (TNC) and KNO_3, has been found to exhibit negligible absorption at near-infrared wavelengths [15], whereas others show high to moderate absorption over the same wavelength band.

Pyrotechnics are more readily ignitable than other energetic materials. For example, commercial gunpowder (G20) has been ignited at a threshold laser power density of $\approx$30 W cm^{-2}. This is remarkably low – approximately three orders of magnitude lower than that for the ignition of high explosives and propellants, even after optical sensitization.

For unconfined ignition tests in the experimental setup (Figure 11.13a and b), materials, usually in the powder form, and sometimes in the form of compacted slabs, are placed at the focus of the laser beam. It was experimentally established that the hose of a domestic vacuum cleaner placed within 5 cm away from the target area was able to satisfactorily remove the copious smoke and soot produced by the ignition process. This removed the need for extra procedures for the protection of optical components. The beam-directing mirror above the target was at a sufficient distance that the high-temperature flash did not cause any damage to it. The use of small quantities of powder also ensured that the flash did

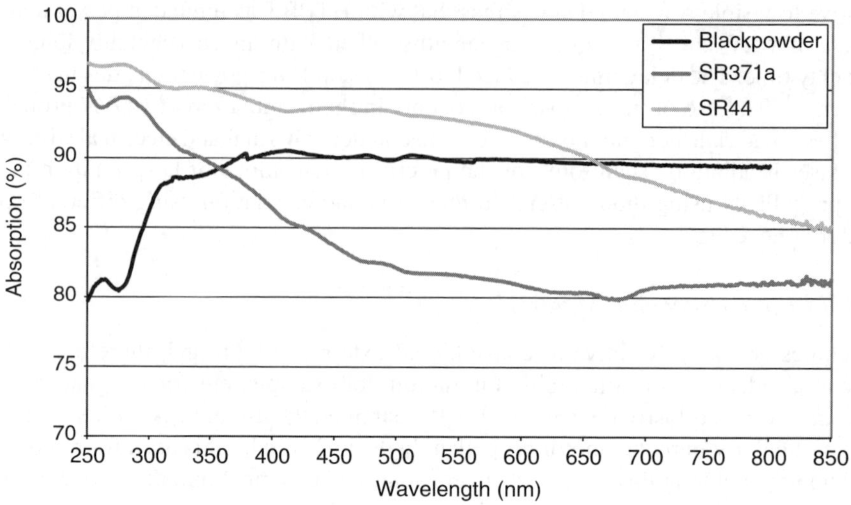

Figure 11.10 *Absorption spectra of some pyrotechnics with different compositions.*

not cause any damage to the neighbouring equipment. Typical oscilloscope traces of the ignition flash, along with the temporal history of the incident laser power, detected by fast photodiodes, are shown in Figure 11.11. Note that the initiation of ignition starts at about 500 ms after the laser is incident on to the sample. The duration of the ignition event will depend upon the quantity and the particular constituents of the sample. Some constituents may exhibit some delayed burning bursts, as is the case in this example.

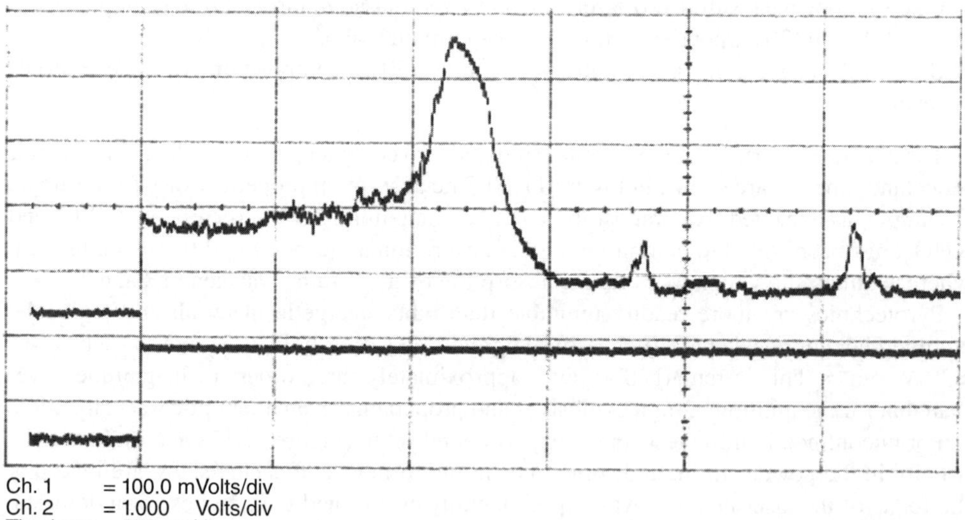

Figure 11.11 *Oscilloscope trace of an ignition event in gunpowder following Ar-ion laser beam irradiation (beam intensity at target ≈75 W cm^{-2}; laser wavelength @500 nm). Reproduced from [16]. Copyright © 2005 WILEY-VCH Verlag GmbH & Co. KGaA, Weinheim.*

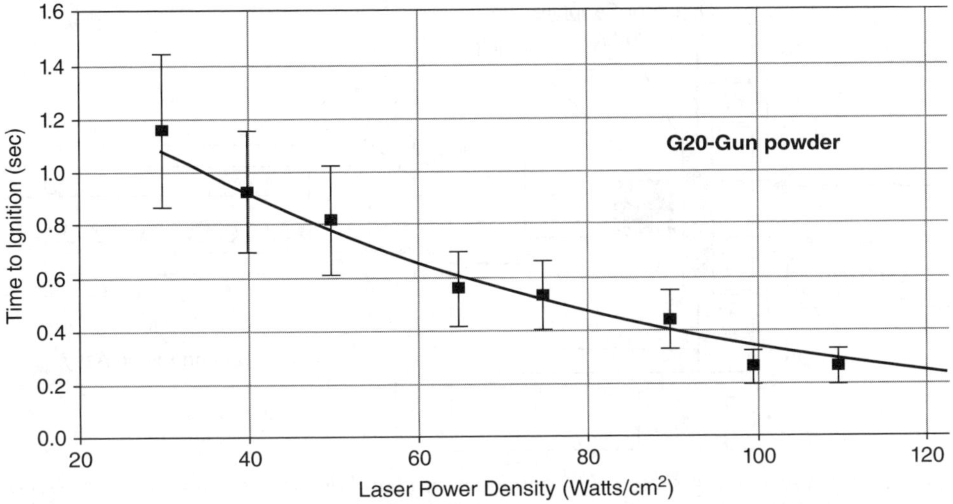

Figure 11.12 Example of an ignition map for a pyrotechnic using Ar-ion laser (@500 nm). Reproduced from [16]. Copyright © 2005 WILEY-VCH Verlag GmbH & Co. KGaA, Weinheim.

Parameters that will affect the ignitibility of pyrotechnics to a smaller or greater extent are:

- laser power density and wavelength;
- focused beam area onto the target;
- confinement condition;
- and of course, the constituents of the material.

It has been observed [16] that the ignitibility of pyrotechnics is more efficient for excitation at the visible wavelength than that at the near-IR wavelength (≈800 nm). An example of a typical ignition map for gunpowder is shown in Figure 11.12. A threshold laser power density of ≈30 W cm^{-2} was recorded for excitation at the visible (@500 nm) wavelength of an Ar-ion laser. It is noted that the speed of the ignition process is faster for near-IR wavelength (ignition delay of ≈500 ms), compared with that for the visible wavelength (ignition delay time ≈1100 ms). At threshold ignition power densities, integrated optical energy density (power density × ignition delay time) required for ignition was found to be similar for both visible and near-IR excitations. Note that the composition of pyrotechnics has no significant effect on laser ignitibility, and also that the difference in absorptivity does not seem to play any significant role in the laser ignitibility of the pyrotechnic materials tested so far.

Most reported research on laser ignition of pyrotechnics had been conducted in an open-air environment. In practical applications, particularly in rocket motors, ignition will take place under confinement. Within confinement, the ignition process is followed – almost in parallel – by a rise in pressure. This dynamic process affects the ignition mechanism by regulating the heat transfer (both convection and conduction) within the sample, and it also regulates laser beam transmission through the vapour onto the sample. Preliminary

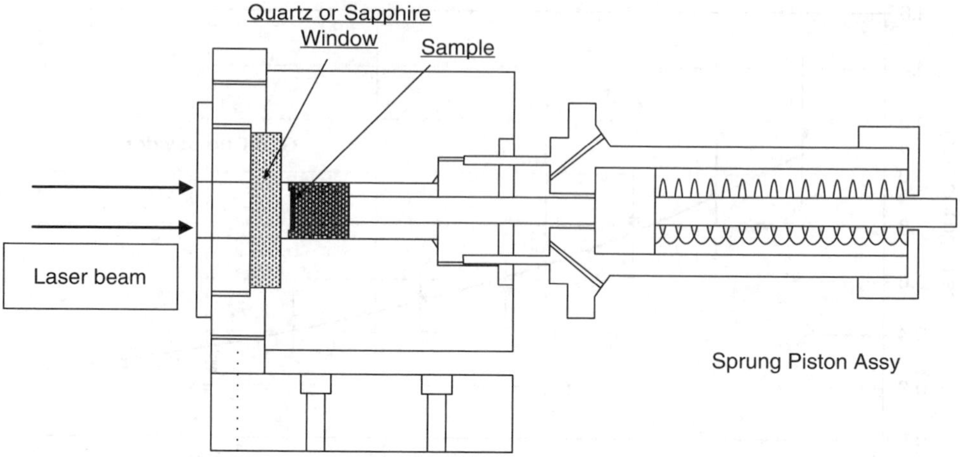

Figure 11.13 Engineering sketch of a confinement chamber for laser ignition tests. Reproduced from [17]. Copyright © 2008 WILEY-VCH Verlag GmbH & Co. KGaA, Weinheim.

results at the author's laboratory have shown that under confinement, the ignition process is more efficient than that for open-air environment. For example, for a gunpowder sample, the ignition delay time at threshold laser power was estimated to be between 30–40 ms whereas, in an open-air environment, it was 200–250 ms. Note that under confinement, and at laser intensities in excess of a few kW cm^{-2}, ignition delay times of a few milliseconds have been achieved [17].

The effects of confinement on the ignitibility of pyrotechnics or primary explosives are more pronounced than that of propellants and high explosives. A primary explosive, nickel hydrazine nitrate (NiHN), also considered as a pyrotechnic, depending upon the specific application, has undergone preliminary laser ignition tests at the authors' laboratories. This material, having the chemical formula Ni $(N_2H_4)_3(NO_3)_2$, is a metal oxyanion salt, complexed with bidentate fuel ligand (MOS-Bid). It exhibits a high absorption peak at ≈500 nm, which is at the centre of the principal output wavelengths (@500 nm) of the argon ion laser. It also shows quite high absorbance at wavelengths above ≈760 nm.

This material was chosen for laser ignition tests because it exhibits a high absorption peak at ≈808 nm, corresponding to the output wavelength of low-power and low-cost commercial NIR diode lasers. Although NiHN is particularly sensitive to flame, it is relatively insensitive to other safety assessment stimuli (e.g. impact, friction and electrostatic charge tests) compared with other primary explosives. Considering its high absorption at wavelengths in the near-IR band, tests were conducted using an output from a near infra-red (NIR) diode laser, operating at a wavelength of 808 nm and providing maximum output power of ≈1.5 W. The test jig is the same as that shown in Figure 11.3, except that the open air metal cubicle was replaced by a confinement chamber. An engineering sketch of the chamber, shown in Figure 11.13 had an aluminium casing and a tough quartz window. Both the laser beam and the light from ignition event (at an angle with the direction of the beam) were detected through the window. A temporal history of the light emission from the detonation event, as measured by a photodiode connected to a transient recorder, is shown

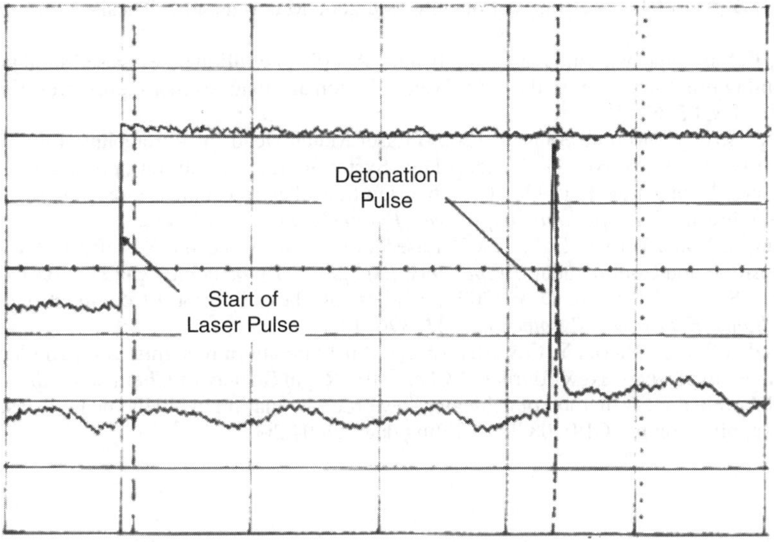

Figure 11.14 *Example of temporal history of low order detonation event in NiHN. Time to detonation = 33.3 ms, laser energy density = 7.21 J cm^{-2}. Reproduced from Ref [19].*

in Figure 11.14. The unusually fast light pulse from the confined ignition event is a low order detonation, as also witnessed by indentation in an aluminium witness plate.

References

(1) Briely, H.R., Williamson, D.M. and Vine, T.A. (2012) Improving laser-driven flyer efficiency with high absorptance layer. *AIP Conference Proceedings on Condensed Matter*, **1426**, 315–318
(2) Goveas, S.G. (1997) The Laser Ignition of Energetic Materials. Ph D Thesis, Univ. of Cambridge.
(3) Ahmad, S.R., Russe, D.A. and Golding, P. (2009) Laser induced deflagration of unconfined HMX. *Propellants, Explosives, Pyrotechnics*, **34**, 513–519.
(4) Traver, C.M. (2004) Chemical kinetic modelling of HMX and TATB laser ignition tests. *Journal of Energetic Materials*, **22**(2), 93–107.
(5) Ostermark, H. and Grans, R. (1990) Laser ignition of explosives: effects of gas pressure on the threshold ignition energy. *Journal of Energetic Materials*, **8**(4), 308–322.
(6) Ostermark, H., Carison, M. and Ekvall, K. (1994) Laser ignition of explosives: effect of laser wavelength on the threshold ignition energy. *Journal of Energetic Materials*, **12**, 63–83.
(7) Bourne, N.K. (2010) On the laser ignition and initiation of explosives. *Proceedings of the Royal Society of London*, **457**, 1401–1426.
(8) Patterson, A., Patterson, J. and Roman, A. (1998) Diode laser ignition of unconfined secondary explosives: A parametric study. 11th detonation symposium, Snowmass, Colorado, USA, Aug. 30, 1998.
(9) Ahmad, S.R., Russell, D.A. and Golding, P. (2009) Laser-induced deflagration of un-confined HMX – the effect of energetic binders. *Propellants, Explosives, Pyrotechnics*, **34**, 513–519.
(10) Ahmad, S.R., Anthony, D.A. and Leach, C.J. (2001) Studies into laser ignition of unconfined propellants. *Propellants, Explosives, Pyrotechnics*, **26**, 235–240.
(11) Zanotti, C. and Giuliani, P. (1998) Composite propellant ignition and excitation by CO_2 laser at sub atmospheric pressure. *Propellants, Explosives, Pyrotechnics*, **23**, 254–259.

(12) Russel, D.A. (2004) Laser Interaction with Energetic Materials. PhD Thesis, Cranfield University.
(13) Saito, T., Shimoda, M., Yamaya, T. and Iwama, A. (1991) Ignition of AP based solid propellants containing nitramines exposed to CO_2 laser radiation at subatmospheric pressures. *Combustion and Flames*, **85**, 68–76.
(14) Lee, A., Stringer, M. and Smit, K. (2003) Laser Match Head for Pyrotechnic Ignition. Report No. DSTO-TR-1448, Weapon Systems Div., Systm Sc. Lab. (Australia), pp. 1–20.
(15) de Young, L. and Lui, F. (1998) Radiative ignition of pyrotechnics: effect of wavelength on ignition threshold. *Propellants, Explosives, Pyrotechnics*, **23**, 328–332.
(16) Ahmad, S.R. and Russel, D.A. (2005) Laser ignition of pyrotechnics – effect of wavelength, composition and Confinement. *Propellants, Explosives, Pyrotechnics*, **30**(2), 131–139.
(17) Ahmad, S.R. and Russell, D.A. (2008) Studies into laser ignition of confined pyrotechnics. *Propellants, Explosives, Pyrotechnics*, **33**, 396–402.
(18) Aravindan Ponnu, Nicola Y. Edwards, Eric V. Anslyn, Pattern recognition based identification of nitrated explosives, New Journal of Chemistry, Royal Society of Chemistry, Apr 7, 2008
(19) Patent Applicaton: A method of achieving low order detonation in NiHN by diode laser, British patent application no. GB0708297.7, Filing date: 28/04.2007

12

Conclusions and Future Prospect

12.1 Introduction

The use of optical energy from a strong flash lamp for the initiation of explosives had been conceived [1] a decade prior to the invention of the working laser. One of the first few areas of research on the application of lasers was in this field, due primarily to the support from the defence industries. The first report on this topic was published within six years. This had opened up a flood gate for publications on a plethora of relevant topics in the field of laser interaction with energetic materials. Half a century on, research and development, albeit still quite sketchy, is still going on in full swing across the industrial world.

Prior to the launch [2] of the LIGHT (Light Ignition in Guns, Howitzers and Tank) project in 1993, the research was almost exclusively aimed at parametric analyses and understanding the interaction mechanisms. This project, funded by the US army, aimed to conduct applied R&D on the use of a Nd-YAG laser operating at @ 1064 nm to ignite black powder bags in the tank ammunition system to eliminate the use of primers. Although no details are available, the united defence Armament System division reported the use of a laser ignition system in their CRUSADER SPX XM2002 Howitzer system in late 2001. Within a short span of time, the project was made classified, and most applied research was shrouded in secrecy. The cutting-edge research and developments in this field, particularly those funded by defence organizations and industries, were not published in open literature. However, since the news of the first use of laser ignition in a Howitzer ammunition system for field use all research reports published in open literature are of fundamental nature, and conclusions presented below are categorized in terms of both laser parameters and material types.

12.2 Theoretical Considerations

Mathematical analyses [3] of the non-steady state generation and propagation of heat in explosive materials due to external stimuli started as early as in 1955, long before the publication of any report on the use of lasers for this purpose. Subsequent analyses of

Laser Ignition of Energetic Materials, First Edition. S Rafi Ahmad and Michael Cartwright.

the interaction processes by many authors have used well-established basic equations of optical energy transfer, conversion of energy into heat and its propagation along the depth (one dimensional) of the sample and the rise in temperature. To set up such equations, assumptions of a zeroth-order Arrhenius reaction kinetic – initiated by the generated heat within the sample – and a homogenous solid media had to be made. The theory, therefore, may not be applicable to pyrotechnics. However, if materials under test are converted to solid pellets, the theory may be applicable with some degree of error margin. Pyrotechnics are inherently inhomogeneous materials, and consist of different combinations of mixtures of different materials. Although models had been put forward to take account of such variability, theories still fail to explain [4] many experimental results.

A simple assumption of the absorption of laser energy and the subsequent generation and propagation of heat is not quite adequate to explain the overall interaction process. At high power densities (photon densities), the generation of heat in the visible and near-IR excitation may be due primarily to bond-breaking due to a single or multi-photon absorption process. Besides, the interaction process involves both the solid target and the ensuing vapour during the interaction process, a dynamic situation for the absorption of energy by the solid target. Therefore, a generalized formula to explain the interaction process is not possible, due, primarily, to the large number of variables involved. In most applications, particularly using low-power diode lasers, the beam needs to be focused to achieve the required threshold power density for reproducible ignition initiation. Due to large beam divergence of diode lasers, diffraction-limited focusing can not be achieved. Also, there exist threshold values of the beam area below which ignition will not commence, despite very high power density. This limit will depend on the type and composition of the sample.

The theory presented in this book can be applicable to evaluate the general pattern of the interaction process within the justifiable validity of assumptions made in the solutions to the equations. The theory may be refined and adjusted to specific interaction parameters, should it be necessary. However, it is to be noted that an analytical solution to any theory based on laser absorption and heat transfer and propagation in the media is not possible without drastic simplifying assumptions. Therefore, the theory developed so far is considered to provide a guideline and, perhaps, to estimate an order of magnitude for the interaction parameters.

12.3 Lasers

In this book, a laser, for all intent and purposes, has been considered to be a tool for safely delivering stimuli (energy) for the generation of heat to effect ignition and subsequent deflagration or detonation, depending upon the type of the energetic material targets and their environments. The field of lasers is vast and ever-expanding. There are hundreds of books on many aspects of lasers, and the market is almost saturated with lasers. Also, the manufactures are capable of tailoring lasers to suit almost any applications. For applications in laser ignition, the lasers have to be small, compact, ruggedized and, most importantly, cost-effective. It is generally accepted that the new generation of diode lasers fits the bills stated above.

However, to study the optical interaction process, it is important to know the properties of such lasers and how beams from them can be delivered to the target effectively. The chapter on the lasers has provided information needed by anybody for choosing a particular laser type for a specific application need, be it for basic research or for use in rocket motor igniters. The chapter gives definitions and dimensions of various laser parameters, which will be necessary for any theoretical analyses. For further understanding of lasers, a few relevant references are cited. Any other information, should it be necessary, is available on the internet and manufacturer's websites. Since the laser is an effective tool, like others, it has a sharp edge. It has to be used with respect, and the chapter includes some safety information.

12.4 Optical and Thermal Properties of Energetic Materials

The theory of laser ignition process involves, almost exclusively, parameters characterizing the optical and thermal properties of the materials. For completeness, however, general chemical and physical characteristics of energetic materials and how these are affected by the environment and processing procedures are given in details in Chapter 4. A high value of absorbance at the wavelength of excitation is a prerequisite for efficient energy transfer from the laser to the target. If the material has natural chromospheres which allow strong absorption of the incident light, the laser ignition process will be efficient.

Different materials do have different absorbing species, with different wavelength selectivity. Therefore, a match between the wavelength of the absorption peak and that of the incident radiation is prerequisite for efficient absorption of the light energy. This match can be made either by locating a laser operating at a wavelength corresponding to the peak of an optical absorption band of samples, or by formulating samples with an appropriate optical sensitizer (e.g. dye or carbon black), so that its absorption band is modified to encompass the wavelength of the selected laser radiation field. The sensitization procedure has been found to be practically indispensible for propellants and most HE materials, particularly for excitation at the diode laser wavelengths in the near-IR wavelength band.

The use of carbon black has been found to be more desirable than other solids, like carbon nanotube or liquid dyes. This is primarily because of its inert properties, which do not alter chemical properties of the host materials. Carbon, however, changes the colour of the materials, and can not be easily formulated to form chemical bonding with the host matrices. It is a general conclusion that if there is no requirement for chemical bonding, wavelength selectivity of absorption or for the maintenance of the colour of the sample, carbon black at concentrations of 1–3% by weight would alter the optical properties of the material so that these are suitable for sensitizing explosive and propellant materials for efficient laser ignition.

12.5 State of the Art: Laser Ignition

Laser initiation has provided a reliable method of moving away from explosives systems with high sensitivity to accidental initiation, and also to reduce the dependency on environmentally hazardous compounds. However, total elimination of primary igniters has not appeared possible for low vulnerability propellant systems. Whilst many of the LOVA ammunition systems have significantly reduced the hazard of accidental DDT, which was a major problem with double base propellants, the laser power required to initiate them without a primer is prohibitive except where high power generator systems are available.

Typically armoured fighting vehicles or expensive missiles may have the necessary power sources available but these may add an additional weight and safety demands require that the vehicles occupants can not be accidentally exposed to such intense laser beams.

In many civilian uses, including in quarries and in civil engineering applications such as building works, demolition so on, that require the production of detonations, lasers may provide a safer option. Instead of using detonation cord, which requires the transport of additional explosive material as well as the detonation charges, or in linking detonating systems between the arrays of charges, the simple deployment of optical fibres and a number of junction devices feeding the laser into the explosive charge is inherently safer. Safety requirements also demand that the laser beam can not be accidentally exposed to operatives and nearby personnel.

Despite the fact that thousands of papers have been published over the last five decades, the subject still remain sketchy, and its application has been very limited as far as we know. Because of both military and commercial interests, the outcome of most applied research had been shrouded in secrecy. The research, so far, indicates that laser ignition could be a commercially viable and practicable solution to overcome the vulnerability problem of conventional ignition initiation by electronic triggering. It has been established that pyrotechnics are most easily ignitable. Since these materials constitute the first stage of a deflagration or detonation bed, it would be possible to ignite this using a laser, by first igniting a pyrotechnic. Examples of such applications are the use of laser triggering in Howitzer gun system [5] and some laser-initiated detonators.

Pyrotechnics, being quite sensitive also to non-laser stimuli, would make ammunitions prone to accidental or deliberate (by enemy) ignition in conventional electronic stimulation (e.g. bridge wire). The recent drive for the use of insensitive ammunitions (i.e. LOVA materials) requires the elimination of pyrotechnics and primary explosives in weapon systems. Direct ignition of explosives (for detonation) and propellants (for gas generation) using diode lasers operating at the near-IR wavelength band cannot be achieved effectively without the use of optical synthesizers. Research on laser ignitibility of LOVA materials have indicated that the variability of their compositions, particularly since these are produced by different manufacturers and for different applications, does not allow any meaningful qualitative assessment on their ignitibility by lasers. However, in general, it may be concluded that these materials will require comparatively high laser outputs from diode lasers for reproducible ignition, albeit the ignition delay times (reaction times) still remain quite long for applications in any practical scenario.

12.6 Future Prospect

Research and development on the use of lasers for the ignition of energetic materials were primarily dictated by the defence industry's needs in operational safety considerations. However, such research and development is not published in open literature. No technical details of systems developed for the Howitzer gun system and for the matchstick device are available in open literature. Even the results of basic research sponsored by defence establishments are not available in open literature. Results are often shrouded in commercial or security classification blankets. Despite such secrecy, much advancement in the knowledge of laser interaction with energetic materials has taken place in recent years.

It is to be noted that while research on laser ignition of energetic materials started in earnest through the sponsorship of defence organizations, the applications in many civilian sectors had been considered. The prospect of remote initiation of detonation of explosives in mining industries, the ignition of flares in oil rigs without the use of command wires, use in air bag inflation in the event of accidents in automobiles, ejection of pilot's seat in emergency for abandoning the aircraft during flights and so on are a few of the civilian applications which laser ignition may find useful. The future prospect of finding effective applications of laser ignition will be dictated mainly by safety concerns. It is also considered that an ignition system without pyrotechnics and primary explosives will eliminate the carcinogenic components and heavy metals used in a conventional ignition system and will be the answer to environmental issues in the manufacture, use and disposal of energetic materials.

Currently, most of the laser initiators rely on heat generation within a carefully controlled region of the sample. The heat transfer mechanism is through lattice and molecular vibrations, and ignition occurs when sufficient energy has been deposited to rupture a number of molecular bonds. Materials containing some relatively weak bonds, which are easily broken, are inherently sensitive to accidental initiation, since impact and friction deliver localized heating. The Arrhenius activation energy is a measure of the energy required to break the weakest chemical bond. Primary explosives have low activation energies. Most of these bonds absorb in the IR region of the spectrum and, hence, this is a restriction on the choice of laser source. As was discussed in the Chapter 8 on novel initiators, considerable effort has been expended on moving away from lead azide and styphnate. Most success has been achieved by considering high nitrogen-containing molecules. These molecules have low activation energies and high explosive output, because of the large number of nitrogen atoms in connecting groups. Nitrogen prefers to be in the diatomic nitrogen molecule, so additional bonds to further nitrogen atoms provide weakly bonded trigger linkages. This means that a number of these molecules show high sensitivity to impact and friction stimuli. Trying to stabilize these molecules against spontaneous decomposition, and reducing their sensitivity, represent major challenges for synthetic organic chemistry. One option is to attach a suitable chromophore to a fairly insensitive high-nitrogen compound, so that the chromophore will not respond to the accidental stimulus but will show high sensitivity to absorption of the laser radiation. This would be a molecular variation on the current addition of highly absorbing dyes to the explosive mixture, enhancing its laser response without increasing its sensitivity to accidental initiation.

An alternative option is to find materials that respond to high-energy photons in the violet and UV regions of the spectrum. In this case, absorption is by excitation of electrons in the outermost electron energy levels of the molecule. This is a much more precise mechanism for the initiation. Some of the high-nitrogen molecules, mentioned again in the new primers chapter (Chapter 8), respond to output from a camera flash. This excites electrons in bonding orbitals, thus destabilizing the molecule. Several coordination complexes between transition metals and high-nitrogen molecular ligands, such a nitrotetrazoles, also show high sensitivity to UV photons. A typical example is the complex with disodium tetra nitrotetrazolate ferrate, which contains environmentally friendly iron atoms. Unfortunately, they also exhibit high sensitivity to friction and impact stimuli, sensitivity that increases with the number of tetrazolate molecules coordinated into the iron. This sensitivity arises from the weak coordinative linkages between the transition metal and the sensitive ligands.

However, experiments with copper chlorotetrazolate show high sensitivity to photon excitation while showing limited sensitivity to conventional stimuli. If the conventional sensitivity can be reduced by coordinating in species such as ethylene diamine, the conventional sensitivity will be reduced and the photon sensitivity will be maintained, although there may be a shift in absorption wavelength for the exciting laser irradiation. An indication of the potential development for such systems is shown by the copper complex of 3(5) hydrazino4-amino-1,2,4 triazole (HATr) with the perchlorate or nitrate anion. This material shows standard impact and friction sensitivities, comparable with lead azide, but its laser sensitivity is incredibly low, requiring less than 0.5 mJ of laser energy per cm^2 for initiation, with an almost instantaneous transition to detonation.

Further work, examining the particular ligand structures, should allow both the conventional initiation threshold and also the laser initiation threshold to be reduced so that a primer will only be sensitive to a specific laser wavelength and will be almost insensitive to any other stimulus, making for an ideal safer explosive primer. The modified primer (explosive) should be able to initiate materials such as RDX and CL20. Perhaps, with the evolution of high nitrogen content and high-performance explosives, with reduced accidental sensitivity and the use of high sensitive laser primers, the explosive train can be simplified and rendered much safer.

References

(1) Bowden, F.P. and Yoffe, Y.D. (1985) Initiation and growth of explosion in liquids and solids; Press syndicate of the Unv. of Cambridge (Pub.), Science Classic series; ISBN 0 521312337.
(2) Forch, B.E. and Beyer, R.A. (1993) Cohen, A., Newbury, J.E., report: Army Research Laboratory (USA), No. AD-A261049, pp. 1–39.
(3) Kemenetiskii, D.A. (1955) *Diffusion and Heat Exchange in Chemical Kinetics*, Princeton university Press, Princeton.
(4) Ostermark, H. (1985) *Laser as a Tool in Sensitivity Testing of Explosives* in Proc. Eighth Sym. On Detonation, NSWC, Silver Spring, MD, MP 86-194, p. 473.
(5) Barrows, A.W., Forch, B.E., Beyer, R.A., Cohen, A. and Newberry, J.E. (1993) *Laser Ignition in Guns, Howitzers and Tanks: The LIGHT programme*, ARL-TR-62, Aberdeen Proving Ground, USA.

Index

Laser Ignition of Energetic Materials, First Edition. S Rafi Ahmad and Michael Cartwright.